Equilibrium, Markets and Dynamics

Springer-Verlag Berlin Heidelberg GmbH

Claus Weddepohl

Cars H. Hommes · Roald Ramer
Cees A. Withagen
Editors

Equilibrium, Markets and Dynamics

Essays in Honour of Claus Weddepohl

With 26 Figures
and 6 Tables

Springer

Prof. Dr. Cars H. Hommes
University of Amsterdam
Center for Nonlinear Dynamics in Economics and
Finance (CeNDEF), Department of Economics
Roetersstraat 11
1018 WB Amsterdam
The Netherlands
hommes@fee.uva.nl

Dr. Roald Ramer
University of Amsterdam
Department of Quantitative Economics
Roetersstraat 11
1018 WB Amsterdam
The Netherlands
roald@fee.uva.nl

Prof. Dr. Cees A. Withagen
Tilburg University
Department of Economics
P.O. Box 90153
5000 LE Tilburg
The Netherlands
c.withagen@kub.nl

ISBN 978-3-540-43470-2 ISBN 978-3-642-56131-3 (eBook)
DOI 10.1007/978-3-642-56131-3

Die Deutsche Bibliothek - CIP-Einheitsaufnahme
Equilibrium, markets and dynamics: essays in honour of Claus Weddepohl / ed.: Cars H. Hommes ... -

http://www.springer.de

Hardcover-Design: Erich Kirchner, Heidelberg

SPIN 10875512 42/2202-5 4 3 2 1 0 - Printed on acid-free paper

Preface

This book contains essays in honour of Claus Weddepohl who, after 22 years, is retiring as professor of mathematical economics at the Department of Quantitative Economics of the University of Amsterdam. Claus Weddepohl may be viewed as the first Dutch mathematical economist in the general equilibrium tradition of Arrow, Debreu and Hahn. The essays in this book are centered around the themes *Equilibrium, Markets and Dynamics*, that have been at the heart of Weddepohl's work on mathematical economics for more than three decades. The essays have been classified according to these three themes. Admittedly such a classification always is somewhat arbitrary, and most essays would in fact fit into two or even all three themes. The essays have been written by international as well as Dutch friends and colleagues including Weddepohl's former Ph.D. students.

The book starts with a review of Claus Weddepohl's work by Roald Ramer, who has been working with him in Amsterdam for all those years. The review describes how Weddepohl became fascinated by general equilibrium theory in the early stages of his career, how he has been working on the theory of markets throughout his career, and how he turned to applications of nonlinear dynamics to price adjustment processes in a later stage of his career.

The first part of the book, *Equilibrium*, collects essays with general equilibrium theory as the main theme. This part of the book starts with a contribution by Jacques Drèze, who presents extensions of the Arrow-Debreu model, motivated by macroeconomic interpretations, concerning money and price rigidities. Drèze discusses problems of indeterminateness of equilibria and argues that macroeconomic interpretations call for an incomplete markets framework, hence for expectations, another source of indeterminacy.

The second paper by Rein Haagsma revisits models of status seeking and conformism, allowing for heterogeneity of agents. In the status seeking model reputation is built up by outperforming others and in the conformist model by being close to the norm (or the median of actions taken). It is shown that multiple equilibria may occur, even when individuals are homogeneous. If individuals are heterogeneous the number of equilibrium actions is shown to be exactly one or two.

Arnold Heertje's contribution fits into the history of economic thought. He discusses various criticisms of general equilibrium theory, with emphasis on a recent book by Walker. Although the analysis of the development and character of general equilibrium theory is appreciated, Heertje is also defending the relevance of the abstract nature of general equilibrium. Technological change and the analysis of non-reproducible goods are suggested as important directions for future work.

Jean-Jacques Herings discusses coordination failures as a possible explanation of unemployment as opposed to or in addition to the neo-classical and the Keynesian points of view. Three model classes with coordination failure are distinguished: strategic models with or without a coordinating role for prices, and general equilibrium models. It is shown that in all cases coordination failures arise as a consequence of pessimistic expectations.

Hans Keiding studies implementation of competitive allocations in economies over time. A problem with dynamic allocation mechanisms is that the class of allocations which can be implemented in subgame perfect Nash equilibria is very large. Keiding proposes a refinement of the game theoretic solution concept to a renegotiation proof equilibrium and shows that competitive allocations over time may be implemented in renegotiation proof equilibrium.

Michael Keren and David Levhari (the latter being one of Weddepohl's temporary predecessors) analyze a model of the management of firms, in relation to the available communications technology. It is shown, among other things, that a decrease of marginal communication time with subordinates implies a larger span of control and a smaller number of hierarchical levels within the organization. This result explains changes which have been reported in corporate structures during recent phases of 'downsizing'.

Gerard van der Laan and Dolf Talman consider a price and quantity adjustment process with supply rationing. It is well known that the classical tâtonnement process need not converge and may lead to cycles or even chaos. The price-quantity adjustment process proposed by van der Laan and Talman always converges to the Walrasian equilibrium. At any price and rationing system generated by the process all markets clear and only supply rationing occurs. In the short run only rationing schemes are adjusted and in the long run both prices and rationings are adjusted.

Pieter Ruys discusses a service economy in a general equilibrium framework. A governance structure is introduced with agents who obey specific rules of management. An equilibrium for an economy with managed stratified services is defined and shown to exist under rather strong conditions on the service and governance structures. This equilibrium is shown to be a generalization of a neoclassical general equilibrium.

The second part of the book, *Markets*, contains contributions on the theory of markets. It starts with an article by Anton Barten on the theory of consumer demand. He deals with two types of households. Both types have similar preferences for most commodities, but one type is not interested at all in a subset of "undesired" commodities. Anton Barten proposes a methodology in applied demand analysis to exploit the similarity and pool the data for both types of households.

Marcel Canoy's contribution has the remarkable and original form of a fairy tale. Canoy discusses regulation of telecommunication markets in the form of dialogues between a fox and an owl. The traditional communication in the animal world by banging sticks has become outdated and a new technology 'Booming Buzzy' has been invented by the rats. The price of the monopolistic rats is much too high though and an auction is organized to sell the rights of the new technology to the highest bidder. But how to provide an optimal regulatory design and will all animals be happy after the auction?

In his essay, Eric van Damme, addresses a theme that Claus Weddepohl has been working on some thirty years ago: Bertrand modeling of price competition. In particular van Damme addresses the important question, both from a theory and a policy viewpoint, whether price competition will produce competitive outcomes. Based on a survey of the existing literature his analysis offers several insights, setting the stage for a discussion of the desirability of a unified framework.

Chris Elbers and Cees Withagen highlight the potential differences in outcomes of a general equilibrium versus a partial equilibrium analysis in a two-country trade model with pollution externalities and Bertrand competition on the world market. The policy implications in terms of optimal emissions taxation of the exposed sector may be opposite.

Jacob Goeree's contribution fits into the theory of auctions. In particular, he discusses what would be the best strategy to auction Claus' handwritten version of his classical 1979 *Econometrica* paper to raise money for Claus' retirement. Jacob Goeree shows that the standard winner-pay auctions yield surprisingly low revenues in this context. It is better to require losers to pay their bids, and in particular a third price all-pay auction generates the highest revenue.

Paul van den Heuvel, who was Weddepohl's first Ph.D. student, compares two pension schemes: 'defined benefit', with no risk for the participant, and 'defined contribution', where the premium is fixed and there is some risk with regard to the payment after retirement. Presently the former system is predominant in the Netherlands but in the paper he argues that the latter also deserves serious consideration.

The third part of the book, *Dynamics*, deals with dynamic economic models. This part starts with a contribution by Frank Hahn, discussing the possibilities and impossibilities of dynamic economic analysis. Hahn discusses both equilibrium and disequilibrium dynamics. Rationality and market clearing put some discipline on equilibrium dynamics, but out of equilibrium adjustment processes with learning agents and heterogeneous expectations leave many possibilities open. Evolutionary game theory, with rationality replaced by an evolutionary theory of learning, is seen as a potentially useful approach in dynamic economic analysis.

Jan Bogers discusses the dynamics of a Diamond type overlapping generations model. In the case of Cobb-Douglas utility and production functions it is shown that a saddle node bifurcation can occur. More precisely, it is shown that if the production coefficient of capital goods is large enough, the number of perfect foresight steady state equilibria changes from zero to two if the growth rate of the population increases.

Buz Brock and Cars Hommes discuss an evolutionary heterogeneous agents asset pricing model, extending an earlier version by including price contingent contracts. This type of nonlinear evolutionary models with boundedly rational agents lead to complicated, chaotic fluctuations, a topic Claus Weddepohl has been working on in the past 15 years. Brock an Hommes' analysis suggests that the introduction of futures markets may destabilize financial markets.

Dave Furth reviews some of the applications of the theory of matrices to the stability of equilibria of both exchange economies and Cournot oligopolies, and explains why this type of work 'faded away'. The 'anything goes' result in the seventies lead to a shift of interest from linear(ized) models towards non–linear ones. As a result interest of economists is shifting from local stability analysis to global dynamic analysis.

In his contribution Sanjeev Goyal argues that the way communication and social interaction is organized in society can have a decisive effect on the outcome of economic behaviour. Consequently Goyal studies a model of network formation and discusses the issue of efficient architectures.

Maarten Janssen touches upon a subject that is dear to Claus Weddepohl, cooperation and competition. It is commonly believed that with larger populations cooperation becomes less likely, because of increasing anonymity. Maarten Janssen shows in an evolutionary dynamic framework that this is not correct in general. In a simple dynamic model with imitative behavior, he shows that if the population size is larger than a certain threshold, cooperation will emerge.

The paper by Kirman and Teyssière is a theoretical as well as an empirical contribution to interacting agent modeling. They present a model for a financial asset with two types of agents, differing in their beliefs about future prices of the asset. The proportions of both types change over time according to a stochastic process describing the interaction of agents. Agents are boundedly rational, in the sense that "wrong" beliefs do not persist over time. Bubble like phenomena and long range dependence in asset price volatility occur in their model. Empirical tests detect regime switches as the cause of long memory.

The final contribution is by Jan Tuinstra, who (until now) is Weddepohl's last Ph.D. student. Tuinstra reviews the literature on the dynamics of tâtonnement processes from a nonlinear dynamics perspective. Tuinstra investigates both the stability and instability of the process and presents simple examples with different kinds of bifurcation routes to chaotic dynamical behaviour.

Finally, we would like to thank all authors for their contributions and kind cooperation. Their efforts made our editorial work easy. We also thank all referees for their prompt and careful reviews of the papers. Special thanks are due to Monique Vis for her help with translating some Word documents into LaTEX Springer style. Without her help this would have been a much more painful experience. We also would like to thank Cees Diks for his help in our struggle with some of the electronic figures in this book, and Peter Heemeijer for redrawing one of the figures. Many thanks also to Werner Mueller for the opportunity to publish this book with Springer Verlag, and to Ruth Milewski for editing and correcting the manuscript. Finally, we thank Ank Weddepohl for kindly providing the photo of Claus. We hope that the readers will enjoy this book as much as we did when editing it.

Amsterdam,
January 2002

Cars Hommes
Roald Ramer
Cees Withagen

Contents

Claus Weddepohl: the First Dutch Mathematical Economist

Roald Ramer*

Department of Quantitative Economics, University of Amsterdam

Keywords: equilibrium, disequilibrium, dynamics, markets.

1 Introduction

In 1955 young Hubertus Nicolaus Weddepohl, or simply Claus for his friends and colleagues, enrolled in the Economische Hogeschool in Tilburg which later was to become Tilburg University. At that time Dutch society was still rather patriarchal. The Dutch economy could be best characterized as corporatist. However it was not one big happy family. Rather there were four of them, so called *zuilen*. Protestants, Catholics, Socialists and Liberals all had their own separate schools, radio and TV stations, sport clubs, and everything else you can think of. There was little direct cooperation and mixing. Luckily, Dutch politics was based on consensus and compromise. Socialists and liberals alike agreed that firms should make profits and workers would not go on strike. Wages and prices were kept low. The macroeconomic growth and stability on the one hand, and an equitable income distribution on the other hand, were major shared concerns of the economists of all denominations. Healthy Dutch pragmatism proved to be a fertile ground for the development of econometric methods but did not produce significant contributions to economic theory in the Netherlands before the seventies.

Weddepohl completed his master studies with specialization in econometrics in 1962 and was appointed as a research assistant in Tilburg. A topic in stability, growth or welfare could provide a suitable Ph.D. research project for a young scholar interested in either economic policy or empirical research. For a student of economic theory however, all that seemed rather traditional and did not promise much intellectual excitement. Mathematical methods employed by Dutch economists did not go beyond simple calculus and linear algebra. Convexity, topology, non-linear dynamics and game theory were still virtually unknown. For a proper education in new mathematical economics Weddepohl had to go to Paris. He spent there one year. Back home in Tilburg he undertook an extensive study of choice theory which resulted in his Ph.D.

* I would like to thank Dave Furth, Cars Hommes, Jan Tuinstra and Cees Withagen for their comments on an earlier draft of this paper.

thesis *Axiomatic Choice Models,* defended in 1970 and published by Wolters-Noordhoff in Groningen. A traditional Dutch thesis was supposed to be a substantial scholar work comparable with a German *Habilitation* or a French *Doctorat d'Etat.* Whereas such prominent economists of Dutch origin like Houthakker, Koopmans and Theil have spent the main part of their academic careers in the US, Weddepohl's thesis appears to be the first significant original Dutch publication in mathematical economics.

The next decade Weddepohl taught mathematical economics and economic theory at his home base at the Tilburg Hogeschool. The only interruption was a research year at CORE in Leuven in 1973/4. The appointment to the prestigious chair of mathematical economics at the University of Amsterdam (UvA) followed in 1980. This chair, alongside those for econometrics, operations research and actuarial studies, was founded as a joint venture of two faculties, economics and mathematics in 1964. The first full professor of mathematical economics was the unforgettable Pieter de Wolff, who combined academic research with the post of the director of the CPB, a government supported institute for economic research and policy. After his retirement in 1976, the vacancy was fulfilled for short periods of time by prominent scholars, first by Carl Christian von Weizsäcker and then by David Levhari. Claus Weddepohl has remained in Amsterdam until his retirement 22 years later.

All published papers of Weddepohl strictly belong to the field of mathematical economics. He saw his task as developing and perfectioning mathematical tools for economic analysis on the one hand, and making them accessible to the practitioners of economic theory on the other hand. Two methodological principles can be found consistently throughout his scientific work. Over the years he would insist on a strictly microeconomic interpretation of the general equilibrium theory and he would mistrust the use of a Nash equilibrium as a game theoretical solution concept. The main theme of Weddepohl's research has always been general equilibrium theory with all its extensions and ramifications. His favourite secondary theme was the theory of the market. His dissatisfaction with the standard textbook treatment of markets proves that Claus Weddepohl has never forgotten his economic studies. He would repeatedly return to this theme in his papers, but he would refrain from commenting in public on matters of economic policy. He would point at various unrealistic assumptions underlying theorems of mathematical economics, in particular those requiring full rationality on the part of economic agents, rendering all predictions quite unreliable. In private conversations he would point out theoretical oversimplifications plagueing political debates on such important yet highly controversial current issues like cartels and free trade, privatization and auctions. His modesty and scientific integrity would dissuade him from arguing on inefficiencies of particular markets without a detailed and specialized knowledge of their functioning. An academic mathematical economist should avoid any suggestion that a mere abstract reasoning could lead to a sensible advice on real life problems. Only recently Weddepohl published a short note ([34] on the list of publications) in ESB, a weekly journal for the audience of Dutch economists, where he argued that in the case of decreasing average costs a public monopoly still appears to be the most efficient solution and warned against easy, popular arguments in favour of privatization of public utilities.

Eight Ph.D. students successfully completed their theses under Weddepohl's supervision. In chronological order they were: Paul van den Heuvel (1981), Cees Withagen (1984), Jan Bogers (1992), Jos Verbeek (1993), Marcel Canoy (1993), Rein Haagsma (1995), Jacob Goeree (1997) and Jan Tuinstra (1999). Seven of them have contributed a paper to this bundle. Their present scientific interests can be judged from their respective papers. Verbeek, on a World Bank mission in Poland, unfortunately was not available. Weddepohl's ninth and final Ph.D. student is expected to graduate soon. Weddepohl produced also two interesting publications ([15] and [23]) with his two other former students, Chris Elbers and Mehmet Yildirim.

Throughout his career Weddepohl has studied a wide variety of mathematical models in economic theory. His scientific publications can be divided into the following categories. The numbers in brackets refer to the list of publications at the end of this overview.

Choice and General Equilibrium	1, 3, 5, 8, 9, 12, 14, 32.
Non-Walrasian Equilibria	7, 10, 11, 16, 23, 27.
Overlapping Generations	13, 15, 17, 19, 23, 32.
Price Dynamics	19, 24, 26, 27, 28, 29, 30, 31.
Markets	2, 4, 6, 18, 20, 21, 25, 33.
Expository and Occasional Papers	1, 8, 12, 14, 17, 22, 28, 33, 34.

Papers which combine two different themes appear on the list twice. The same holds for the expository and occasional papers which also are classified according to their topics. I shall briefly discuss Weddepohl's contributions to all fields listed above. Short characterizations and interpretations of these subjects given in this overview reflect perhaps closer my own present day ideas, than Weddepohl's original views and intentions. This, to some degree, is unavoidable. To my defence I can add that I have tried to stay as close as possible to the spirit of Weddepohl's remarks expressed in countless private conversations over those 22 years. I shall begin with Weddepohl's Ph.D. thesis.

2 Axiomatic Choice Theory

In his Ph.D. thesis [1] Weddepohl discussed relations between various axiomatizations of the theory of choice. *Choice* in the theory of choice is a generalization of demand and supply functions introduced in the nineteenth century. A consumer would choose so as to equate the ratios of his marginal utilities with relative prices. *Preferences* evolved from utilities considered in the welfare theory already in the eighteenth century. The theory of *choice under risk* has its roots in gambling and in insurance. In his thesis Weddepohl, following Arrow, Samuelson and Houthakker, introduces abstract formulations which obliterates the origins of the separate concepts. The virtue of this approach is a possibility of a direct comparison of various axiomatic systems.

Following [1], the set of all non empty subsets of a choice set X is denoted by $\mathcal{A}$. A *choice model* is a triple $(X,\ \mathcal{P},\ C)$, with the domain $\mathcal{P} \subset \mathcal{A}$ and the choice correspondence $C : \mathcal{P} \to \mathcal{A}$ mapping every $P \in \mathcal{P}$ to a non empty subset $C(P) \subset P$. Of special interest are the following domains, in my own notation:

$$X(2) = X \times X,\ X(\#) = \text{finite subsets of } X,\ \mathcal{B} = \text{budget subsets of } X \subset \mathbb{R}^n_+.$$

A choice correspondence C defined on $X(2)$ is in fact the same as a *complete preference* relation R over X. We say that x is not worse than y and write xRy if and only if $x \in C(\{x,y\})$. A single-valued choice function $D : \mathcal{B} \to X$ defined on budget sets, is the well known demand function.

A complete and transitive preference R over X has always a natural extension to a choice model $(X,\ X(\#),\ C)$ over all finite subsets of X. A complete, transitive and continuous preference over $\mathbb{R}^n_+$ determines a demand correspondence, or in our terminology a choice model $\left(\mathbb{R}^n_+,\ \mathcal{B},\ D\right)$. The classical problem of axiomatic choice theory was the search for properties of choice models which would guarantee that they can be obtained from binary preference relations. In the case of $X(\#)$, studied by Arrow, it appears that every C satisfying the axiom of *independence of irrelevant alternatives* is consistent with the revealed transitive preference relation. Weddepohl has generalized this result to the case of general domains $\mathcal{P}$ of subsets of X, closed under the union operation.

The question if a given demand function D defined on budget sets can be obtained from a preference relation R defined on $\mathbb{R}^n_+$ by a concave utility function, has been of major interest for economic theory. At first it did not seem to be much of a problem. It took a long time until economists came to realize that marginal utilities can not always be *integrated* to one utility function. In order to deal with this case the *weak and strong axioms of revealed preference* were introduced. The weak axiom guarantees that revealed preference is well defined. The strong axiom implies that it is transitive. Ville and Houthakker independently have shown that a Lipschitz-continuous, single valued demand function satisfying weak and strong axioms of revealed preference can be obtained from a convex, continuous and transitive preference relation. Weddepohl has extended this result to demand correspondences. This is a rather subtle result as it is not at all obvious how to define Lipschitz-continuity for correspondences. The proof makes use of dual correspondences defined on the price simplex.

The significant conclusion from all these technical results is that it really does not matter if one takes the preference or the choice correspondence as the *primitive concept* of choice theory. Unfortunately the axiomatic approach underlying this work does not say much about plausibility of the axioms of transitivity, completeness and convexity with respect to preferences, and of the axioms of revealed preference and independence of irrelevant alternatives with respect to choices. They are usually interpreted normatively as the essential prerequisites of rationality. Here they are foisted upon us by the consistency of preferences with choice correspondences defined over large domains. Weddepohl was fully aware that in economic theory axioms function primarily as conditions for interesting mathematical theorems. Their significance and relation to reality is strongly suggested by the terminology. Descriptive and/or normative corroboration of the axioms is left for the scientists engaged in the applied work.

Weddepohl did not come back to choice theory in his scientific papers after 1970. During the next decade he devoted a large part of his research time to the study of various axiomatic systems and topological methods in general equilibrium theory. However, he kept teaching individual and social choice theory both in Tilburg and Amsterdam throughout his whole academic life.

3 General Competitive Equilibrium

General competitive equilibrium is undoubtedly the centerpiece of mathematical economics. It belongs to the realm of *pure economic theory* where perfect competition rules unchallenged. Walras' own interpretation of these concepts was clear. For an economy with capital goods, credit and money it would have to be an equilibrium steady state. A real existing economy is never exactly in an equilibrium state, Walras has argued, but oscillates around one. Walras himself considered the tâtonnement process as an argument for the hypothesis that such an equilibrium, which he called a 'scientific solution', would indeed be an approximately correct description of a market economy where prices are formed by the forces of free competition. However, such important real life phenomena like monopolies and taxation are likely to cause permanent and significant deviations from any competitive equilibrium. Their study would be taken up by Walras in his *applied economic theory*, which unfortunately has not progressed beyond the preparatory stage.

This brilliant idea did not make much impact on the economic profession for some 80 years. It is often said that this neglect was due to the fact that it was far ahead of its time and that a proof of existence of such an equilibrium was lacking. There is some truth in it but the two main reasons may be quite different. To begin with, Walras' own description of an equilibrium steady state was not altogether correct and had to be completed by Wicksell. Moreover, Walras did not succeed in solving the most important problem of pure economic theory of the nineteenth century, that of the determination of the interest rate in the long run. On the more practical side, the progressing large scale industrialization, the development of the institutions of the capitalist world economy and finally the great depression of the 1930's made it increasingly less likely that any main economic issue could be sensibly studied as if the economy were at or near an equilibrium (quasi)-steady state.

The rapid formalization and mathematization of economic theory between 1940 and 1950 following the lead of Von Neumann, Samuelson and Arrow have led to a radical reformulation and a spectacular revival of general competitive equilibrium theory. Some time later it was to be complemented by theories of imperfect competition and asymmetric information founded on the game theoretical solution concept of a perfect Bayesian equilibrium. The formal treatment of time, uncertainty, information and money, introduced by Arrow, has turned general equilibrium theory into a powerful mathematical tool of economy theory. Or, one could say that it has transformed pure economic theory, which anyhow does not make more sense nowadays than it made to Keynes, into a beautiful mathematical toy.

Debreu's proof of the existence of a general equilibrium under so called 'reasonable assumptions', was both cheered and ridiculed. When Weddepohl fell under the spell of general equilibrium theory the radical students were throwing tomatoes at Debreu in California. At that turbulent time general equilibrium was yet little known in the Dutch economic faculties, and considered far too abstract to be taught to the students. Weddepohl became the first Dutch economist working in the Arrow-Debreu tradition. For Weddepohl the search for existence theorems for equilibria under various conditions and more realistic assumptions, became one of his main academic activities for twenty years. From a mathematical point of view the problem of existence is a natural one. For an economist much less so. After all the economy which we try to understand, does really exist and the question of the adequacy of the proposed model or the solution concept can not be solved by mere pointing to an existence theorem. Still, this mathematical problem of existence of equilibria was certainly beneficial for mathematical economics as it spurred a lot of important theoretical work and has produced many beautiful, though perhaps not always very useful, theorems.

Around 1975 Gale and Mas-Collel and also Shafer and Sonnenschein have extended existence theorems to economies where consumers have nontransitive and possibly also incomplete preferences. At about the same time Weddepohl, working at CORE, obtained independently similar results, combining Michael's selection theorem with the techniques applied by Aumann and Schmeidler to economies with a continuum of agents ([5]). He continued his study of topological methods in general equilibrium theory at the Tilburg University in collaboration with his friend and colleague, Pieter Ruys. Their original results including an exposition of convexity and duality, are presented in a sizable overview [8] which appeared in 1979. There they also show how to incorporate Lindahl equilibria for economies with public goods and turnpike-like equilibrium paths for economies with capital goods within the framework of general equilibrium theory. Five years later in [14] Weddepohl published another exposition of topological methods in general equilibrium theory.

These developments demonstrate conclusively that the technical improvements were after all motivated by economic considerations. The general idea was, roughly speaking, to provide a unifying mathematical general equilibrium approach to concepts already known and studied in various and seemingly unrelated economic theories. In his inaugural lecture at the University of Amsterdam, delivered in 1982 and published as [12], Weddepohl has indicated the most important developments in this field and discussed some problems he was currently working on.

To conclude this section I want to devote a few words to an important methodological issue, namely that of the basic interpretation of general equilibrium theory. We were taught and we teach our students that it is a model of individual agents, possibly millions of them, and of their detailed consumption and production plans, concerning all possible commodities with exactly specified quality, time, place and other contingencies, possibly billions of them altogether. Apart from being nonoperational, the main objection against this, as it were a nano-economic interpretation of general equilibrium, is its incompatibility with the postulate of perfect competition. This was perfectly well understood by Walras who was discussing large markets and Jevons who was explicitly considering 'trading bodies' rather than individual consumers and producers. It is highly significant that Hicks in his famous Capital and Value, the first original English contribution to general equilibrium theory, presents the IS-LM model as the prime application of the general equilibrium method.

Weddepohl's inclination to a micro-economic interpretation of general equilibrium is on the one hand rooted in the doctrine of optimizing, rational economic agents, which lies at the basis of mathematical economics, and is on the other hand the consequence of his dissatisfaction with the use of the notion of a representative agent in macro-economics. Thus in [12] he comments on the Barro-Grossman-Malinvaud model, with three commodities, one producer and a single consumer, to be discussed in the next section, as follows.

> *"This is a very simple micro-economic model. It may be hoped that it also gives an idea of the behaviour of an economy with many agents and many goods and types of labour. It may also be interpreted as a macro-economic model with aggregated agents and aggregated commodities. The present author rather prefers a micro-economic interpretation. (As a macro-economic model it is consistent with micro-economic theory. Therefore it plays a role in the discussion on the micro foundations of macro-economics.)"*([12], p.387/8.)

This micro-economic interpretation is well in accordance with the current practice where the foundations and the exposition of general equilibrium theory are presented in textbooks on micro-economic theory, while the applications are found in textbooks on macro-economic theory. For example, in

Chapter 20 of the magnificent Microeconomic Theory by Mas-Collel, Whinston and Green, the authors give, in their own words *"an extremely succinct account of overlapping-generations economies, a model of central importance in modern macroeconomic theory."*

4 Non-Walrasian Equilibria

Weddepohl's first probe beyond the traditional Walrasian framework is his elegant paper [7], published in *Econometrica* in 1979. In the first part he assumes that labour time t, is fixed and identical for all possible employments of all kinds of labour. In the second part t must still be the same for everyone but it is not fixed in advance. This common value of t is therefore a kind of *public good* and the appropriate solution concept is the Lindhal equilibrium. The situation is relatively simple if there is only one kind of labour and everyone must work. In general complications are bound to occur because the set of feasible allocations need not be convex.

General equilibrium theory defines and describes equilibrium states, but it does not provide sufficient reasons for thinking that an economy as described in terms of preferences, production possibilities and available resources must always be at or in the vicinity of a Walrasian equilibrium where all markets clear. Even if we believe that the forces of free competition should quickly bring an economy, which is temporarily out of equilibrium after an exogenous shock, close to a new equilibrium, we would like to be able to say something definite about those temporary disequilibrium states and possibly also to predict the dynamic paths running through such states.

In the early seventies a number of economists independently began to study such market disequilibrium states in terms of efficient demand and supply functions, which take into account spillover effects from one market to another. All that originated from formal interpretations of Keynesian economics and Hicksian short term equilibria where, at any rate for the sake of simplicity, prices and wages were assumed be fixed. This has led to the idea of the so called *fixed price equilibria.*

In the simplest and most influential macro-economic version due to Barro-Grossman and Malinvaud there is some given supply of money (M) held by consumers/workers and a fixed real wage rate (w) but the markets for labour and goods need not clear. Under some standard simplifying assumptions there is a unique Walrasian equilibrium (M^*, w^*). For all other combinations of (M, w) there exists a non-Walrasian equilibrium of one of three possible kinds. A *classical unemployment* occurs at relatively high values of both M and w. Consumers are constrained on both markets. At low values of M producers can not sell all they would like to produce, while the consumers are partially unemployed. This is so called *Keynesian unemployment.* If, on the other hand, there is an ample supply of money and wages are not too high, producers will not be able to find all labour they could profitably employ and consumers can not buy all goods they demand and are therefore spending less money than they would have liked. Such a state is called *repressed inflation.*

In subsequent years Drèze, Benassy, Grandmont and others have provided truly micro-economic foundations for such equilibria. The difficulty they had to cope with was the necessity of a specification of *rationing schemes* which would distribute excess demand and supply among individual agents. There are many such schemes and it is not a priori clear how to choose one above another.

Weddepohl's three important contributions to this field of research, [10], [11] and [16], were published in the *Journal of Economic Theory* between 1982 and 1987. In the first two papers he considered elaborations of the production sector of the standard Barro-Grosman-Malinvaud model. In [10] production takes place under increasing returns to scale. In this case the firms must be rationed so that both Walrasian equilibrium and the classical unemployment regime must disappear. For wages not too high equilibria of two other kinds - Keynesian and inflationary - are shown to exist. In [11] Weddepohl examined rationing schemes for firms with decreasing returns to scale allowing for configurations where some firms are rationed on the labour market while others are rationed on the commodity market. There exist many such schemes. This non-uniqueness leads to the appearance of a fourth regime on the boundary between the Keynesian and inflationary domains. It has been called *underconsumption* by Muelbauer and Portes in their model with firms holding inventories. In [16], in a fully micro-economic model, Weddepohl studied supply-constrained equilibria in economies with prices which are not necessarily rigidly fixed but rather linked to one another by some index functions which determine lower and upper limits for various price ratios. Important applications of this idea are wages for various kinds of labour or wages indexed by prices of consumer goods.

5 Overlapping Generations

The chair of mathematical economics at the University of Amsterdam belongs to the Department of Quantitative Economics, which includes actuarial sciences, econometrics and operation research. The formal and informal contacts with his new colleagues and students have spurred Weddepohl's interests in economic aspects of savings for old age consumption. They can be realized by individuals through operations on the capital market, or alternatively they can be provided by a pension scheme where workers pay a uniform premium to a pension fund and retired receive a uniform pension as long as they live. A natural framework to study relative advantages and disadvantages of both systems is provided by *overlapping generations models.* Weddepohl has devoted three papers [13], [15] and [19] to these problems.

In [15] only steady state equilibria are considered. I quote from the conclusions. *"In an economy without productive capital the golden rule equilibrium is the only possibility under the pension scheme and the no credit equilibrium is the most natural result under individual optimization. ... This difference does not exist in an economy with productive capital. ... For realizing the golden rule equilibrium in both cases an institution is needed for ensuring that exactly the necessary amount of capital is invested as productive capital and that a surplus is invested in a different way (bonds, gold, money). However, under the pension scheme the pension fund itself could do this job, whereas under individual optimization a supplementary institution is indispensable. ... it can be considered as an element of planning in a basically decentralized economy."* In the complementary paper [19] various adjustment rules for pension funds are studied. *"If the system is out of equilibrium in the sense that at the going interest rate the rate of capitalization is not equal to 1, an adjustment of the premium rates can restore the rate of capitalization. The rule setting premiums equal to the rate at which the rate of capitalization is maintained plus or minus a correction term will lead to a steady state with certainty. The other adjustment rules give complications and therefore must be handled with care."*

A few years later Weddepohl published a very readable introduction to overlapping generations models [17]. For simplicity and clarity of exposition he considers there only models with two overlapping generations and a single commodity and assumes *perfect foresight* or *rational expectations* throughout. It appears that almost all general phenomena can be obtained already in this simplest case. They include stable and unstable, stationary and periodic equilibria alongside chaotic equilibrium paths and sunspot equilibria.

In [23] and [32], written in collaboration with his Ph.D. students, Weddepohl has shown how to combine models of different kinds with each other. In the first paper, fixed price equilibria in overlapping generations models with investment are studied. In the second paper, an equivalence of equilibrium cycles in overlapping generation models with asymmetric equilibria of corresponding general equilibrium models, is established. This correspondence is used in order to detect cycles of all periods in some quite natural specifications of overlapping generation models.

6 Markets

Cournot's theory of the market for a single homogeneous commodity has been hailed as the first major achievement of mathematical economics. At the one end he considered a monopolistic market with a single large producer, and at the other end, a perfectly competitive market with many small producers. An oligopoly was to provide a bridge between these two extreme cases. Cournot's purpose was to show that the oligopolistic price would converge to the competitive price as the number of producers increases beyond any limit. Bertrand's idea that prices rather than quantities should be taken as the strategic variables of market games, has led him to the conclusion that two independent suppliers are enough in order to bring the price down to the level of marginal costs. However, Cournot's and Bertrand's oligopolists were merely vendors of mineral water, each of whom could satisfy the whole demand at no cost.

It was Edgeworth who first argued that Cournot's oligopoly solution, which was nothing else than a Nash equilibrium for a single period market game, does not adequately represent strategic competition within a small group of intelligent and clever producers. He considered the idea of sequential production decisions, and pondered over an outcome which later was to become known as the von Stackelberg equilibrium. He rejected it as well, because he could not think of an argument for which one producer rather than another should become the leader. He also pointed out that in the case of price competition, the producers will not be willing to supply more than their profit maximizing output. Eventually Edgeworth came to the conclusion that in general oligopolistic markets are likely to be unstable, and that an oligopolistic equilibrium, if it at all exists, is *indeterminate.*

Edgeworth's critique of Cournot and Bertrand was very influential. So much so, that in a similar vein, von Neumann has reasoned that for non-cooperative games no generally valid solution can exist. In modern economic theory the predominance of the Nash equilibrium as a solution concept is not questioned and need not be discussed here. Both Cournot's and Bertrand's solutions are Nash equilibria of single period games (though unnoticed for several years by Shubik and Nash himself). Innumerable modifications of Cournot and Bertrand oligopoly games have been proposed in the economic literature of the twentieth century. Well known are two stage games, one due to Hotelling where locations, or rather qualities, are set in the first stage, and another examined by Kreps and Scheinkman where capacities are chosen in the first stage. In both models price competition takes place in the second stage of the game. On its own Bertrand's price competition for a given homogeneous commodity is an oversimplification rather than the benchmark case and is quite unwieldy, as it is known that in general there are no equilibria in pure strategies.

Already a long time ago Bowley has argued that not only quantities, but also their actual and potential changes, so called *conjectural variations,* play a role in market games. With two producers and prices as variables, the argument is as follows. Whenever producer i examines if he should be satisfied with the pair of prices (p_i, p_j), he may not assume that p_j will remain unchanged after he has shifted from p_i to p_i'. Therefore, he should compare his profit $\pi_i\,(p_i, p_j)$ at unchanged prices with $\pi_i\,(p_i', p_j')$ for $p_j' = r_{ij}\,(p_i, p_i', p_j)$ where the function r_{ij} indicates how he expects producer j to react to such a shift. This looks like a dynamical reasoning within a static framework of a single period market. One should be suspicious of introducing *ad hoc* reaction functions, and in fact one would prefer to obtain such effects as solutions of explicitly dynamic games, rather than to assume their existence *a priori.* A way to deal with such considerations is to treat (p_i, p_j) as an initial proposal rather than an outright commitment and try to model an ensuing bargaining process. Another possibility is to study explicitly multi-period market games. Unfortunately, in either case general solutions are not easy to obtain, in particular if the number of producers is larger than two.

This idea of conjectural variations captured Weddepohl's imagination already in the beginning of his academic career. In his early papers [4] and [6] he examined the consequences of a very simple form of such conjectures, namely that firms with decreasing average costs would match any price reduction of their rivals and satisfy their fixed share of demand. In [25] he considered more complicated assumptions, called *alternative conjectures,* in the case of firms with increasing average costs. On the whole Weddepohl's approach to oligopolistic markets has been remarkably consistent over a period of thirty years. Its characteristic tenets are as follows.

- All producers supply one homogeneous commodity although their cost functions need not be identical. Market shares need not be equal. Firms can incur selling expenses.
- Strategic variables are not quantities but prices. In addition to prices, also price changes must play a strategic role. Mixed strategies are not considered.
- A straightforward Nash equilibrium of a single period simultaneous moves game is not an adequate solution concept for market games in the case of homogeneous commodities.
- The cases of increasing returns and decreasing returns are quite different and must be treated separately. U-shaped marginal and average costs functions do not merit a separate treatment.

In his simplest model of a market for a single homogeneous commodity, Weddepohl considered n, possibly a large number, of identical firms who all have the same decreasing average cost function $c(x)$. Firm i can announce price p_i or remain inactive. Suppose that firms 1, ..., m have announced prices $(p_1, ..., p_m)$, while the remaining $n - m$ firms have quit the market. Then $p = \min(p_1, ..., p_m)$ becomes the market price and every active firm is committed to supply its share of the demand $d(p)$, and obtains profit $\pi = \frac{1}{m} d(p)(p - c(d(p)/m))$, which can be negative, while the profit of nonactive firms is 0. This is a well defined non-cooperative game with a continuum of Nash equilibria. The given rules make undercutting unprofitable, and at the same time allow for a profitable entry. At the lowest possible equilibrium price $\underline{p}$ satisfying the equation $\underline{p} = c(d(\underline{p})/2)$ there are two active firms breaking just even (although it appears that Weddepohl thought that $\underline{p} = c(d(\underline{p}))$ holds and that single firm equilibria are possible). The number of active firms can not be bigger than the maximal integer k such that $\overline{p}$ satisfying $\overline{p} = c(d(\overline{p})/k)$ exists. This $\overline{p}$ is the highest possible equilibrium price. Notice that Weddepohl's assumptions effectively prevent *price wars* and thus allow for solutions with quite a large number of small inefficient firms. This is due to the special assumption that active firms are not allowed to withdraw after they have discovered that they are bound to make losses. That is, producer i who in the first instance has posted price p_i is committed to deliver commodities demanded at the market price p even if it is lower than his announced p_i. If we allow such firms to quit, the game becomes more complicated. Later Weddepohl also examined an extension of this simple model where the market shares of the firms are not given in advance and are not necessarily equal, but are determined by the relative selling expenses of active firms. The results obtained in [18] and [20] are qualitatively similar. As can be expected, the maximal number of active firms is lower than in the simple case with fixed market shares.

In the case of increasing average costs entry on a small scale is always profitable. Therefore the number of suppliers must be fixed in advance. For a given decreasing demand function there exists in this case a unique competitive equilibrium price p^*. Unfortunately, this price is not a Bertrand equilibrium price as observed by Arrow and Shubik. Namely, every individual producer could raise his price slightly above the competitive price and capture the residual demand as other firms would not be willing to extend their production. Only Nash equilibria in mixed strategies can exist in this case. According to Weddepohl, alternative conjectures are called for. However, the need to consider residual demand causes complications with respect to such conjectures as well. In [21] Weddepohl discussed one-price solutions which under alternative conjectures of similar kind as in the case of decreasing costs, need not exist. The next step was to allow for solutions where prices asked by different firms need not be equal. In [25] there are two firms only, and by definition, their prices p_i and p_j must be either equal or their absolute difference $|p_i - p_j|$ must be exactly α, for some fixed constant α known in advance. Alternative conjectures which are natural in this setting are quite complicated, but the positive result is, that at any rate such two-price solutions always exist.

7 Price Dynamics

Another important, so far unmentioned, issue is modelling of economic dynamics, both in the short and in the long run. In his teaching and occasional papers, like his inaugural lecture [12], Weddepohl would often point out the importance of the problem of stability of Walrasian equilibria. General conditions, guaranteeing convergence of various tâtonnement processes to Walrasian equilibria in continuous time were well known. There were however two major problems. After Debreu, Mantel and Sonnenschein demonstrated thirty years ago that every homogeneous and continuous function of prices $z(p)$ satisfying Walras' law could be obtained as an excess demand function, it became clear that price adjustment processes can not in general be expected to converge to a Walrasian equilibrium. Moreover, it was argued that tâtonnement processes are not very realistic and do not describe adequately observed economic dynamics. In [12] Weddepohl discussed extensively price adjustment processes driven by effective excess demand taking into account spillover effects, in the context of sequences of short term fixed price equilibria. This was also the subject of the thesis of his first Ph.D. student van den Heuvel. Weddepohl's own study of economic dynamics was initially centered around the problem of long term stability of interest rates. Results presented in [17] and [19] were already presented in the section on overlapping generations models.

The discovery of the phenomenon of deterministic chaos in mathematics and physics in the sixties and seventies, and the first applications of this new mathematics to economics in the eighties, have reinforced Weddepohl's interest in dynamic economic processes. Weddepohl, following Saari, has argued in [24] that the use of systems of differential equations to describe a tâtonnement process has little economic motivation and that the correct way to model a price adjustment process is likely to be an iterative one. Such a process can be interpreted as a computational method carried out by an auctioneer. Alternatively it can be seen as the evolution of actual prices in an economy with perishable goods only, where the data remain constant for all periods of time so that the trades out of equilibrium have no real effects on future periods. Stability or at least approximate stability is a very desirable property for tâtonnement processes under both interpretations. Unfortunately, Day and Saari have discovered that the discrete version of a tâtonnement process can be unstable and generate erratic price fluctuations, even in the case of gross substitutability for which the continuous version always converges to the unique equilibrium.

While experimenting with simple computer models Weddepohl has observed in [24] that by introducing *a priori* upper and lower bounds for relative or absolute price changes in consecutive periods, interesting stability results can be obtained. He has called such discrete tâtonnement processes *cautious* and has noted two phenomena. Firstly, for some equilibria there exist small regions around them, such that dynamic paths beginning in such a region remain there for ever although they do not always seem to converge to an equilibrium. Secondly, every path beginning outside the union of these regions eventually enters one such region and remains there forever. This is a very encouraging result for economists who often take the stability of general equilibria for granted. In [27] and [29] Weddepohl presented computer simulations of such cautious adjustment processes, respectively for disequilibrium dynamics driven by effective excess demand functions, and for tâtonnement processes in economies with a production sector.

Theoretical results on cautious tâtonnement processes are presented in [26] and [30]. In the first paper Weddepohl considered generic economies with two goods only. With the price of one good fixed as a numeraire at 1, there is an odd number of equilibrium prices of the other good in increasing order: $p_1 < p_2 < ... < p_k$. Let also a cautious tâtonnement process be fixed. It appears that for every odd-numbered equilibrium p_j there is an interval S_j around it such that all S_j's are disjoint and that the following holds.

- A dynamic path starting in S_j remains in S_j for ever.
- A path starting outside $\cup S_j$ enters S_j for some odd j.

In [30] this result is extended to economies with many goods which all are gross substitutes. In this case there is a unique equilibrium price-vector p^* and there exists a cube C around p^* such that the properties (i) and (ii) hold for the region C. Within the region C the character of such a price path can not be predicted, it can converge to the unique equilibrium p^*, but periodic or chaotic fluctuations are also possible.

8 Final Remarks

Weddepohl's interest for discrete dynamical processes in economic theory proved to be very contagious and fruitful. At the heart of the research he has initiated lie such modern themes like bounded rationality and heterogeneous expectations. His last two Ph.D. students, Goeree and Tuinstra, have been working in this area. As a senior researcher Weddepohl prompted his younger colleague Cars Hommes in setting up a new research group, the Center for Nonlinear Dynamics and Economics and Finance (CeNDEF) at the University of Amsterdam. The articles in this Festschrift *Equilibrium, Markets and Dynamics* reflect Claus' broad interests and the stimulating role he has played for mathematical economics over a period of thirty years.

Publications of H.N. Weddepohl

1. Weddepohl, H.N., (1970), *Axiomatic Choice Models and Duality.* Rotterdam University Press, Rotterdam.
2. Weddepohl, H.N., (1971), "Concurrentie-evenwicht en de core van een directe markt." *Maandschrift Economie* 35: 427-448.
3. Weddepohl, H.N., (1972), "Duality and Equilibrium." *Zeitschrift für Nationalökonomie* 32: 163-187.
4. Weddepohl, H.N., (1975), "Concurrentie in geval van afnemende gemiddelde kosten." *Maandschrift Economie* 40: 87-102.
5. Weddepohl, H.N., (1978), "Equilibrium in a Market with Incomplete Preferences where the Number of Consumers May Be Finite." In: *Equilibrium and Disequilibrium in Economic Theory*, ed. G. Schwödiauer, Reidel Publishing Company, Dordrecht-Boston, pp. 13-26.
6. Weddepohl, H.N., (1978), "Increasing Returns and Fixed Market Shares." *International Economic Review* 19: 405-414.
7. Weddepohl, H.N., (1979), "An Equilibrium Model with Fixed Labor Time." *Econometrica* 47: 921-938.
8. Ruys P.H.M. and Weddepohl, H.N., (1979), "Economic Theory and Duality." In: *Convex Analysis and Mathematical Economics*, ed. J. Kriens, Lecture Notes in Economics and Mathematical Systems 168, Springer, Berlin, pp. 1-72.
9. Ruys P.H.M. and Weddepohl, H.N., (1979), "Economic Theory and Duality: an Outline." *Operations Research Verfahren* 34: 261-273.
10. Weddepohl, H.N., (1982), "Equilibria with Rationing in an Economy with Increasing Returns." *Journal of Economic Theory* 26: 143-163.
11. Weddepohl, H.N., (1983), "Fixed Price Equilibria in a Multifirm Model." *Journal of Economic Theory* 29: 373-399.
12. Weddepohl, H.N., (1983), "Developments in the Theory of General Equilibrium." *De Economist* 131: 95-108.
13. Weddepohl, H.N., (1984), "Sparen in modellen met overlappende generaties." In: *De praktijk van de econometrie*, ed. P.A. Verheyen, T.M.A. Bemelmans, A. Kapteyn, J. Kriens, P.H.M. Ruys and G.J.C.Th. van Schijndel, Stenfert Kroese, Leiden, pp. 109-120.
14. Weddepohl, H.N., (1984), "General Equilibrium Theory." In: *Mathematical Methods in Economics*, ed. F.van der Ploeg, John Wiley & Sons, pp. 331-369.
15. Elbers, C. and Weddepohl, H.N., (1986), "Steady State Equilibria with Saving for Retirement Purposes in a Continuous Time Overlapping Generations Model." *Zeitschrift für Nationalökonomie* 46: 253-282.
16. Weddepohl, H.N., (1986), "Supply Constrained Equilibria in Economies with Indexed Prices." *Journal of Economic Theory* 43: 203-224.
17. Weddepohl, H.N., (1990), "Overlapping Generations Models, an Introduction." In: *Advanced Lectures in Quantitative Economics,* ed. F. van der Ploeg, Academic Press, London, pp. 249-313.
18. Weddepohl, H.N., (1990), "Increasing Returns and Selling Expenses." In: *Optimal Decisions in Market and Planned Economies*, ed. R.E. Quandt and D. Triska, Westview Press, Boulder, pp. 130-148.
19. Baneke, J., Ramer, R. and Weddepohl, H.N., (1990), "Stability in a Growth Model where Savings are Provided by a Pension Scheme." In *Liber Amicorum Walther Kok and Jaap van Klinken*, ed. H. Wolthuis, R. Kaas, M.J. Goovaerts and B.J.J. Alting von Geusau, University of Amsterdam, pp. 11-49.

20. Weddepohl, H.N., (1990), "Market Equilibria under Increasing Returns." *Nieuw Archief voor Wiskunde*, 8: 413-430.
21. Weddepohl, H.N., (1991), "Op zoek naar evenwicht in een Bertrand model." In *Samenspel (Opstellen aangeboden aan G.F.W.M. Pikkemaat)*, ed. J.L. Bouma, Th. Junius, S.K. Kuipers and L. Schoonbeek, Wolters-Noordhoff, Groningen, pp. 85-102.
22. Weddepohl, H.N., (1992), "Wiskundige Economie." In: *Samenleving en Economie in de Twintigste Eeuw,* ed. M.M.G. Fase en I. van der Zijpp, Stenfert Kroese, Leiden, pp. 355-379.
23. Weddepohl, H.N. and Yildirim, M., (1993), "Fixed Price Equilibria in an Overlapping Generations Model with Investment." *Journal of Economics* 57: 37-68.
24. Weddepohl, H.N., (1994), "Erratic Dynamics in a Restricted Tâtonnement Process with Two and Three Goods." In: *Predictability and Nonlineair Modelling in Natural Sciences and Economics*, ed. J. Grasman and G. van Straten, Kluwer, Dordrecht, pp. 609-620.
25. Canoy, M. and Weddepohl, H.N., (1995), "Alternative Conjectures in Bertrand-Edgeworth Oligopoly." *European Journal of Political Economy* 11: 577-598.
26. Weddepohl, H.N., (1995), "A Cautious Price Adjustment Mechanism: Chaotic Behavior." *Journal of Economic Behavior and Organization* 27: 293-300.
27. Weddepohl, H.N., (1995), "A Simulation Study of Disequilibrium Price Dynamics." In: *Dynamic Disequilibrium Dynamics*, ed. W.A. Barnett, G-C. Gandolfo, and C.Hillinger, Cambridge University Press, Cambridge, pp. 421-443.
28. Weddepohl, H.N., (1995), "Chaos in prijzen." In: *Liber Amicorum voor Arnold Heertje*, ed. W. Kanning, Stenfert Kroese, Leiden, pp. 144-149.
29. Weddepohl, H.N., (1997), "Simulating Tâtonnement in a Production Economy." *Economic Theory* 10: 551-558.
30. Goeree, J.K., Hommes, C.H. and Weddepohl, H.N., (1998), "Stability and Complex Dynamics in a Discrete Tâtonnement Model.", *Journal of Economic Behavior and Organization* 33: 395-410.
31. Weddepohl, H.N., (1999), "Price Adjustment Models in a Production Economy." *Annals of Operations Research* 89: 149-164.
32. Tuinstra, J. and Weddepohl, H.N., (1999), "On the Equivalence Between Overlapping Generations Models and Symmetric General Equilibrium Models." *Journal of Economics* 70: 187-207.
33. Weddepohl, H.N., (1999), "Models of the Market." In: *Economics, Welfare Policy and the History of Economic Thought, Essays in the Honour of Arnold Heertje*, ed. M.G. Fase, Edward Elgar, pp. 90-103.
34. Weddepohl, H.N., (2001), "Publiek monopolie blijft het goedkoopst." *Economisch Statistische Berichten* 86: 200-202.

Equilibrium

Indeterminateness of Equilibria and Macroeconomics*

Jacques H. Drèze

CORE, Université catholique de Louvain

Abstract. The paper reviews results on indeterminateness of equilibria in two extensions of the standard (Arrow-Debreu) model of general equilibrium. These extensions, motivated by macroeconomic interpretations, concern money and price rigidities. In a natural extension to money (held for transaction purposes), if monetary policy fixes either nominal interest rates or money supply (but not both), the variability of inflation rates is unrestricted, at equilibrium. In the absence of initial nominal asset positions, the indeterminateness of inflation rates is harmless, in the complete-markets framework of Arrow-Debreu. When some relative prices are predetermined, there exists generically a continuum of real equilibria, indexed by the overall degree of rationing. In a model combining money and nominal price rigidities, the fixed nominal prices limit the indeterminateness of inflation rates, but the real indeterminateness subsits. When one introduces in addition a tâtonnement process of nominal price formation, incorporating some downward nominal rigidities, both the nominal and the real indeterminateness may be eliminated (through the initial conditions), in the complete-markets framework. It is argued that macroeconomic interpretations call for an incomplete-markets framework, hence for expectations, another source of indeterminateness. A concluding section offers some heuristic remarks on the open problems associated with market incompleteness.

Keywords: indeterminateness of equilibria, inflation, price rigidities, incomplete markets.

1 Introduction and Preview

Macroeconomic theory is seen nowadays as a study in "dynamic general equilibrium theory under uncertainty, with incomplete (and possibly imperfect) markets".[1] The *general equilibrium* theory initiated by Walras (1902) and often referred to as "the Arrow-Debreu (1954) model" has acquired canonical status for microeconomics. It is a natural candidate to set the standards of "general equilibrium" in macroeconomics as well, with the prospect of methodological reunification of micro- and macrotheories.

The Arrow-Debreu model is built on simple primitives: commodities, firms described by productions sets, households or individuals described by consumption sets, preferences and endowments. The interaction of many agents

* The author thanks Jean-Jacques Herings and David de la Croix for stimulating discussions on the contents of this paper.

[1] Quoted from Blanchard and Fisher (1989, p.29).

through decentralised markets raises the *general equilibrium* issue. *Uncertainty* means that the primitives evolve over time under the influence of exogenous circumstances, as captured by an event-tree. A finite tree, extending from date 0 to date T, contains nodes or date-events labeled s_t at date t. Commodities, and prices, are identified by a pair of subscripts (ℓ, s_t) where $\ell = 1 \cdots L$ refers to physical characteristics and s_t to the date-event contingently on which the commodity is available. Markets for contingent commodities are in the nature of insurance or futures markets.

An equilibrium with complete markets is defined by the property that markets clear at some initial date for delivery of any physical commodity contingent on any date event. This is clearly an idealisation: describing fully every conceivable date-event, and transacting on so many markets, lies beyond our capabilities; besides, as yet unborn or inactive agents have a role to play in transactions at future dates. *Incomplete markets* is the only relevant framework. At each date-event, active economic agents trade on a limited set of markets (spot markets for commodities, markets for assets and some futures or options), knowing that similar markets will open at future date-events. Trading at a date-event is thus of necessity based on *expectations* (about prices and possibly quantities on markets to be opened later). Two strands of incomplete-markets theory have emerged: temporary general equilibrium theory (TGE) and general equilibrium with incomplete markets (GEI). They rest on different assumptions about expectations: GEI typically assumes common, single-valued, exact expectations, whereas TGE only requires continuity with respect to current realisations.[2]

The event-tree provides a natural support for *dynamic* Arrow-Debreu general equilibrium theory. Tracing properties of a sequence of GEI's or TGE's along a path of events would seem like a natural pursuit. If the properties were of a macroeconomic nature, the link to macroeconomics would be established. A primary concern of macroeconomics is the short-run evolution of real national income and of the price level. The two variables are linked by an "output-inflation" or "unemployment-inflation" tradeoff, popularised under the name "Phillips curve" - an umbrella that covers a variety of interpretations, from wage bargaining to staggered price setting. As for national income, the popular tool for a discussion of its comparative-statics adjustments is still the IS-LM diagram - with modern presentations deriving the IS curve from intertemporal optimisation by savers and investors, thereby bringing in expectations under incomplete markets.[3]

If Arrow-Debreu general equilibrium theory is to relate to that kind of macroeconomics, the model should be extended in at least three directions: monetary exchange, price rigidities and incomplete markets. I review concisely in section 2.1 the extension to *money* developed in my joint work with

---

[2] Cf. Grandmont (1982), Magill and Shafer (1991); on the scope for reconciling the two approaches, at least partly, see also Drèze (1999b).

[3] See Clarida *et al.* (1999) for a recent exposition.

Heracles Polemarchakis (1999, 2001a, 2001b). The conclusion of that work, related to those of some earlier authors[4] (but not of others),[5] is clear: in a natural extension to monetary exchange of the Arrow-Debreu economy, if monetary policy sets interest rates (or alternatively money supplies), the price level and its evolution are essentially indeterminate. However, in the absence of side-conditions (like pre-existing nominal contracts), the indeterminateness is purely nominal and devoid of real consequences. But nominal interest rates have real consequences, of a microeconomic nature.

Next, the model should be extended to allow some form of *price and wage rigidities*, apt to sustain temporary underutilisation of productive capacities or labour, and to introduce a measure of nominal stickiness in the model.[6] I have argued elsewhere (Drèze, 2001) that incomplete markets account for downward rigidity of real wages and of nominal markups, in the sense that allocations sustained by such rigidities dominate, at a second-best level, allocations sustained by flexible prices and wages. The argument, still heuristic, cannot be repeated here. But I shall review concisely in section 2.2 recent work on equilibria in real economies with some price rigidities, see Herings (1996), Drèze (1997), Herings and Drèze (1998), Citanna *et al.* (2001). That work concludes to generic multiplicity of equilibria (a continuum), i.e. real indeterminateness.

As a modest step in the direction of Arrow-Debreu macroeconomics, I outline in section 3 how the models with money and with price rigidities can be combined into a model of nominal rigidities. Nominal indeterminateness disappears at those date-events where some nominal prices are predetermined, but monetary policy now has real macroeconomic consequences.

To go from fixed prices to downward rigidities, and to explore the selection of a unique equilibrium, I review concisely in section 4 a tâtonnement process in prices and quantity constraints for an exchange economy, where the Phillips curve property holds (under the additional assumption of fixed budget shares).

Sections 2 - 4 deal with complete-markets models. Greater realism calls for *incomplete-markets* modeling. The difficulties associated with modeling expectations are here compounded by the existence of a continuum of potential equilibria. A few heuristic remarks on these difficulties conclude the paper.

---

[4] E.g. Sargent and Wallace (1975), Sims (1994), Woodford (1994).

[5] E.g. Lucas and Stokey (1987).

[6] This statement is not meant to exclude altogether other approaches to macroeconomics; rather it reflects the view that approaches involving price and wage rigidities should not be excluded from a general equilibrium formulation.

2 Two Extensions of the Arrow-Debreu Model

2.1 Money

The simplest extension of the Arrow -Debreu model, that captures the essentials of monetary exchange, proceeds from the following premises.[7]

1. Non-interest-bearing fiat money is dominated by interest-bearing nominal assets as a store of value; the demand for money accordingly results from its role in facilitating transactions: "money buys goods."
2. Fiat money is created at no cost by banks that lend it to individuals and firms at non-negative nominal interest rates.
3. Banks keep balanced accounts, so that outstanding money is the counterpart of assets, claims on individuals and firms; it is "inside money."
4. Banks are owned; profits, equal to interest earned on assets, accrue to shareholders.[8]

These features are readily introduced in the Arrow-Debreu model. A bank is brought in as an additional agent. It lends money to individuals and firms, collecting interest at non-negative interest rates. Either the bank sets interest rates and accommodates money demand; or the bank sets the money supply, and interest rates adjust to clear the money market. I shall privilege the first interpretation, at no loss in generality. The bank profits (interest earned) are redistributed to the agents. *The transactions technology is represented by abstract correspondences*, one for each agent; they define, for each price system, which combinations of trades and money balances are feasible. For instance, in the cash-in-advance model of Clower (1967), beginning-of-period money balances must cover the value of the period's purchases of consumption goods or inputs. The abstract specification covers this and other more complex specifications.[9] The crucial specific assumption is that all transactions correspondences are homogeneous of degree zero in prices and money balances, i.e. they are free of "money illusion".

The analytical work consists in replacing the consumption set of a consumer by an "exchange set correspondence" defined as the intersection of the consumption set and the transactions correspondence. Similarly, a firm's production set is intersected with its transactions correspondence to define a "production set correspondence". Optimising behavior of consumers and firms, constrained by these correspondences, imply well defined and well behaved money demands.

This simple and general specification proves fruitful. Under natural assumptions, competitive equilibria exist, for arbitrary (bounded, non-negative) interest rates. The interesting property concerns nominal indeterminateness.

[7] Drèze and Polemarchakis (2001b, section 1).

[8] Households, in the model under review. If a government were added as a separate agent, it could own the bank or part thereof.

[9] Including for instance the inventory-demand model of Baumol (1952) and Tobin (1956).

Consider an event-tree extending over $T+1$ periods numbered $t = 0, 1, \cdots T$.[10] Let there be S_t possible events s_t at date t. Fix *arbitrarily* the price levels (e.g. sum over commodities of nominal spot prices) at each of these events. This operation entails S_t degrees of freedom, which come in addition to those underlying the independent choice of interest rates by the bank.

Proposition: There exists a competitive equilibrium, with the price level at each date-t event equal to the arbitrary predetermined value.[11]

In other words, setting interest rates (or alternatively money supplies), in a natural extension to monetary exchange of the Arrow-Debreu economy, leaves the price level and its evolution essentially indeterminate.

In order to understand this basic result, let $t = T = 1$. There are L physical commodities indexed $\ell = 1 \cdots L$ with nominal spot prices $p_{\ell 0}$ at time 0, $p_{\ell s}$ at time 1 in event s, $s = 1 \cdots S_1$, $\ell = 1 \cdots L$. Write $P_0 = \Sigma_\ell p_{\ell 0}, P_s = \Sigma_\ell p_{\ell s}$ for the price levels respectively at time 0, and at time 1 under event s, $s = 1 \cdots S_1$. P_s is also the spot nominal price of a composite index commodity, $s = 0, 1, \cdots S_1$. Under complete markets, claims to one unit of money at time 1 contingent on state s have a well-defined price in terms of money at time 0, say q_s, $s = 1 \cdots S_1$. A collection of such claims, one for each s, corresponds to a one-period default-free nominal bond. Its money price at time 0 is equal to $\frac{1}{1+r_0}$, where r_0 is the one-period nominal rate of interest at time 0, set by the bank. Hence, $\Sigma_s q_s = \frac{1}{1+r_0}$. This is a no-arbitrage condition.

The relative price of one unit of the index commodity contingent on state s to one similar unit at time 0 is $\frac{q_s P_s}{P_0} := \Pi_s$. That relative price is a real price (marginal rate of substitution or transformation) associated with the real equilibrium. Note that it is obtained as the ratio to P_0 of the product $q_s P_s$. Accordingly, both P_s and q_s are variation-free, subject to a single condition on their product. If, for each $s = 1 \cdots S_1, P_s$ is set exogenously, then $q_s = P_0 \frac{\Pi_s}{P_s}$, and P_0 is determined by the constraint $\Sigma_s q_s = P_0 \Sigma_s \frac{\Pi_s}{P_s} = \frac{1}{1+r_0}$. Given $P_s, s = 1 \cdots S_1$, there always exist a P_0 and a vector q such that $\frac{q_s P_s}{P_0} = \Pi_s$, $s = 1 \cdots S_1$. Alternatively, if P_0 is set exogenously (for instance, inherited from yesterday), the products $q_s P_s$ are uniquely defined, and there remain $S_1 - 1$ degrees of freedom in splitting the products under the condition $\Sigma_s q_s = \frac{1}{1+r_0}$.

One may also express the ratios $\frac{P_s}{P_0}$ in terms of the contingent rates of inflation $\phi_s = \frac{P_s - P_0}{P_0}$, so as to retrieve the uncertainty-adjusted Irving Fisher formulation of no-arbitrage. If $S_1 = 1$, i.e. certainty, then $\Pi_1 = \frac{1}{1+r_0}(1+\phi_1)$ is equal to the real discount factor $\frac{1}{1+\rho_0}$, where ρ_0 is a real discount rate; this yields the standard Fisher condition $1+r_0 = (1+\phi_1)(1+\rho_0)$. Defining the real

[10] An infinite-horizon version of the model is presented in Bloise *et al.* (2001); qualitative properties are unchanged.

[11] This is proposition 2 in Drèze and Polemarchakis (2001b).

discount factor under uncertainty by $\frac{1}{1+\rho_0} = \Sigma_s \Pi_s = \Sigma_s q_s(1+\phi_s)$ and the q-expected rate of inflation by $\overline{\phi}_q = \frac{\Sigma_s q_s \phi_s}{\Sigma_s q_s}$, we obtain $1+r_0 = (1+\overline{\phi}_q)(1+\rho_0)$.

This relation has a natural interpretation, which also provides the clue to understanding the indeterminateness property: *at an equilibrium with complete markets, a monetary policy that sets nominal interest rates controls at each node the one-period q-expected rate of inflation, but leaves the variability of inflation across successor nodes unrestricted.* Indeed, the single instrument r_0 imposes one constraint on the S_1 variables $\phi_s, s = 1 \cdots S_1$ - neither more nor less. Additional constraints may come either from additional *initial conditions* - as in the so-called "fiscal theory of the price level" (e.g. Woodford 1994) - or from a *substantive theory of nominal price formation* (e.g. a Phillips curve); more on this below.

There is a saving grace, however.

Proposition: In the absence of additional initial conditions, the allocations of commodities associated with alternative price levels at the date-t events are equivalent.[12]

In other words, inflation uncertainty is harmless, *under complete markets*. Alternatively stated, perfectly insurable inflation-uncertainty is harmless. This may be interpreted as an additional motivation to study incomplete-markets models. On the other hand, positive nominal interest rates have real consequences: (i) they drive a wedge between effective buying prices (inclusive of a "liquidity cost") and selling prices; (ii) they induce agents to incur real costs in order to economise balances, which are costless to society; (iii) they redistribute resources from borrowers to lenders.

2.2 Price Rigidities

In my work, also with others, on equilibria with price rigidities, I have found it helpful to study a *particular* economy where the set of commodities is partitioned *a priori* into two subsets, or "groups", labeled I and II. Group I consists of commodities with flexible prices equating supply and demand at all date-events. Group II consists of commodities with either preset or downward rigid prices, and with markets clearing through *quantity constraints on supply alone* (*not* on demand). This specification is not as general as one would wish, but it has proved usefully tractable. The relevant equilibrium concept has been labeled *supply-constrained equilibrium* (meaning with constraints on supply) in van der Laan (1984), Dehez-Drèze (1984), or Drèze (1997); it is labeled "underemployment equilibria" in Herings (1996), Herings-Drèze (1998) or Citanna *et al* (2001). [13]

[12] See proposition 2 in Drèze and Polemarchakis (2001b).

[13] In van der Laan (1984) and Herings (1996), group I is empty, but it is required that at least one commodity is unconstrained.

For some purposes, like the study of Walrasian price adjustments, group I is naturally restricted to perishable commodities (fish or vegetables) sold through auctions, and assets (oil futures or equities) traded on organised exchanges. For macroeconomics, it is more natural to restrict group II to commodities subject to cyclical underutilisation, namely the non-storable inputs labour (especially unskilled labour) and productive capacities,[14] with all other commodities in group I. The latter interpretation underlies the sequel.

A *supply-constrained equilibrium at fixed prices* $\overline{p}_{II}$ *for group* II is then defined by an allocation, a vector of prices p_I for group I and agent-specific quantity constraints s^a_{II} on the supply of group II commodities, such that the allocation is optimal for each agent on its (supply-constrained) budget set.

Proposition: Under standard assumptions plus differentiability, generically in initial endowments, the set of supply-constrained equilibria owns a connected component, i.e. a continuum, of different (utility-wise) equilibria.[15]

There is thus, generically, real indeterminateness of supply-constrained equilibria.

In order to understand this result, one should first realise that the fixed prices of the L_{II} commodities in group II freeze $L_{II} - 1$ relative prices. But the specification allows for L_{II} quantity constraints. This leaves one "degree of freedom", which corresponds simultaneously: to the price level P_I of group I commodities, relative to the fixed level P_{II} of group II prices, hence to the overall price level; and to the overall level of supply constraints on group II commodities.[16] The ratio $\frac{P_I}{P_{II}}$ affects demand for group II commodities through substitution effects; the severity of supply constraints affects demand through income effects. Through these combined effects, excess demands are eliminated, at all levels of supply. Because demands are unconstrained, equilibrium allocations satisfy first-order conditions from the demand side, and this contains the multiplicity of equilibria. The question also arises whether the multiple equilibria display a positive association between price levels and activity levels. Examples reveal that neither that property, nor its converse, have a claim to generality: "stagflation" is also possible.[17]

---

[14] "Productive capacities" may appear as specific commodities (like machine rentals), or may be priced indirectly through markups on outputs.

[15] This is theorem 3.1(ii) in Herings-Drèze (1998) or Citanna *et al.* (2001). Technically, the theorem is proved by application of Browder's (1960) fixed-point theorem, as extended to correspondences by Mas-Colell (1974).

[16] See Citanna *et al.* (2001, section 5) for a detailed example.

[17] See Drèze (1997, section 4.4).

3 Combining Money and Price Rigidities

This work on supply-constrained equilibria is an extension of the canonical Arrow-Debreu model to price rigidities. Accordingly, it concerns an idealised economy with complete markets (even though I claim that incomplete markets provide the most convincing explanation of price rigidities). Because the economy is real, the price rigidities concern ultimately some real relative prices. Extending the Arrow-Debreu model to money, as outlined in section 2.1, permits studying as well absolute nominal price rigidities, along with the implied real relative rigidities. Nominal rigidities are a standard theme of macroeconomics. Parallely, extending the theory of monetary equilibria to nominal rigidities carries the prospect of lifting the nominal indeterminateness of those equilibria. I state now a simple, straightforward result to that effect, still in the framework of economies with complete markets. The key to that result is the definition of monetary economies in terms of "transactions correspondences", reviewed in section 2.1.

Because the bank of section 2.1 issues loans (balances) at *fixed interest rates* and accommodates money demand, it fits into the model of section 2.2 as a firm supplying the commodity "money" at a fixed price r, subject to the quantity constraints on supply defined by the demand for money of the agents. That is, loans belong to group II. Because the transactions technology is homogeneous of degree 0 in prices and money balances, the fixed prices $\overline{p}_{II}$ of the other group II commodities, and the equilibrium prices p_I of group I commodities, are nominal prices. Accordingly, *the model of section 2.2 applies immediately to a monetary economy with nominal price rigidities* (fixed prices). In particular, there exists a continuum of *different* equilibria, typically characterised by a continuum of different nominal price levels P_I for the group I commodities.

The more interesting implications concern the nominal indeterminateness of monetary equilibria. Consider again an event-tree extending over $T + 1$ periods, with nodes (date-events) s_t. Commodities are indexed (ℓ, s_t), with nominal spot prices p_{ℓ,s_t}.

Proposition: Assume that, for some t, $T \geq t \geq 0$, the set of group II commodities contains, for each s_τ with $\tau \leq t$, at least one element (ℓ, s_τ); then the set of supply-constrained equilibria is free of nominal indeterminateness up to and including date t.

By "free of nominal indeterminateness", I mean that, generically, it will not be the case (as under fully flexible prices) that arbitrary overall price levels support equivalent equilibria.

The result just stated is intuitively transparent. At a node s_t, if some prices are nominally fixed, it is not possible to rescale nominal spot prices and nominal balances: the fixed prices $p_{\ell s_t}$, (ℓ, s_t) in group II, stand in the way! Thus, nominal price rigidities are apt to eliminate nominal indeterminateness.

Of course, the price rigidities bring along real indeterminateness, as explained in section 2.2. But the continuum of different supply-constrained equilibria is indexed by the price level P_I of group I commodities, hence by the general price level (P_I given). If the monetary model of section 2.1 were augmented with a specification of a process of nominal price formation (as in section 4), the resulting overall price level would lift the real indeterminateness of supply-constrained equilibria. Thus the forms of indeterminateness associated respectively with money and with price rigidities need not compound each other; they might instead compensate each other, under complete markets and explicit dynamics.

4 Elementary Dynamics

I have argued in section 2.2 that, for macroeconomic purposes, the more relevant price rigidities concern labour services and productive capacities, i.e. wages and markups. In section 3, I have entertained fixed nominal prices. *Fixed* nominal wages or markups make sense in short-run analysis, but definitely not over longer horizons. On the other hand, *downward rigid* nominal wages make sense in a short to medium run; for some commodities (produced under nominal overhead costs), downward rigid nominal markups similarly make sense. It would thus be desirable to extend the results quoted above to downward rigid, rather than fixed, nominal prices.

Existence of an equilibrium with $p_{II} \geq \overline{p}_{II}$ is proved trivially for monetary economies which own a Walrasian equilibrium with strictly positive finite prices $p^* \gg 0$. Indeed, there will exist $\lambda > 0$ such that $\lambda p^*_{II} \geq \overline{p}_{II}$, and λp^* is also a Walrasian price vector. Supply constraints are not binding, at this equilibrium; so, it has the additional property, usually imposed on supply-constrained equilibria with *downward rigid* prices, that supply constraints are not binding for any good ℓ in group II such that $p_\ell > \overline{p}_\ell$.

In a multi-period setup, downward nominal rigidity of prices $p_{\ell s_t}$ means $p_{\ell s_{t+1}} \geq p_{\ell s_t}$ for all date-events s_{t+1} successor to s_t and for all s_t, $t \leq T-1$. If one imposes the side conditions that $p_{\ell s_{t+1}} = p_{\ell s_t}$ whenever some agent is constrained on the supply of ℓ at s_t, then the trivial existence proof via competitive prices evaporates, unless it could be proved in addition that $p^*_{\ell s_{t+1}} \geq p^*_{\ell s_t}$ whenever required.

Also, nominal interest rates (monetary policy) now have further (macroeconomic) real consequences. Let downward rigidities impose $p_{\ell s_{t+1}} \geq p_{\ell s_t}$. The relative price of commodities (ℓ, s_{t+1}) and (ℓ, s_t) is then $\frac{p_{\ell s_{t+1}}}{(1+r_{s_t})p_{\ell s_t}}$ $:= \pi_{\ell, s_{t+1}}$. If unconstrained equilibrium called for a relative price equal to $\pi_{\ell, s_{t+1}} < 1$, it would also call for a nominal rate satisfying $1+r_{s_t} \geq (\frac{1}{\pi_{\ell, s_{t+1}}}) >$

1. Nominal stickiness breeds monetary policy effectiveness, but inflation may be required to alleviate the distortions induced by the stickiness.[18]

I am not aware of a general existence result for equilibria with downward rigid nominal prices and no price increases under supply rationing. But something close to such a result appears in my paper "on the dynamics of supply constrained equilibria" (Drèze 1999a).[19] That paper deals with an exchange economy, where resources are supplied inelastically to the market but not consumed directly (a natural case for inputs). I study a tâtonnement process under which the agents receive two kinds of signals: prices for all commodities, and quantity constraints on the supply of group II commodities. The process starts from arbitrary vectors of prices and supply constraints, say inherited from yesterday's allocation and today's revised expectations; these signals define the constrained budget sets of the agents. The agents express effective demands, the aggregation of which defines market excess demands. Prices of group I commodities adjust up or down, proportionately to excess demands, as under Walrasian tâtonnement. Supply constraints for group II commodities adjust up or down, proportionately to excess demands. These constraints are bounded below by zero and above by unconstrained supply. In case of excess demand at unconstrained supply, prices of group II commodities adjust upward. That is, prices of group II commodities are not raised until all possible quantity adjustments have been performed. This reflects the idea that "quantities move faster than prices", as suggested by Keynes (1936) and Leijonhufvud (1968). These adjustment rules for the signals, and the reactions of the agents expressing the effective demands corresponding to the prevailing signals, define together a tâtonnement process in continuous time.

Proposition: Under standard assumptions, plus differentiability and extended substitutability, this adjustment process is quasi-stable and every limit point of a trajectory is a supply-constrained equilibrium.[20]

It is thus a built-in property of the tâtonnement process that group II prices do not move, when some agent is supply-constrained. Convergence is established using as a Lyapunov function the sum of inflation and excess supply (output gap). An extension of the model to non-tâtonnement would thus yield a result close to"existence of equilibria with some downward rigid nominal

[18] This is the so-called "grease effect of inflation", underlined e.g. by Tobin (1972).

[19] Dynamic models of price and quantity adjustments with rationing have received surprinsgly little attention; one notable exception is Weddepohl (1996).

[20] This is theorem 4.1 in Drèze (1999a). Substitutability is a standard condition for stability of Walrasian tâtonnement. Combined with non-inferiority, it yields "extended substitutability": the net demand for any commodity does not increase when the prices and/or supply possibilities of other commodities are decreased.

prices and no price increases under supply rationing." It would also provide the explicit dynamics advocated at the end of section 3.

As the paper stands, it does not cover successive periods through non-tâtonnement dynamics, but rather through indexing of commodities by date-events. The intertemporal downward rigidities must be imposed through the additional conditions $p_{\ell s_{t+1}} \geq p_{\ell s_t}$, for all ℓ in group II, all s_t and all s_{t+1} successor to s_t. Under these side conditions, which impose raising $p_{\ell s_{t+1}}$ whenever $p_{\ell s_t}$ is raised (although excess supply of ℓ may prevail at s_{t+1}), my Lyapunov function need not be monotone. That property is retained, however in an example (homethetic preferences, section 5.2 of Drèze 1999a), with the following properties.

Initiation of the process is defined by a price vector and a vector of quotas (rationing coefficients on supply) for group II commodities.[21] Note that these signals are defined for all nodes of the tree, under the complete markets assumption implicit in the analysis. For date 0, the initial prices and quantity constraints might be those prevailing "yesterday"; for dates $t \geq 1$, they are best interpreted as "rational expectations" - though any reasonable interpretation should allow for asymmetric expectations across agents.

Tâtonnement generates (as limit point) a new price vector and a new vector of quotas. In the homothetic example, a revision at date 0 of expectations about rationing affects the new signals monotonically *as functions of one variable*: the present value of quotas at the initial prices.[22] That single variable is reminiscent of "the overall level of supply constraints on group II commodities" mentioned in the penultimate paragraph of section 2.2. It may also be interpreted as an aggregate wealth, or "permanent income", variable. *Both inflation and activity levels are monotone strictly increasing in that macroeconomic variable*, thereby also defining a "generalised Phillips curve".

Indeterminateness has two faces in this model. Taking the complete-markets hypothesis literally, one would associate indeterminateness with the initial conditions - noting from the property mentioned above that initial quantity constraints are reducible to a single macroeconomic variable. Alternatively, one may reason conditionally on initial conditions, and treat future quantity constraints as an expectations variable. The implication would be that the degree of freedom associated with the overall level of supply constraints in the models of sections 2.2 and 3 corresponds to *macroeconomic expectations*. Lifting that degree of freedom thus calls for introducing a *substantive theory of macroeconomic expectation formation*. That elementary dynamics should lead to such a requirement is not surprising. The task, however, is complicated by the existence of a continuum of equilibria over which

[21] Drèze (1999a) relies on proportional rationing, but the analysis should remain valid under general Lipschitz-continuous rationing functions.

[22] A similar property obtains in the homothetic exchange economy discussed in section 5 of Citanna *et al* (2001).

expectations must be defined. How does one define rational expectations when theory is silent?

5 Incomplete-Markets Heuristics

As announced in the introduction, the only interesting intertemporal models are incomplete-markets models. Unfortunately, they raise conceptual and analytical difficulties; accordingly, they have been investigated less extensively. I shall limit myself to a few heuristic remarks, addressed to the question raised at the end of section 3: under price rigidities, how is a particular equilibrium selected from the continuum of different supply-constrained equilibria?

5.1

Under incomplete markets, agents must substitute *expectations* for the missing market signals. That feature alone opens the door to multiple equilibria, indexed by expectations. Under price rigidities and quantity constraints, the levels of the constraints are part of the signals, and future levels must again be "estimated". With a continuum of possible levels of severity for the constraints, the problem of selecting one particular level from the continuum is shifted to expectations. Defining "rational expectations" is problematic, if one wished to be more specific than simply defining a range. On the other hand, *well-defined period* 0 *decisions do not require either common or single-valued expectations.*

5.2

Incomplete markets are apt to compound the indeterminateness problem. Consider, as an extreme example, an economy with $t = T = 1$ and states 0 at date 0, $s = 1 \cdots S_1$, at date 1. Assume that trading across date events is *impossible*. Thus there are $S_1 + 1 = N$ distinct and disjoint economies. Each one of these owns a continuum of different (utility-wise) supply - constrained equilibria, indexed by its own overall level of rationing. There are thus altogether N degrees of indeterminateness.

As a less extreme case, assume that there exist markets at date 0 for monetary claims contingent on states $s = 1 \cdots J < S_1$ at date 1 and only on these. In that case, there exist $S_1 - J + 1$ distinct and disjoint economies, that is 1 plus the "degree of market incompleteness" $S_1 - J$; and the degree of indeterminateness, $S_1 - J + 1$, is increasing in the degree of incompleteness.[23]

[23] This conclusion rests on the premise that fixed prices for group II commodities are the present-value prices of Arrow-Debreu; if spot prices in states $s = 1 \cdots S$ are fixed, but the prices of contingent claims q_s of section 2.1 are flexible, a qualitatively different model emerges, the properties of which are the subject of research in progress with Jean-Jacques Herings.

I *conjecture* that this simple example is illustrative of a general property, namely that the dimension of the continuum of different supply-constrained equilibria should be positively related to the degree of market incompleteness. But this remains a mere conjecture, at this stage.[24]

5.3

What is the meaning of a high-dimensional real indeterminateness, due to incomplete markets? One would expect that the allocation for period 0, the only one which is observable, would still belong to a single-dimensional continuum; the remaining degrees of indeterminateness should concern allocations at future date events, in particular how they relate to each other and to the period-0 allocation. That is, the *future* becomes more indeterminate, less predictable (contingently on a date-event). With a single degree of indeterminateness, observing today's allocation would (at least could) reveal the selection from the continuum, and the indeterminateness would be lifted now and forever. [25] With a higher degree of indeterminateness, associated with incomplete markets, that strong property is lost.

The analogy with nominal indeterminateness is telling. With a single degree of indeterminateness concerning the overall nominal price level (as in the Arrow-Debreu model), observing a single nominal price is enough to infer the overall price level, and there remains no indeterminateness ever. With pervasive indeterminateness, of the kind exhibited in section 2.1, assessing tomorrow's price level conditionally on any event remains an issue all along.

[24] As ground for caution, I note an opposed property in the "fiscal theory of the price level", which postulates the existence of nominal government debt backed by tax receipts or seignorage revenue. To illustrate with seignorage, assume that debt service at date event s_t amounts to $D_{s_t} > 0$ and must be covered through seignorage $r_{s_t} M_{s_t}$. Given a real allocation entailing balances $M_{s_t} > 0$ at the price level P_{s_t}, there exists a unique price level $\hat{P}_{s_t}$ such that $r_{s_t} M_{s_t} \frac{\hat{P}_{s_t}}{P_{s_t}} = D_{s_t}$. This $\hat{P}_{s_t}$ is the price level at s_t implied by the exogenous nominal debt obligation D_{s_t}. If there existed such an equation for each s_t at date t, the indeterminateness of price levels through date t would disappear, according to the "fiscal theory"; see e.g. Woodford (1994). Note however that the reasoning ignores trades in contingent nominal claims. The present value at 0 of D_{s_t} is $q_{s_t} D_{s_t}$, so the present value of all debt obligations is $\Sigma_{s_t} q_{s_t} D_{s_t} := D$. Similarly, the present value of seignorage is $\Sigma_{s_t} q_{s_t} r_{s_t} M_{s_t} := V_M$. Provided $V_M \geq D$, all debt obligations can be met from seignorage revenue through trades in contingent claims. Thus, with complete markets, there is a single constraint on the overall price level, and indeterminateness is reduced by a single unit. It is a surprise that, under the fiscal theory of the price level, indeterminateness increases with the extent of market completeness, and not the other way around. (This conundrum is avoided by Woodford, who rules out trading in contingent claims by the government.)

[25] At least for equilibrium allocations, and in the sense of forecasting the future allocation at a date-event; events themselves of course remain nature's choices.

The upshot is that, under high-dimensional real indeterminateness, we may expect repeated surprises (like the 1990 recession, for instance); and we must face the fact that allocations given a date-event are not uniquely defined. This is in contradiction to the premise of "General Equilibrium Theory with Incomplete Markets" (GEI), and more in line with the premise of "Temporary General Equilibrium Theory" (TGE).

5.4

What can we conclude, regarding the selection of today's allocation from the continuum of supply-constrained equilibria? *Heuristically*, I would say that the selection results from the interaction of three elements:

(i) *Initial conditions* , as inherited from the past and/or implied by today's prices and event; say, prices and quantity constraints inherited, technology or natural resources given by the event, etc.
(ii) *Expectations*, conditioned by today's information and possibly influenced by today's emerging allocation;[26] this includes expectations about future constraints, as also influenced by today's prices and constraints.
(iii) *The market mechanism*, which generates the allocation (prices and constraints) from the given initial conditions and expectations.

The elementary dynamic process of section 4 provides an illustration of such a market mechanism, initiated by prices and quantity constraints inherited from yesterday, plus initial expectations regarding future quantity constraints, or their overall level in the special (homethetic) formulation. Under incomplete markets, there would remain to specify how these expectations respond to current adjustments.

In macroeconomic theory of Keynesian inspiration, there also exists a continuum of feasible supply-constrained allocations, among which selection is entrusted to the "IS-LM diagram". As noted in the introduction, modern versions of the IS curve explicitly aim at combining equality of aggregate demand and supply today with intertemporal optimisation. That is also the problem implicitly addressed by the selection of a specific equilibrium today. There seems to exist no alternative to letting unobserved expectations enter the process. Justifying, from microeconomic reasoning, the concentration of expectations on macroeconomic variables (income, inflation, ...) remains an unmet challenge, outside of representative agents models or of specific models like those of section 4 above.[27] The development of a theory of expectation formation for macroeconomic variables is another incompletely met challenge,

[26] Radner (1979) discusses conditions under which today's prices "reveal" the information of individual agents.

[27] See, however, Hahn-Solow (1995, Chap.6) for an example. My own interest in coordination failures explains why I am not content with representative-agents models, though I do not deny their usefulness.

compounded by indeterminateness. There we stand - with some well-defined research avenues, but a long way to go.[28]

References

1. Arrow, K.J. and G. Debreu (1954), "Existence of an equilibrium for a competitive economy", *Econometrica*, **22**, 265-290.
2. Baumol, W. J. (1952): "The transactions demand for cash: an inventory theoretic approach", *Quarterly Journal of Economics* 66, 545-556.
3. Blanchard, O. and S. Fisher (1989), *Lectures on Macroeconomics*, MIT Press, Cambridge (Mass).
4. Bloise, G., Drèze, J.H. and H. Polemarchakis (2001), "Monetary Equilibria over an Infinite Horizon", mimeo, CORE, Louvain-la-Neuve.
5. Browder, F.E. (1960), "On Continuity of Fixed Points under Deformations of Continuous Mappings", *Summa Brasiliensis Mathematicae*, **4**, 183-91.
6. Citanna, A., H. Crs, J.H. Drèze, P.J.J. Herings and A. Villanacci (2001), "Continua of Underemployment Equilibria Reflecting Coordination Failures, Also at Walrasian Prices", forthcoming in *Journal of Mathematical Economics.*
7. Clarida, R., Gali, J. and M. Gertler (1999), "The Science of Monetary Policy: A New Keynesian Perspective", *Journal of Economic Literature*, XXXVII, **4**, 1161-1707.
8. Clower, R. (1967), "A Reconsideration of the Microfoundations of Monetary Theory", *Western Economic Journal* **6**, 1-8.
9. Dehez, P. and J.H. Drèze (1984), "On Supply-Constrained Equilibria", *Journal of Economic Theory*, **33**, 1, 172-182.
10. Drèze, J.H. (1997), "Walras-Keynes Equilibria, Coordination and Macroeconomics", *European Economic Review*, **41**, 1735-1762.
11. Drèze, J.H. (1999a), "On the Dynamics of Supply-constrained Equilibria" in P.J.J. Herings, G. van der Laan and A.J.J. Talman, Eds, *Theory of Markets*, pp. 7-25, North-Holland, Amsterdam.
12. Drèze, J.H. (1999b), "The Formulation of Uncertainty: Prices and States", Chap. 3 in G. Chichilnisky, Ed. *Markets, Information, and Uncertainty,* Essays in Economic Theory in Honor of K.J. Arrow, Cambridge University Press, Cambridge, UK.
13. Drèze, J.H. (2001), "On the Macroeconomics of Uncertainty and Incomplete Markets", *Recherches Economiques de Louvain*, **67**, 1, 3-28; Chap. 3 in J.H. Drèze, Ed., *Advances in Macroeconomics*, Palgrave, London.
14. Drèze, J.H. and H.M. Polemarchakis (1999), "Money and Monetary Policy in General Equilibrium" in: L.-A. Gérard-Varet, A. P. Kirman and M. Ruggiero, Eds, *Economics, the Next Ten Years*, Oxford University Press, Oxford.
15. Drèze, J.H. and H.M. Polemarchakis (2001a), "Intertemporal General Equilibrium and Monetary Theory", Chap. 2 in A. Leijonhufvud, Ed., *Monetary Policy as a Basis for Monetary Policy*, Palgrave, London.
16. Drèze, J.H. and H.M. Polemarchakis (2001b), "Monetary Equilibria", Chap. 5 in G. Debreu, W. Neuefeind and W. Trockel, Eds, *Economics Essays, A Festschrift for Werner Hildenbrand* , Springer, Berlin-Heidelberg.

[28] Assessing the promises of recent work on expectation formation would require an independent paper; there is indeed a substantial emerging literature.

17. Grandmont, J.M. (1982) "Temporary General Equilibrium Theory", Chap.19 in K.J. Arrow and M.D. Intriligator, Eds, *Handbook of Mathematical Economics*, North-Holland, Amsterdam.
18. Hahn, F. and R. Solow (1995), *A Critical Essay on Modern Macroeconomic Theory*, MIT Press, Cambridge (Mass).
19. Herings, P.J.J. (1996), *Static and Dynamic Aspects of General Disequilibrium Theory*, Kluwer, Boston.
20. Herings, P.J.J. and J.H. Drèze (1998), "Continua of Underemployment Equilibria", CenTER DP 9805, Tilburg and CORE DP 9845, Louvain-La-Neuve.
21. Keynes, J.M. (1936), *The General Theory of Employment Interest and Money*, Macmillan, London.
22. Leijonhufvud, A. (1968), *On Keynesian Economics and the Economics of Keynes*, Oxford University Press, Oxford.
23. Lucas, R. and N. Stokey (1987), "Money and Rates of Interest in a Cash-in-Advance Economy", *Econometrica*, **55**, 491-513.
24. Magill, M and W. Shafer (1991), "Incomplete Markets", Chap. 30 in K. Arrow and M. Intriligator, Eds, *Handbook of Mathematical Economics*, North-Holland, Amsterdam.
25. Mas-Colell, A. (1974), "A Note on a Theorem of F. Browder", *Mathematical Programming*,**6**, 229-33.
26. Radner, R. (1979), "Rational Expectations Equilbrium: Generic Existence and the Information Revealed by Prices, *Econometrica*, **44**, 655-78.
27. Sargent, T.N. and N. Wallace (1975), "Rational Expectations, the Optimal Monetary Instrument, and the Optimal Money Supply Rule", *Jounral of Political Economy* , **83** 241-254.
28. Sims, C.A. (1994), "A Simple Model of the Determination of the Price Level and the Interaction of Monetary and Fiscal Policy", *Economic Theory* **4**, 381-399.
29. Tobin, J. (1956): "The interest-elasticity of transactions demand for cash", *Review of Economics and Statistics* **38**, 241-247.
30. Tobin, J. (1972), "Inflation and Unemployment", *American Economic Review*, **62**, 1, 1-18.
31. Van der Laan, G. (1984), "Supply-Constrained Fixed Price Equilibria in Monetary Economies", *Journal of Mathematical Economics*, **13**, 2, 171-187.
32. Walras, L. (1902) (4th Edition), *Elements d'Economie Politique Pure*, F. Rouge.
33. Weddepohl, C. (1996), "A Simulation Study of Disequilibrium Price Dynamics", Chap 15 in W.A. Barnett, G. Gandolfo and C. Hillinger, Eds, *Dynamic Disequilibrium Modelling*, Cambridge University Press, Cambridge, UK.
34. Woodford, M. (1994), "Monetary Policy and Price Level Determinacy in a Cash-in Advance Economy", *Economic Theory*, **4**, 345-380.

Equilibrium Properties of Status Seeking and Conformism

Rein Haagsma*

Department of Social Sciences, Wageningen University and Research Centre, Wageningen

Abstract. The article examines the relation between the heterogeneity of individuals and the distribution of their actions, for the case where individuals compare themselves to others. Two types of social comparison are studied: status seeking and conformism. Both are modelled as an implicit contest where individuals get higher reputation only if they surpass other people. In the status model, reputation increases with the fraction of people who take a lower action. In the conformist model, reputation increases with the fraction of people who take actions with a larger distance to the norm. Both types of social comparison may generate multiple equilibria with action distributions, even if individuals are homogeneous. If individuals are heterogeneous, in general only equilibria with one or two actions exist. Such clustering probably agrees with empirical observations of status seeking and conformism.

Keywords: status seeking, conformism, norm(s), reputation.

1 Introduction

The desire of people to distinguish themselves from the "common herd" or, alternatively, to fit in and to follow others seems to underlie many decisions of everyday life. Examples are the clothes we wear, the size and brand of our car, our work pace on the shop-floor, where we go on holiday, the house and neighbourhood we live, our number of children, the education we aspire, and so on. Of course, individual considerations play a part too. But analysing such decisions just like the consumer's choice between apples and oranges in any microeconomic textbook would be misleading. It would obscure the inherent externalities involved: the impact one's choice has on the other members of his social network, i.e., his friends, relatives, and neighbours.

After the first formal approaches of Morgenstern [9] and Leibenstein [7], the literature has expanded with a diversity of rational-choice models dealing with social comparison.[1] Akerlof [1] discusses two rudimentary models that

* The author would like to thank Maarten Vendrik, Hans-Peter Weikard, and an anonymous referee for help. In particular, he benefited from discussions with Pierre van Mouche.

[1] Classics on status seeking and conformism are Frank [3] and Jones [5], respectively. These books also contain surveys. For more recent literature, see e.g. [1] and [2].

illustrate the particular features most clearly. The *status model* relates to people who seek status by trying to distance themselves from the actions of others. The *conformist model* relates to people who wish to conform to other people's actions. As an introduction to our analysis, let us briefly examine these models.

In each model, agent i chooses an action $x_i \geq 0$ of some economic or social kind to maximize utility. The action yields not only direct or *intrinsic* utility but also indirect or (what I call) *reputational* utility, by altering the relative social position of the agent . Reputational utility is gained by performing better than others in the status model, and by doing the same as others in the conformist model. All agents have identical utility functions. As usual in this literature, the focus is on Nash-equilibria in which agents choose their utility-maximizing action, given what others are doing. An equilibrium is a set of mutually consistent actions.

In the *status model*, utility is given by

$$U := d(x_i - x_{-i}) - ax_i^2 + bx_i + c \tag{1}$$

with strictly positive parameters. Besides intrinsic benefits in amount $-ax_i^2 + bx_i + c$, agent i gains reputation in amount $d(x_i - x_{-i})$ if he surpasses the actions of others, denoted by x_{-i}. Note that x_{-i} is well defined (in equilibrium), since agents are identical and, by construction of U, the utility-maximizing action does not depend on x_{-i}. The equilibrium allocation is

$$x_i = x_{-i} = \frac{b+d}{2a}. \tag{2}$$

Hence, the competitive race to outdo other people results in individual actions that exceed the socially efficient level by $d/2a$. The reason for the overactivity is that by striving for a higher social position, individuals disregard the fact that their actions affect the relative standing of other people. Attaining a high rank clearly implies a low rank for some others.

In the *conformist model*, utility is given by

$$U := -d|\, x_i - x_{-i}| - \ ax_i^2 + bx_i + c \tag{3}$$

with strictly positive parameters and $b \geq d$. Now agent i wants to conform to the actions of others, and he loses reputation in amount $d|x_i - x_{-i}|$ if he fails. Although x_{-i} may not be correctly defined here,[2] the symmetrical equilibrium action is not unique but has a range of values:

[2] Akerlof's specification assumes that the other agents take the same action, but this is not evident. An alternative approach is to replace x_{-i} in (3) by an arbitrary function $\varphi^i(\mathbf{x}_{-i})$, where $\mathbf{x}_{-i}$ is a vector of the actions of other agents. Assume φ^i has five properties: *(i)* $\varphi^i(\mathbf{x}_{-i}) \geq 0$ for $\mathbf{x}_{-i} \geq 0$; *(ii)* φ^i is increasing in all vector elements; *(iii)* $\varphi^i(\alpha, \ldots, \alpha) = \alpha$ for $\alpha \geq 0$; *(iv)* $\varphi^i = \varphi^j$ for any pair of agents i

$$x_i = x_{-i} \text{ with } \frac{b-d}{2a} \leq x_i \leq \frac{b+d}{2a}. \tag{4}$$

The multiple equilibria are such that conformism generally entails either overactivity or underactivity. Individual actions exceed or fall below the socially efficient level by at most $d/2a$, and maximize welfare only in a borderline case (i.e., if $x_i = b/2a$).

What causes the multiple equilibria is the discontinuity in the marginal utility of x_i. A little more x_i increases reputational utility with d as long as x_i is less than x_{-i}, but decreases reputational utility with d if x_i is greater than x_{-i}. It implies that the choice of x_i at x_{-i} is no longer a marginal decision. Over a range of parameter values, the agent would set x_i equal to x_{-i}. This implication is important, for it concurs with social-psychological findings that norms generally prescribe clear codes of behaviour that can be either obeyed or violated. Conforming to a certain norm then reflects a dichotomous choice.[3]

These two simple models cogently demonstrate the negative externalities of social comparison. Yet their construction of reality has some basic flaws, which seem characteristic for much of the formal literature in this field. First, both models implicitly assume that each individual belongs to his preferred reference group, or that it is too costly or simply impossible for him to switch to another group. However, people are often free to choose their own associates.[4] Status seekers with a low social position in one group may wish to move to a high social position in another group, even if it has lower intrinsic benefits. Conformists who find a specific norm too costly to follow may disobey the norm, but may also choose to avoid the implied dissonance by switching to some group with a more friendly norm. For example, if I consider the publication norm in my university department as too strict, I may move to a more flexible department, even at the expense of some salary or travelling time. In short, a full analysis should include both the choice of action and the choice of reference group.

Second, the two models ignore the relation between status seeking and conformism. Granting that people are to some extent both status seekers and

---

and j; and *(v)* $\varphi^i(\mathbf{x}^a_{-i}) = \varphi^i(\mathbf{x}^b_{-i})$ if $\mathbf{x}^a_{-i}$ and $\mathbf{x}^b_{-i}$ contain the same elements. The meaning of properties *(i)-(iii)* is obvious. Properties *(iv)* and *(v)* imply that a given set of other people's actions affects all individuals in the same way. Under these conditions, it can be shown that only symmetrical Nash-equilibria exist, and they are given by (4). Examples of φ^i are the arithmetic mean and the geometric mean.

3 For this reason, Akerlof rejects a quadratic specification of reputational utility: $-d(x_i - x_{-i})^2$, which he considers to be a special case [1:1010].

4 That people can freely choose their reference group is indeed one of the three postulates of Frank's *Choosing the Right Pond* [3]; chapter three offers an illuminating treatment of the choice of reference group.

conformists, what is the interaction between both intentions and how does it affect behaviour? One interpretation that comes to mind is conformists get higher reputation only if they surpass others in norm compliance. Our houses all must have a flowery front garden, but the roses I grow are bigger than yours. Collective measures that eliminate inefficient norms then may also curtail the intemperance of status seeking.

Third, although the two models assume that *ex ante* people wish to perform better or just the same as others, *ex post* their actions turn out to be identical. Whether people's actions result from status seeking or conformism (or neither of them) is, therefore, difficult to ascertain. In everyday life, however, status seeking and conformism are often identified by the simultaneous existence of individuals whose behaviour seems to ignore the actions of others. Probably the following observations hold. Status seeking typically applies to a minority of people who exuberantly consume (or invest) by maximizing especially reputational utility, while a majority bases consumption mainly on intrinsic utility. Conformism typically applies to a majority of people who conform to a certain level of consumption by maximizing especially reputational utility, while a minority consumes less or more mainly for intrinsic reasons. More general, we should know more about the relation between the heterogeneity of consumers and the distribution of their actions, and how this relation is affected by social comparison.

The article deals with the second and third limitation. The next section examines two alternative models of status seeking and conformism, which among other things can explain the mentioned stylized observations. The final section gives some remarks.

2 Status Seeking and Conformism Revisited

It is not obvious how to express reputational utility when a reference group has more than two members with potentially different actions. Consider the case of status seeking and assume there are three agents. A straightforward generalization of the reputation term of (1) for agent $i = 1$ is $d((x_1 - x_2) + (x_1 - x_3))$. This can be rewritten as $2d(x_1 - \bar{x}_{-1})$, where $\bar{x}_{-1} := (x_2 + x_3)/2$ denotes the average action of the others. Now suppose the first agent increases his action from $\bar{x}_{-1}$ to $\bar{x}_{-1} + \Delta$, so he gains reputation in amount $2d\Delta$. The gain is clearly independent of the spread of the other actions. For example, if he increases his action from 10 to 12, his reputation gain in case $x_2 = 5$ and $x_3 = 15$ is the same as the gain in case $x_2 = 9$ and $x_3 = 11$. But this seems unrealistic, because in the latter case he surpasses the third agent.[5]

[5] Insofar as the literature deals with a distribution of actions, the role of social differences is often simplified by assuming that reputation depends on the difference between (or the ratio of) the agent's action and some average of all the actions (see e.g. [2], [6], and [10]). As indicated here, the simplification obscures the relevance of other distribution measures.

If reputational utility is given by a *non*-linear combination of social differences, gains in reputation generally do depend on the spread of other people's actions. In addition, we want the specification to be such that reputation gains increase with the number of surpassed agents. For example, suppose the reputation term of the first agent is $d(\sin(\frac{x_1-x_2}{e}) + \sin(\frac{x_1-x_3}{e}))$, with each $\frac{x_1-x_i}{e} \in [-\frac{1}{2}\pi, \frac{1}{2}\pi]$ and $e > 0$. Using the numerical example above with $e = 21/\pi$, if the first agent increases his action from 10 to 12, his reputation gain in the first case is $d(\sin(\pi/3) - \sin(\pi/7)) \approx .43d$, and in the second case, where he overtakes the third agent, $d(\sin(\pi/7) + \sin(\pi/21)) \approx .58d$.

This approach has two key characteristics. First, moving closer to the action of a higher-status agent, without matching his action, still increases reputational utility. Second, whether or not to match the action of a higher-status agent is a marginal decision. Both properties are probably too special. In general, people's status seems related to a non-differentiable ranking of actions, as exemplified by the metaphor "social ladder". The status model below exhibits this ladder.

Next consider the case of conformism. With n agents, a similar generalization of the reputation term of (3) for the first agent is $-d(|\,x_1 - x_2|+\cdots+|\,x_1 - x_n|)$.[6] It implies that moving closer to the location of maximum reputation (here a single action or a set $[x_j, x_{j+1}]$), without matching the action of a better conformist, increases reputational utility. Second, matching his action is not a marginal decision: going beyond a better conformist involves a discontinuity in the marginal utility of reputation (lowering it with $2d$). Both properties are in line with Akerlof's basic model. The conformist model below maintains the second property, but there the discontinuity in marginal utility is caused by a jump in reputation. In other words, as with status seeking, conformist behaviour is seen as a contest, where only overtaking a better conformist yields higher reputation. The conformist model thus incorporates status effects.

Therefore, suppose an individual's reputation depends on a single factor: the number of people he leaves behind. In the status model, reputational utility increases with the fraction of agents who take a lower action. In the conformist model, reputational utility increases with the fraction of agents who take actions with a larger distance to the norm. We will examine Nash-equilibria in which each agent chooses his utility-maximizing action, given the group distribution and the median of *all* actions (thus including his own action). This provision is plausible in case of a large reference group, where the individual impact on overall fractions is negligible and perceived as such by individual agents.

Let F denote the cumulative distribution function indicating how individual actions are distributed in the reference group and let m denote the median

[6] Note that this generalization simplifies to $-d\,|\,x_1 - \bar{x}_{-1}|$, where $\bar{x}_{-1} := (x_2+\cdots+x_n)/(n-1)$, only if x_1 is the lowest or the highest action, which further disputes the use of averages (see the previous note).

of the actions. Distribution function F is increasing and continuous from the right. Median m is defined as the smallest x satisfying $F(x) \geq \frac{1}{2}$ (see, e.g., [8]). For simplicity, assume that actions have negative intrinsic benefits. This leaves the qualitative results of the status model intact and particularly eases the analysis of the conformist model, although it obscures the possibility of underactivity equilibria.

With *status seeking*, agent i chooses an action $x_i \geq 0$ that maximizes utility:

$$U^s := d(F(x_i) - F(m)) - ax_i, \tag{5}$$

with strictly positive parameters.[7] Although he loses intrinsic benefits of ax_i, the agent may enjoy reputational utility in amount $d(F(x_i) - F(m))$ if he surpasses the median of the actions. For example, suppose F is continuous, so $F(m) = \frac{1}{2}$. Then his reputational utility is positive when he out-performs more than fifty percent of the others.

With *conformism*, agent i chooses an action $x_i \geq 0$ that maximizes utility:

$$U^c := \begin{cases} -d\,|F(x_i) - F(2m - x_i)\,| - ax_i & \text{if } 0 \leq x_i \leq 2m \\ -dF(x_i) - ax_i & \text{if } x_i > 2m\ , \end{cases} \tag{6}$$

with strictly positive parameters. The norm is represented by the median of the actions m. His reputational utility depends negatively on the fraction of agents with actions x closer to the norm, or those with $|x - m| < |x_i - m|$. If he chooses some $x^\star \in [0, 2m]$, their fraction equals $|F(x^*) - F(2m - x^*)|$. If he chooses an action $x^\star > 2m$, the fraction of agents closer to the norm simply equals $F(x^*)$.

Hereafter the analysis focuses on the case of status seeking; the results on conformism are mentioned in passing. For convenience, we usear the following *reputation functions*:

$$r^s(x_i; m) := F(x_i) - F(m) \tag{7}$$

for status seeking, and

$$r^c(x_i; m) := \begin{cases} -\,|F(x_i) - F(2m - x_i)\,| & \text{if } 0 \leq x_i \leq 2m \\ -F(x_i) & \text{if } x_i > 2m \end{cases} \tag{8}$$

for conformism. As will be seen, both types of social comparison may generate an equilibrium characterized by a distribution of actions. Actually there are

[7] Frank [4] also assumes that the reputation of status seekers depends on $F(x_i)$, but holds the distribution of actions fixed. Such a partial-equilibrium approach is likely to suffer from inconsistencies: the generated set of utility-maximizing actions may form a distribution that differs from the hypothesized one.

multiple equilibria with action distributions, and this result holds even if individuals have the same utility function.[8]

Equilibrium with a Single Action

Suppose all agents choose action x_0. Does such a single-action equilibrium exist? Since the median equals x_0, we find $F(m) = F(x_0) = 1$. Thus, in case of status seeking, the reputation function of agent i is given by

$$r^s(x_i; x_0) = \begin{cases} -1 & \text{if } x_i < x_0 \\ 0 & \text{if } x_i \geq x_0 \,. \end{cases} \tag{9}$$

It is seen that quantity x_0 acts as a norm prescribing a certain minimum level of action. If the agent chooses a lower level, he loses reputational utility in amount d, though he can avoid intrinsic losses by setting $x_i = 0$. If he chooses the minimum or higher level, he saves his reputation but incurs intrinsic losses of at least ax_0. For a single-action equilibrium, everyone must strictly prefer the latter action, which holds if and only if $-d < -ax_0$. It follows that the equilibrium action has a range of values:

$$0 \leq x_0 < \frac{d}{a}. \tag{10}$$

The multiple equilibria are generally not socially efficient. Status seeking causes overactivity, except in the borderline case where $x_0 = 0$. Note that individuals may be homogeneous or heterogeneous. It is only required that their personal relative taste for reputation, as measured by d/a, is large enough.

In case of *conformism*, the reputation function is given by

$$r^c(x_i; x_0) = \begin{cases} -1 & \text{if } x_i \neq x_0 \\ 0 & \text{if } x_i = x_0 \,. \end{cases} \tag{11}$$

Now any deviation from quantity x_0 entails a loss of reputation. Proceeding as before, we find the same set of equilibria (10).[9]

Equilibrium with Two Actions

Suppose a fraction $(1 - \pi)$ of the agents chooses action x_0 and the remaining fraction π takes a strictly greater action x_1. To examine the possibility of such

[8] Formally, the analysis is inspired by Stiglitz's [11] exposition on equilibrium wage distributions.

[9] With quadratic intrinsic utility ($-ax^2 + bx + c$, with $b \geq 2\sqrt{ad}$), the set of equilibria contains states with underactivity: $[\frac{b}{2a} - \sqrt{d/a}, \frac{b}{2a} + \sqrt{d/a}]$, whereas status seeking produces equilibria with only overactivity: $[\frac{b}{2a}, \frac{b}{2a} + \sqrt{d/a}]$.

a two-actions equilibrium, two cases must be distinguished. If $\pi \leq \frac{1}{2}$, most agents prefer x_0, so the median of the actions is x_0, and $F(m) = F(x_0) = (1-\pi)$. If $\pi > \frac{1}{2}$, most agents prefer x_1, so the median equals x_1, and $F(m) = F(x_1) = 1$. Hence, the two cases produce different reputation functions.

First consider the case where $\pi \leq \frac{1}{2}$, with reputation function:

$$r^s(x_i; x_0) = \begin{cases} -(1-\pi) & \text{if } 0 \leq x_i < x_0 \\ 0 & \text{if } x_0 \leq x_i < x_1 \\ \pi & \text{if } \quad x_i \geq x_1. \end{cases} \tag{12}$$

If agent i chooses a low level of action, he loses reputational utility in amount $(1-\pi)d$, but he can avoid intrinsic losses by setting $x_i = 0$. If the agent chooses a medium level, he saves his reputation and can restrict intrinsic losses by setting $x_i = x_0$. If the agent chooses a high level, he gains reputational utility in amount πd but incurs high intrinsic losses of at least ax_1. For a two-actions equilibrium, actions x_0 and x_1 must yield the same utility, which must strictly exceed the utility of the best alternative. It follows that for any given fraction π, the pair of equilibrium actions (x_0, x_1) has a range of values:

$$\frac{x_1 - x_0}{\pi} = \frac{d}{a} \tag{13}$$

provided

$$0 \leq x_0 < x_1 < \frac{d}{a}. \tag{14}$$

In fact, since fraction π is arbitrary, any tuple (x_0, x_1, π) satisfying (13), (14), and $\pi \leq \frac{1}{2}$ is an equilibrium. Of course, if the number of agents is an integer, π must be a rational number.

Each equilibrium may also be sustained if individuals differ in their relative taste for reputation (e.g., because of income differences). To see this, let δ^- and δ^+ denote the fractions of agents with relative tastes strictly less (but not too small) and strictly greater than d/a, so $1- \delta^- - \delta^+$ (> 0) is the fraction of agents with d/a. Then action x_0 is chosen by those with low relative tastes plus a fraction $(1-\pi- \delta^-)$ of the agents, and action x_1 is taken by those with high relative tastes plus a fraction $(\pi- \delta^+)$ of the agents. If there are no agents with d/a, so if $\delta^- + \delta^+ = 1$, the equilibrium fractions choosing x_0 and x_1 are fixed by $\pi = \delta^+$. Now d/a in (13) has a certain range of values bounded by the highest relative taste of the δ^--subgroup and the lowest relative taste of the δ^+-subgroup.

Next consider the case where $\pi > \frac{1}{2}$, with reputation function:

$$r^s(x_i; x_1) = \begin{cases} -1 & \text{if } 0 \leq x_i < x_0 \\ -\pi & \text{if } x_0 \leq x_i < x_1 \\ 0 & \text{if } \quad x_i \geq x_1\,. \end{cases} \tag{15}$$

Proceeding as before, we get the same result: any tuple (x_0, x_1, π) satisfying (13), (14), and $\pi > \frac{1}{2}$ is an equilibrium. The two cases yield similar outcomes because utility U^s is assumed to be linear in $F(m)$.

All two-actions equilibria are socially inefficient. Status seeking leads to overactivity, at least by the subgroup of individuals choosing amount x_1. Particularly the possibility of two-actions equilibria with $x_0 = 0$ and $\pi < \frac{1}{2}$ nicely illustrates the earlier observation (see Introduction), that status seeking often relates to a minority of people who overconsume to enhance their reputation, while a majority bases consumption on intrinsic utility. When more people engage in overconsumption (π rises), either one's reputation from consuming x_0 falls or that from consuming x_1 rises, so the ex post status seeker anyhow increases his consumption (x_1 rises).

Finally, consider the case of conformism. If $\pi \leq \frac{1}{2}$, action x_0 represents the norm, so the reputation function is decreasing for $x_i > x_0$. It implies that additional intrinsic losses from actions beyond x_0 are not compensated by any gains in reputation. Therefore, two-actions equilibria do not exist. If $\pi > \frac{1}{2}$, action x_1 represents the norm. The reputation function is given by

$$r^c(x_i; x_1) = \begin{cases} -1 \text{ if } & 0 \leq x_i < x_0 \\ -\pi \text{ if } & x_0 \leq x_i < x_1 \\ 0 \text{ if } & x_i = x_1 \\ -\pi \text{ if } & x_1 < x_i \leq 2x_1 - x_0 \\ -1 \text{ if } & x_i > 2x_1 - x_0 \,. \end{cases} \tag{16}$$

Choosing less action than the norm costs reputation, but restricts intrinsic losses. Now multiple two-actions equilibria do exist, and are indicated by any tuple (x_0, x_1, π) satisfying (13), (14), and $\pi > \frac{1}{2}$. With $x_0 = 0$, this result illustrates the earlier observation, that conformism often applies to a majority of people who conform to a certain consumption norm to secure their reputation, while a minority of nonconformists bases consumption on intrinsic considerations.

Equilibrium with Three Actions

Suppose a fraction π_1 of the agents takes action x_1, a fraction π_2 takes a strictly greater action x_2, and the other agents set the smallest action x_0. As before, depending on the size of each subgroup, the median of the actions can be x_0, x_1, or x_2. Each case generates a specific reputation function but, because of the linearity of U^s in $F(m)$, yields the same result. Let us examine the case where $\pi_2 > \frac{1}{2}$. If a majority prefers x_2, then $F(m) = F(x_2) = 1$, so the reputation function becomes

$$r^s(x_i;x_2)=\begin{cases}-1 & \text{if } 0\le x_i<x_0\\ -(\pi_1+\pi_2) & \text{if } x_0\le x_i<x_1\\ -\pi_2 & \text{if } x_1\le x_i<x_2\\ 0 & \text{if } \quad x_i\ge x_2\,.\end{cases} \tag{17}$$

By increasing his action, agent i gains reputation step by step, but loses intrinsic utility all the time. A three-actions equilibrium requires that x_0, x_1, and x_2 yield the same utility, which must be strictly higher than the utility of the best alternative. For given fractions π_1 and π_2, the tuple of equilibrium actions (x_0, x_1, x_2) has a range of values:

$$\frac{x_1-x_0}{\pi_1}=\frac{x_2-x_1}{\pi_2}=\frac{d}{a} \tag{18}$$

provided

$$0\le x_0<x_1<x_2<\frac{d}{a}\,. \tag{19}$$

Since fraction π_1 and π_2 can be freely chosen, any tuple $(x_0, x_1, x_2, \pi_1, \pi_2)$ satisfying (18), (19), and $\pi_1+\pi_2<1$ is an equilibrium.[10] As before, none of the equilibria is socially efficient. Status seeking causes too much activity, at least by the subgroups choosing x_1 and x_2.

Each equilibrium may also hold if individuals have different relative tastes for reputation. Using earlier notation, action x_0 is chosen by those with low relative tastes plus a fraction $(1-\pi_1-\pi_2-\delta^-)$ of the agents, action x_1 is taken by fraction π_1, and action x_2 is chosen by those with high relative tastes plus a fraction $(\pi_2-\delta^+)$ of the agents. However, if none of the agents has relative taste d/a (i.e., if $\delta^-+\delta^+=1$), then no one prefers the medium action x_1, and all three-actions equilibria collapse and switch over to two-actions equilibria. This result clearly underscores the relevance of two-actions equilibria.

As for conformism, if $\pi_2\le\frac{1}{2}$, the reputation function is decreasing for $x_i>x_0$ or $x_i>x_1$, so three-actions equilibria do not exist here. If $\pi_2>\frac{1}{2}$, where action x_2 represents the norm, the reputation function is given by

$$r^c(x_i;x_2)=\begin{cases}-1 & \text{if } \quad 0\le x_i<x_0\\ -(\pi_1+\pi_2) & \text{if } \quad x_0\le x_i<x_1\\ -\pi_2 & \text{if } \quad x_1\le x_i<x_2\\ 0 & \text{if } \quad x_i=x_2\\ -\pi_2 & \text{if } \quad x_2<x_i\le 2x_2-x_1\\ -(\pi_1+\pi_2) & \text{if } 2x_2-x_1<x_i\le 2x_2-x_0\\ -1 & \text{if } \quad x_i>2x_2-x_0\,.\end{cases} \tag{20}$$

[10] This result corrects an error in [11: 604].

As before, taking less action than the norm reduces reputational utility, but also restricts intrinsic losses. Multiple three-actions equilibria exist, and are indicated by any tuple $(x_0, x_1, x_2, \pi_1, \pi_2)$ satisfying (18), (19), and $\pi_2 > \frac{1}{2}$. The equilibria are also sustained if agents differ in relative taste for reputation, but again note that if no one has taste d/a, all three-actions equilibria switch over to two-actions equilibria.

Equilibrium with Many Actions

The foregoing implies that equilibria with any finite number of actions can be constructed, provided the number of agents is large enough. If π_j $(j = 1, 2, \cdots, n)$ is the fraction of agents taking action x_j, any tuple $(x_0, x_1, \ldots, x_n, \pi_1, \ldots, \pi_n)$ satisfying

$$\frac{x_1 - x_0}{\pi_1} = \cdots = \frac{x_n - x_{n-1}}{\pi_n} = \frac{d}{a} \tag{21}$$

is an n-actions equilibrium, subject to

$$0 \leq x_0 < x_1 < \cdots < x_n < \frac{d}{a} \quad \text{and} \quad \pi_1 + \cdots + \pi_n < 1 \,. \tag{22}$$

Conformism generates n-actions equilibria only if action x_n is the norm. The equilibria are the same as those with status seeking, provided $\pi_n > \frac{1}{2}$.

For both types of social comparison, it holds that if agents are heterogeneous, x_0 is chosen by those with low relative tastes for reputation plus a fraction $(1 - \pi_1 - \cdots - \pi_n - \delta^-)$ of the agents, x_j is chosen by fraction π_j $(j = 1, \cdots, n-1)$, and x_n is chosen by those with high relative tastes plus a fraction $(\pi_n - \delta^+)$ of the agents. If none of the agents has taste d/a, the equilibria fall apart and become two-actions equilibria.

Equilibrium with a Continuum of Actions

Finally, it may be instructive to look for equilibria with a continuous distribution of actions. Evaluating $U^s(x) = U^s(x + \Delta x)$ for status seeking, we find

$$\frac{F(x + \Delta x) - F(x)}{\Delta x} = \frac{a}{d} \,. \tag{23}$$

For Δx approaching zero, it says equilibrium actions have a continuous uniform density of a/d. Now integrate over $x \geq 0$ and impose interval $[0, 1]$ as range. The corresponding distribution function then follows as $F = \frac{a}{d}(x - \underline{x})$ with domain $[\underline{x}, \overline{x}]$, where $\underline{x} \geq 0$ and $\overline{x} = \underline{x} + \frac{d}{a}$. Granting that actions lower than or equal to $\underline{x}$ affect reputation the same, as do actions greater than or equal to $\overline{x}$, the reputation function of agent i becomes

$$r^s(x_i; \frac{d}{2a} - \underline{x}) = \begin{cases} -\frac{1}{2} & \text{if } 0 \leq x_i < \underline{x} \\ \frac{a}{d}(x_i - \underline{x}) - \frac{1}{2} & \text{if } \underline{x} \leq x_i \leq \overline{x} \\ \frac{1}{2} & \text{if } \quad x_i > \overline{x}\,. \end{cases} \tag{24}$$

Actions $x_i \in [\underline{x}, \overline{x}]$ yield equal utility, and are strictly preferred to the other actions if and only if $\underline{x} = 0$. Hence, the set of equilibrium actions is given by $[0, d/a]$.

In case of conformism, no equilibrium exists with a continuum of actions. The reason is simple. A continuous distribution of actions would necessarily imply that some agents take actions beyond the median, yet these actions generate lower reputation and lower intrinsic value and, therefore, have lower utility than the median action.

3 Final Remarks

With two simple models, we have studied the relation between the heterogeneity of individuals and the distribution of their actions, for the case where individuals compare themselves to others. Two types of social comparison, status seeking and conformism, have been modelled as an implicit contest where individuals gain reputation only if they surpass other people. In the status model, reputation increases with the fraction of people who take a lower action. In the conformist model, reputation increases with the fraction of people who take actions with a larger distance to the norm.

The incorporation of status seeking and conformism into the decision framework of economic agents alters some basic results of standard economic theory. If consumers have similar preferences and face similar constraints (prices, income, time), text book economics predicts they will demand the same amount of consumption. If consumers are status seekers or conformists, however, this outcome is only one of many possibilities. Moreover, consumption patterns are generally characterized by under- or overconsumption.

If consumers diverge in their preferences or constraints, text book economics predicts they will typically choose different amounts of consumption. Again our analysis has pointed in the opposite direction. If consumers are status seekers or conformists, their choices converge on one or two consumption quantities. Such clustering probably agrees with empirical observations of status seeking and conformism.

Finally, the possibility of clustered actions also suggests an explanation of the existence of social subgroups. Akerlof [1] focuses on this question, using a conformist type of model. Essentially he explains the formation and reproduction of social subgroups, or clusters of actions, by assuming a certain degree of heterogeneity among individuals. This article provides a similar result on the basis of identical individuals.

References

1. Akerlof G. (1997) Social Distance and Social Decisions, *Econometrica* 65: 1005-1027.
2. Clark A.E., Oswald A.J. (1998) Comparison-Concave Utility and Following Behaviour in Social and Economic Settings, *Journal of Public Economics* 70: 133-155.
3. Frank R.H. (1985) Choosing the Right Pond: Human Behavior and the Quest for Status, Oxford University Press, New York.
4. Frank R.H. (1985) The Demand for Unobservable and Other Nonpositional Goods, *American Economic Review* 75: 101-116.
5. Jones S.R.G. (1984) The Economics of Conformism. Basil Blackwell, Oxford.
6. Layard R. (1980) Human Satisfactions and Public Policy, *Economic Journal* 90: 737-750.
7. Leibenstein H. (1957) Bandwagon, Snob, and Veblen Effects in the Theory of Consumers' Demand, *Quarterly Journal of Economics* 72: 183-207.
8. Mood A.M., Graybill F.A., Boes D.C. (1974) Introduction to the Theory of Statistics, 3rd edn. McGraw-Hill Tokyo.
9. Morgenstern O. (1948) Demand Theory Reconsidered, *Quarterly Journal of Economics* 63: 165-201
10. Romer D. (1984) The Theory of Social Custom: a Modification and Some Extensions, *Quarterly Journal of Economics* 99: 717-727
11. Stiglitz J.E. (1985) Equilibrium Wage Distributions, *Economic Journal* 95: 595-618

Recent Criticism of General Equilibrium Theory

Arnold Heertje

Department of Economics, University of Amsterdam

Abstract. This article, meant to honour Claus Weddepohl, who made important contributions to general equilibrium theory, discusses recent criticism of this theory. In particular, attention is paid to a critical book by Donald Walker. Although high regard is expressed for his analysis of the development and character of general equilibrium theory, the author is defending the relevance of the abstract nature of the theory of general equilibrium.

Keywords: General equilibrium, existence, uniqueness, technical change, non-reproducible goods.

1 Introduction

In a number of papers Claus Weddepohl devoted himself to the study of specific aspects of general equilibrium theory. In particular, he looked at choice models (Weddepohl, 1970), duality and equilibrium (Weddepohl, 1972, 1973), incomplete preferences (Weddepohl, 1974), equilibrium and fixed labour time (Weddepohl, 1979), supply-constrained equilibria (Weddepohl, 1987) and models of the market (Weddepohl, 1999). It seems therefore fitting to discuss recent criticism of general equilibrium theory as it has been developed by K. Arrow, G. Debreu and F. Hahn. In particular we concentrate our observations on a book by Donald Walker (Walker, 1997), but we shall not leave aside a recent critical article by Makowski and Ostroy (Makowski and Ostroy, 2001).

I shall not restrict myself to a critical evaluation of the critique of general equilibrium theory, but also suggest some topics for further research within the realm of general equilibrium theory.

2 Walker and General Equilibrium

Walker is well known for his scholarly work on Leon Walras (Walker, 1996). On the basis of his knowledge of Walras's thinking, he presented a critical analysis of general equilibrium theory in his Hennipman Lecture of 1995 at the University of Amsterdam. This lecture has been developed into a book entitled *"Advances in General Equilibrium Theory"* (Walker, 1997). It is Walker's intention to lay the groundwork for a new direction of general equilibrium theory. He is very critical of general equilibrium theory as it has been developed in the hands of Walras (1874), Wald (1936), Von Neumann (1937), Arrow and Debreu (1954), Hahn (1962) and Negishi (1962). According to Walker, these authors, in fact, constructed equilibrium equation systems that may have a solution. Their equation systems have to be distinguished from models. According to Walker equations can only describe conditions or behaviour or results of the functioning of a model that are established verbally or result from verbal specifications, and therefore equations cannot specify or provide or establish those results (Walker 1997, page 41). Models may reflect the existence of equilibrium, if the structural and behavioural features of the model have been established. If these features are carefully specified, the model is a functioning system. In economics as an empirical science, equations can be accepted only if they reflect the features of a functioning and realistic model. Equilibrium is not the outcome of the characteristics of an equation system, but is achieved by a model. The Arrow-Debreu-Hahn constructions are not supported by the features of a reasonably functioning model and have no relation to the real economy. The existence of a solution of an equation system is in Walker's view no more than a mathematical property. Equilibrium refers to solid economic content. This is - in Walker's words - "*... a question of whether or not effective assumptions have been made on the basic level of such features as institutions, technology, physical structures, procedures, laws, rules, and practices of the participants in the model. Those features create the dynamic characteristics of the model, without the knowledge of which the ascertainment of the existence of equilibrium cannot be undertaken*" (Walker, 1997, page 106). It is important to note the properties a model must exhibit in order to meet Walker's standard. I, therefore, quote him in full on this issue.

"*A model should contain firms, consumers, and resource suppliers, all with specified characteristics. It should contain general economic institutions, by which is meant institutions other than those which are within markets; and it should contain technology, productive process, and systems of communication that link economic agents with markets. The model should have markets with well-defined physical features that enable the participants to make purchases and sales, and with marketplace institutions, technology, rules and pricing conventions and procedures. The model should specify concretely how the traders are enabled to come together, how prices are quoted and publicized in each particular market, how and when information about prices and sales in each particular market is transmitted to traders in other markets, by whom and how prices are changed, and how offers to buy and sell are made public*" (Walker, 1997, page 35).

Against this background Walker takes issue with Wald, who in his view only provided a mathematical property of an equation system. In this connection Walker is also very critical of Nash, who in game theory examined the existence of equilibrium in a game or bargaining situation. According to Walker Nash also defined equilibrium as the property of an equation system considered to be just a logical structure without economic content. Walker points out that Arrow and Debreu employed Nash's ideas together with fixed-point theorems, convex sets and topology. According to Walker the Arrow-Debreu model lacks structural and behavioural foundations. Furthermore, he observes that the Arrow-Debreu model is not only purely competitive, but also virtual. In a virtual model there are changes of price and of desired supply and demand quantities in disequilibrium, but there is no actual exchange, production, hiring, consumption, saving or any economic activity in disequilibrium (Walker, 1997, page 107). A third drawback in Walker's opinion is that Arrow and Debreu made specific mathematical assumptions because they were convenient from a mathematical point of view. In this sense the Arrow-Debreu equation system is according to Walker methodologically inferior. In view of Walker's interpretation of equilibrium as a characteristic of a functioning system it is natural that existence and stability are no separate issues as they are in the case of formal equation systems.

Within the framework of general equilibrium analysis the role of competition is essential. Walker devotes a lengthy discussion to several interpretations in the literature of pure and perfect competition. His analysis boils down to the observation that perfect competition is not complicated, fanciful or incredible if it refers to an organised market. *"Such a market will generate the competitive results if there is a homogeneous commodity, a few rules of behavior, and the physical and technological conditions that are in fact found in real organized markets, including a means of disseminating information, and a pricing procedure that leads to the quoted price being changed in the same direction as the sign of the known or unknown market excess demand or to the selection of the equilibrium price by an omniscient particular-market authority"* (Walker, 1997, page 89). According to Walker large numbers of participants, knowledge of all consequences of actions, perfect divisibility of the commodity and costless information are not required for such markets. It follows that a properly constructed perfectly competitive general equilibrium model requires the structural and behavioural features that result in markets being organized *à la Walker.*

In a recent article Makowski and Ostroy also discussed the variety of interpretations of perfect competition (Makowski and Ostroy, 2001). The purpose of their reformulation is to regard perfect competition not only as a model of price setting, but also as the focal point for a broader range of incentive-related topics. Their emphasis is on full appropriation. A perfectly competitive equilibrium is one in which every individual fully appropriates his social contribution, when viewed as the marginal individual added to the economy (Makowski and Ostroy, 2001, page 500). Their definition is related to the adding-up problem of marginal productivity theory. Full appropriation is both a zero profits condition for individuals and a maximum profits condition (Makowski and Ostroy, 2001, page 502). Although this statement seems compatible with the outcome of standard theory on perfect competition, the framework still allows the authors to bring in structural features, like transaction costs, incentives, innovations and creativity. From this point of view they are not too far away from Walker, although they do not refer to him. On the other hand they pay far less attention to the consequences of their approach for general equilibrium theory.

3 Critical Evaluation

When reading Walker one can hardly avoid the impression that the whole literature on general equilibrium theory is of no importance. This impression derives its significance from the impressive knowledge by Walker of the relevant literature. It is clear from Walker's sophisticated writing on general equilibrium theory that he is fully aware both of the verbal issues involved and of the mathematics of the theory. Therefore, Walker's fundamental criticism cannot be put aside as the expression of frustration caused by not being able to follow the rather abstract economic literature. In this connection, one should not overlook that Walker concludes his book with a number of ideas on a new general equilibrium model. Walker's book cannot be looked at as a simple attack on mathematical economics.

Against this background of a positive attitude, let me first of all dwell on the question of the importance of general equilibrium theory. What does it mean to interpret Walker as saying that general equilibrium theory as it has been developed up to now has no relevance or importance? It means that general equilibrium models are neither functioning nor realistic. They are not functioning because they lack wholly or partly specifications of institutions, technologies, rules and regulations, firms, households, other participants of various types, financial institutions, ways of doing business, and the other elements that go into the making of markets. They are not realistic because they employ unrealistic or absurd postulates, so that they do not simulate the real economy.

In essence, this raises the much debated question of the relationship between economic theory and economic reality, or broader between theory and reality. It seems that Walker is right in pointing out that general equilibrium theory in its present state is no one-to-one mapping of reality. This state of affairs is there to stay forever. It will always be possible to prove that in a theory certain aspects of reality are left out or with reason not taken into account. The essence of a theory does not consist in the formal system of equations, but in the set of relationships between assumptions and conclusions. This set may be expressed in terms of equations that have a solution. In this sense theory has to be distinguished from mere description of reality. Without discussing the methodological difficulties of describing reality, it may be pointed out that building a theory aims at understanding reality. Theory creates its own world and sometimes a new world, that as such is part of enlarged reality. In other words, there is no one-way avenue between theory and reality. Not all economic theory is meant to solve practical problems, nor should it be. On the other hand, abstract reasoning, which is part of building a theory, is often of more use to solving real-life economic problems than pure description on the basis of experience and past events. I am inclined to suggest that Einstein would never have developed his theory of relativity, if he had carried out his scientific research with a pencil and a piece of paper on the basis of the procedures laid down by Walker (Pais, 1982). By creating his own world of reality, Einstein proved to be able to predict phenomena in the real world that nobody had seen or recognised before and that even came to the fore long after his death. A case in point is the Einstein-Bose condensation (Pais, 1986, page 284).

In order to understand and explain reality, it is justified to simplify in a theory relationships and even concepts and definitions of phenomena. This opens the door to imagination, speculation and even mathematical beauty. It is not only justified, but it may even be considered the ultimate aim of all science. To redefine reality so that it is reduced to a simple structure. Or to put it in the words of Weintraub: *"If all knowledge is structurally organized, is developed out of some basic and unchanging mental structures, the connection between mathematics and the growth of knowledge is not accidental"* (Weintraub, 1985, page 179).

General economic theory is the core of economics as a science. As such the development of the theory is a never-ending story. Even if we can admit that it will never end up with a complete description of reality, as Walker would like to see, along the way it provides us with useful insights and educational profitability. By way of illustration I refer to the connection between Pareto efficiency and competitive equilibrium, between bargaining and equilibrium, the introduction of future markets, missing markets and incomplete markets (Starr, 1997, Hahn, 1989, Magill and Quinzii, 1996). To work with equations and even with equations that are introduced for the sake of mathematical convenience is a step forward compared with a verbal treatment, in which the relationship between necessary assumptions and derived statements and consistency of reasoning are missing links. This insight also applies to the concept of equilibrium. To study whether a general equilibrium system does have an equilibrium solution is a necessary building block of the theory, even if the concept of equilibrium does not have a direct counterpart in reality. It may be argued that it is just the other way round. The development and further analysis of equilibrium in a theory, existence, uniqueness and stability are intriguing guidelines for the discussion of how our economy works in everyday life. In this respect, it seems appropriate to refer to the history of the Cournot-Nash concept of equilibrium in the case of oligopoly and strategic games. When Cournot wrote his book in 1838 he did not think of reality when in his analysis he moved from one seller on the market to the case of two sellers (Cournot, 1838). His assumptions with regard to suppliers of mineral water had nothing to do with reality - nevertheless he developed a concept of equilibrium for this situation, which up to the present time has proved to be the best in the case of profit maximisation. As is well known, Nash generalised this concept, which is now the basis for all non-cooperative game theory (Nash, 1951, 1996). In a certain sense to establish the conditions under which an equilibrium does exist is a necessary completion of a theory. This statement does not depend on the concrete specification of the concept of equilibrium. Both Cournot and Nash show their brilliance in defining equilibrium out of the blue and in an abstract world, such that many unforeseen practical illustrations arise (Dougherty, 1988, Nasar, 1998).

A word has to be said about the relationship of Walker's book to the impressive study by Ingrao and Israel on economic equilibrium in the history of science (Ingrao and Israel, 1990). It is clear that Walker knows that book as he quotes it several times in a critical sense. However, Ingrao and Israel after a very thorough study reach a conclusion which is not too far away from Walker's opinion. Let me quote in full their conclusion:

"We feel it is possible to conclude that the mathematical analysis of general equilibrium theory in the context of the classical hypotheses as codified by Debreu's axiomatization has led to one clear result. There is a contradiction between the theory's aims and the consequences derived from the system of hypotheses constituting its structure, and hence a contradiction between aims and hypotheses within the theory. The only way out of this situation is to jettison explicitly the programmatic central core that has been so carefully preserved throughout the many paradigmatic shifts. And *"The only way possible is a thorough reexamination of the theory's basic hypotheses, i.e. a true paradigmatic revolution"* (Ingrao and Israel, 1990, page 362).

It is remarkable that Walker did not notice the resemblance between his view and that of Ingrao and Israel. In this respect, I must draw the attention of the reader to the discussion of multiple solutions of a systems of equations and the existence of only one equilibrium in the functioning model, by Walker. This aspect of uniqueness is also dealt with by Ingrao and Israel. They draw the conclusion that *"uniqueness theorems can only be obtained on assumptions so restrictive as to appear unacceptable"* (Ingrao and Israel, 1990, page 360).

In a certain sense this is the other side of Walker's coin. The main question is whether Walker is right in pointing out that a theory, a model, a system of equations, is incomplete or even wrong from the point of view of explaining reality as long as there is no unique equilibrium, the implication being that in reality there is always at a certain moment of time only one, unique equilibrium that is relevant. In my view, the possibility that in reality multiple equilibria may emerge cannot be excluded. It seems, therefore, inappropriate to downgrade economic analysis, which opens up the possibility of non-unique equilibria.

All this being said does not imply that general equilibrium theory cannot be improved in accordance with the suggestions of Donald Walker. In this respect, it should be pointed out that several issues are already taken up.

Not always within the framework of a full-fledged general equilibrium model, but in separate studies of specific aspects: let me refer to work on financial institutions, information theory and product innovation. To introduce these studies in a general equilibrium model is not always easy, but also in the development of our science patience is often beneficial. Of course, the analysis of general equilibrium over time remains a challenge, just as the introduction of imperfect competition. Of course, perfect competition is the exception and imperfect competition the rule. However, in theory it is not forbidden to gain insights on the basis of the simple case. We cannot do without learning all the time. And let us not forget that important attempts have been carried out by Negishi, Novshek and Sonnenschein, Mas-Colell and Dierker and Grodal to combine monopolistic competition and Cournot equilibrium, with general equilibrium theory (Negishi (1961, 1962), Novshek and Sonnenschein (1978), Mas-Colell (1982), Dierker and Grodal (1986)). Walker is not satisfied with the attempts so far as their " .. constructors have made simplifying assumptions" (Walker, 1997, page 145), but this criticism does not exclude that progress will be made in the future. This observation is also applicable to the analysis of Costa on general equilibrium and the theory of markets, who to a large extent agrees with Walker (Costa, 1998). Let me, therefore, suggest two important topics for the development of general equilibrium theory in the years to come.

4 Further Development of General Equilibrium Theory

Technical change has not been looked at from a welfare theory point of view in a systematic way. The confrontation of the analysis of technical change and welfare economics is lacking in economic theory. Several years ago I devoted my Hennipman lecture to this interesting topic (Heertje, 1995). The main issue is to link the development of new methods of production and new products to the individual welfare of consumers of present and future generations. It implies enlarging the production set with new goods and even virtual goods of which there is not the slightest idea in the research and development area. I am convinced that a general equilibrium approach is needed to study Pareto-optimality over time in view of the development, diffusion and application of new technology. I am even more convinced that this is a very complicated task from the point of view of both economic and mathematical analysis.

The second topic I would like to bring to the fore is the analysis of non-reproducible goods. General equilibrium theory so far is based on the production and consumption of reproducible goods. Ricardo made for the first time the distinction between reproducible and non-reproducible goods (Ricardo, 1817). Economic theory did not take up this distinction. A few years ago I devoted my farewell lecture from the Faculty of Law at the University of Amsterdam to this topic. I tried to illustrate the importance of non-reproducible goods for economic theory and in reality. Goods like old paintings, historical monuments and rare books are non-reproducible. Nature too is in a fundamental sense non-reproducible. To enlarge the consumption set with non-reproducible goods may be considered a first step in order to develop a general equilibrium theory that contains reproducible and non-reproducible goods. Here too I can only repeat how difficult the mathematical description of the generalisation of general equilibrium theory appears to be. This is even more the case if we combine the two suggestions for a further development of general equilibrium theory.

However, one can hardly deny the relevance of a general equilibrium approach in order to study the welfare theoretical aspects of new goods, virtual goods and non-reproducible goods in a dynamic setting. Even if several items of Walker's shopping list for a complete and so-called realistic general equilibrium model are missing, the analysis undoubtedly brings into the open many new questions. And perhaps the progress of advanced economic theory is first of all characterised by new important questions and less so by answers given at a certain moment of time.

References

1. Arrow, K.J. and Debreu, G. (1954), Existence of an equilibrium for a competitive economy, *Econometrica*, Volume 22, pages 256-290.
2. Arrow, K.J. and Hahn, F.H. (1971), *General Competitive Analysis*, Holden-Day, Inc, San Francisco.
3. Costa. M.L. (1998), *General Equilibrium Analysis and the Theory of Market*, Edward Elgar, Cheltenham.
4. Cournot, A.A., (1838), *Recherches sur les principes mathématiques de la théorie des richesses*, Hachette, Paris.
5. Debreu, G. (1983), *Mathematical Economics: Twenty Papers of Gerard Debreu*, Cambridge University Press, Cambridge.
6. Dierker, H. and Grodal, B. (1986), Non-existence of Cournot-Walras equilibrium in a General Equilibrium Model with Two Oligopolists, in *Contributions to Mathematical Economics, In Honor of Gerard Debreu*, edited by W. Hildenbrand and A. Mas-Colell, North Holland, Amsterdam, pp. 167-186.
7. Dougherty, A.F., (1988), (ed.), *Cournot Oligopoly*, Cambridge University Press, Cambridge.
8. Hahn, F. (1982), A stable adjustment process for a competitive economy, *Review of Economic Studies*, Volume 29, pages 62-65.
9. Hahn, F. (1989), (ed), *The Economics of Missing Markets, Information and Games*, Clarendon Press, Oxford.
10. Heertje, A. (1995), Observations on technical change and Paretian economics, *De Economist*, Volume 143, pages 433-456.
11. Hildenbrand, W. and Mas-Colell, A. (1986) (ed), *Contribution to Mathematical Economics, In Honor of Gerard Debreu*, North-Holland, Amsterdam.
12. Ingrao, B. and Israel, G. (1990), *The Invisible Hand*, MIT Press, Cambridge, Mass.

13. Makowski, L. and Ostroy (2001), Perfect Competition and the Creativity of the Market, *Journal of Economic Literature*, Volume XXXIX, number 2, page 479-535.
14. Mas-Colell, A. (1982), The Cournot foundations of Walrasian equilibrium Theory: an exposition of recent theory, in *Advances in Economic Theory*, edited by W. Hildenbrand, Cambridge University Press, Cambridge.
15. Magill, A. and Quinzii, M. (1996), *Theory of Incomplete Markets*, MIT Press, Cambridge, Mass.
16. Nasar, S. (1998), *A Beautiful Mind*, Simon and Schuster, New York.
17. Nash, J. (1951), Non-cooperative games, *Annals of Mathematics*, pages 286-295.
18. Nash, J. (1996), *Essays on Game Theory*, Edward Elgar, Cheltenham.
19. Negishi, T. (1961), Monopolistic Competition and General Equilibrium, *Review of Economic Studies*, Volume 29, pages 196-201.
20. Negishi, T. (1962), The stability of a competitive economy: a survey article, *Econometrica*, Volume 30, pages 635-691.
21. Neumann, J. von (1937), A model of general economic equilibrium, *Review of Economic Studies*, Volume 13, pages 1-9.
22. Novshek, W. and Sonnenschein, H (1978), Cournot and Walras equilibrium, *Journal of Economic Theory*, Volume 19, pages 223-266.
23. Pais, A. (1982), *Subtle is the Lord, the Science and Life of Albert Einstein*, Oxford University Press, Oxford.
24. Pais, A. (1986), *Inward Bound*, Clarendon Press, Oxford.
25. Ricardo, D. (1817), *On the Principles of Political Economy and Taxation*, John Murray, London.
26. Starr, R.M. (1997), *General Equilibrium Theory*, Cambridge University Press, Cambridge.
27. Wald, A. (1936 a), Über die Produktionsgleichungen der ökonomischen Wertlehre, translated by W.J. Baumol as On the Production Equations of Economic Value Theory, in Baumol, W.J. and Goldfeld, S.M. (1968) (ed), *Precursors in Mathematical Economics: An Anthology*, London School of Economics, London.
28. Wald, A. (1936 b), Über einige Gleichungssysteme der mathematischen Ökonomie, *Zeitschrift für Nationalökonomie*, Volume 7, page 637-670, translated by O. Eckstein as On Some Systems of Equations of Mathematical Economics, *Econometrica*, Volume 19, 1951, page 368-403.
29. Walker, D.A. (1996), *Walras's Market Models*, Cambridge University Press, Cambridge.
30. Walker, D.A. (1997), *Advances in General Equilibrium Theory, The Hennipman Lectures*, Edward Elgar, Cheltenham.
31. Walras, L. (1874) *Eléments d'Économie Politique Pure ou Théorie de la Richesse Sociale*, L. Corbaz, Lausanne.
32. Weddepohl, H.N. (1970), *Axiomatic Choice Models*, Universitaire Pers Rotterdam, Rotterdam.
33. Weddepohl, H.N. (1972), Duality and Equilibrium, *Zeitschrift für Nationalökonomie*, Volume 32, pages 160-187.
34. Weddepohl, H.N. (1973), Dual Sets and Dual Correspondences and their Application to Equilibrium Theory, Tilburg Institute of Economics.

35. Weddepohl, H.N. (1974), Equilibrium in a Market with Incomplete Preferences where the number of consumers may be finite, Center for Operations Research and Econometrics, Katholieke Universiteit Leuven.
36. Weddepohl, H.N. (1979), An Equilibrium Model with Fixed Labor Time, *Econometrica*, Volume 47, pages 921-938.
37. Weddepohl, H.N. (1987), Supply-constrained Equilibria in Economies with Indexed Prices, *Journal of Economic Theory*, Volume 43, pages 203-22.
38. Weddepohl, H.N. (1999), Models of the Market, in: *Economics, Welfare Policy and the History of Economic Thought, Essays in Honour of Arnold Heertje*, Edward Elgar, Cheltenham.
39. Weintraub, E.R. (1985), *General Equilibrium Analysis*, Cambridge University Press, Cambridge.

Coordinating Thoughts on Coordination Failures

P. Jean-Jacques Herings*

Department of Economics, Universiteit Maastricht

Abstract. Coordination failures constitute an alternative explanation for underemployment that complements the Keynesian and neo-classical views. The paper proposes to distinguish three classes of models with coordination failures. The classes are formed by strategic models with or without a coordinating role for prices, and general equilibrium models. The main insights resulting for each class of models are exhibited. It is argued that coordination failures are likely to arise in a decentralized economy, even under conditions where perfect competition could prevail. The paper concludes by pointing out several promising directions for future research.

Keywords: coordination failures; unemployment; indeterminacy of equilibrium

1 Introduction

There are two quite distinct views when studying the underemployment of resources. One view is the neo-classical view. Underemployment of resources may occur, but is either a voluntary choice, or a temporary phenomenon that results from unimportant frictions in the process of matching supply and demand. This view is expressed by the well-known quote from Friedman (1968)

> At any moment in time there is some level of employment which has the property that it is consistent with equilibrium in the structure of real wage rates ... The 'natural rate of unemployment', in other words, is the level that would be ground out by the Walrasian system of general equilibrium equations, provided that there is embedded in them the actual structural characteristics of the labor and commodity markets.

The Keynesian tradition on the contrary stresses non-rational expectations and wage and price rigidities in its explanation of underemployment of resources. This lead has been followed in theories on temporary equilibrium, see Grandmont (1987) for a survey, and in the fix-price literature, originating

* I would like to thank Joan Muysken, Erik de Regt and an anonymous referee for helpful comments to an earlier draft of this paper.

from the work of Bénassy (1975), Drèze (1975), and Younès (1975). The fix-price literature allows for a consistent modeling of the consequences of trade at non-Walrasian prices, including the modeling of spill-over effects to other markets caused by price rigidities in one market.

The typical argument made against the fix-price literature is that it fails to explain why prices are rigid, and in particular why prices might deviate from Walrasian values for prolonged periods of time. Within the general equilibrium paradigm, there are at least three possible responses to this question. Blad (1978) stresses that price adjustment processes need not converge to an equilibrium, and that convergence, even if it takes place, can take quite some time. This view is reinforced in Weddepohl (1997), where it is shown that a discrete time tâtonnement process can generate periodic and chaotic paths. Secondly, Herings (1997) and Tuinstra (2000) show that political interference in the market mechanism can be rational from a partisan point of view and might be responsible for sustained deviations from prices that clear markets. Thirdly, Drèze and Gollier (1993) and Drèze (2001) argue that price rigidities are a response to market incompleteness. This argument is particularly valid for the two forms of underemployment most frequently encountered, unemployed labor and excess capacities. These are two clear examples of commodities for which future markets are hardly developed.

Notwithstanding the importance of the models in the fix-price tradition, it is by no means clear that they encompass all that can be said about underemployment of resources. One is lead to wonder whether underemployment of resources is impossible when expectations are rational and prices attain Walrasian values, a conclusion that seems to emerge from the neo-classical literature. Any underemployment resulting under these ideal circumstances is said to be caused by coordination failures.

This paper identifies three classes of models with coordination failures. The first class consists of rather abstract game-theoretic models, following the seminal work of Bryant (1983). This part of the literature makes it very clear that coordination failures are perfectly consistent with rational expectations. They lack, however, a coordinating role for the price mechanism. Strategic models with a coordinating role for prices constitute the second class. It is argued that outcomes in strategic models are often not robust to slightly different specifications of the model. Moreover, strategic models usually only permit a partial analysis, and this only under very specific assumptions with respect to the economic primitives. The third class of models consists of general equilibrium models of coordination failures. This class extends a partial analysis to a global one, and leads to results that are quite general. Section 2 lists the contributions of the game-theoretic literature on coordination failures. The following three sections discuss the three model classes. The final section concludes and highlights some issues that figure prominently on the research agenda.

2 Coordination Games

The analysis of economies where Walrasian prices prevail and where agents make rational decisions has been the centerpiece of economics for a long time. In this setting, the Walrasian allocation is within reach, and, in the absence of externalities, is known to be Pareto optimal. This is the central message of the first welfare theorem. It is clear that any underutilization of a valuable resource, that is a resource with a positive price, is bound to lead to inefficiencies. Underutilization of resources is not even second best in this case. If trade takes place at Walrasian prices, and all agents behave rationally, then moving to the Walrasian allocation leads to a Pareto improvement. This suggests the following as a general definition of a coordination failure: If society fails to reach an outcome that is feasible and that everyone prefers, then the members of society have failed to coordinate in some way.

A mere inspection of the definition of a coordination failure may suggest a coordination failure is far-fetched. If a feasible Pareto improving outcome exists, it seems that rational agents should be able to enforce it in one way or another. A simple game-theoretic insight reveals that this is not the case. Consider the normal form game with two players depicted in Figure 1.

	L	R
T	1, 1	0, 0
B	0, 0	2, 2

Fig. 1. A pure coordination game

This game has two Nash equilibria in pure strategies, (T,L) and (B,R). Moreover, there is one equilibrium in mixed strategies, where the actions T and L are chosen with probability 2/3 by agents 1 and 2, respectively. Despite the extensive literature on refinements of Nash equilibrium, none of the existing equilibrium refinements can exclude any of the three Nash equilibria as being nonsensible. It is clear that the Nash equilibrium (T,L) suffers from severe coordination failures.

It is obvious that a far majority of players will be able to coordinate on (B,R), which has been confirmed in experiments. Moreover, a number of more sophisticated approaches in the literature that attempt to model communication, also reach the conclusion that (B,R) should be played by rational players. Kalai and Samet (1985) show that a multistage unanimity game where non-cooperative players end up at an equilibrium which is symmetric and persistent leads to the play of (B,R). Herings, Mauleon and Vannetelbosch (2001) show that the possibility to make a commitment not to play certain strategies combined with the use of a mediator that produces an arbitrary match within the remaining strategies when possible, leads to the play of (B,R).

A more serious challenge is posed by the stag hunt game depicted in Figure 2 that is taken from Aumann (1990).

	R	D
R	7, 7	8, 0
D	0, 8	9, 9

Fig. 2. The stag hunt game

A player has the choice to hunt rabbit or deer. A player can hunt rabbits by himself, and get a guaranteed payoff of at least 7. To hunt deer, the collaboration of others is needed. The potential payoffs of hunting deer are equal to 9, but are also quite risky. It is only if the other player also chooses to hunt deer that these payoffs are collected. Otherwise the player is left with payoffs equal to 0.

The most frequently applied game-theoretic solution concept, the one of pure strategy Nash equilibrium, specifies two possible solutions. Either both players hunt rabbits, or both players hunt deer. Both players have payoffs equal to 7 in the former equilibrium, versus payoffs of 9 in the latter. None of the existing equilibrium refinements rules out one of the two equilibria. Notice that communication will not be of great help in this game. Both players would like to invoke the other one to choose D, and will announce to do so themselves. But when actually making their decision, they might well choose the safe strategy R. Contrary to the game in Figure 1, experiments reveal a frequent occurrence of the Pareto dominated outcome (R,R).

In fact, it is possible to make a case for the play of (R,R) on at least three different accounts. Harsanyi and Selten (1988) argue that the solution to a game should not change if one adds a constant to the payoff of a player, where the constant depends only on the strategy chosen by the player's opponent. The reason is that addition of such a constant does not change the best response correspondence of the game. Adding the constant -8 to the payoff of a player when the opponent chooses D, shows that the game of Figure 2 is equivalent to the one depicted in Figure 3.

	R	D
R	7, 7	0, 0
D	0, 0	1, 1

Fig. 3. Transformation of the stag hunt game

Any argument that was used to motivate the lower-right cell as the most sensible outcome in the game of Figure 1, can now be used to argue that the Pareto dominated outcome (R,R) results in the game of Figure 3. This equilibrium is called the risk dominant equilibrium by Harsanyi and Selten (1988). For general symmetric 2x2 games with two strict Nash equilibria, the risk dominant Nash equilibrium is the one that consists of the strategies players would choose if they believe each of the actions of the opponents to be played with equal probability. A risk dominant Nash equilibrium need not be Pareto optimal.

A second argument for the play of the risk dominant Nash equilibrium comes from Carlsson and van Damme (1993). They show how a small amount of uncertainty about the coordination game that is actually played, forces players to take a global perspective, in which they can reason themselves to the play of the risk dominant equilibrium.

Finally, Ellison (1993), Kandori, Mailath, and Rob (1993) and Young (1993) take the perspective of repeated play of a coordination game by myopic players. Players choose in each period the action that performed best in the previous period, but do so subject to mistakes. They show that the resulting dynamic system is most frequently in the state where all players choose the risk dominant action.

3 Strategic Models of Coordination Failures

Bryant (1983) provides an economic model with coordination failures. Consider an economy consisting of nN individuals. The individuals are located at $N \geq 2$ separate sites with n individuals at each site. Each site produces an intermediate good, that is transformed into a final output by means of a Leontiev technology. The amounts $I_1, \ldots, I_N$ of intermediate goods are transformed into $N \min\{I_1, \ldots, I_N\}$ units of the final output. The intermediate goods are produced from leisure on a one for one basis. Individuals care about leisure and the consumption of the final output. For simplicity it is assumed that the final output is divided equally among all individuals.

The crux of the model is that each site decides independently from the other sites how much of the intermediary good to produce. This captures the essence of the specialized, multi-staged and decentralized production that characterizes an advanced economy. When a site has to determine the amount of intermediary good to be produced, it is not informed about the amounts produced at the other sites. It is therefore important that there are at least two sites.

Bryant's model admits a straightforward competitive analysis. The constant returns to scale production process fixes the real wage to be equal to one. The first order conditions for utility maximization then require that each household supplies an amount $\overline{I}$ of leisure in return for $\overline{I}$ units of the final output, where $\overline{I}$ is such that the marginal rate of substitution of leisure for consumption equals one.

Bryant's model as formalized above is simply a normal-form game, which is analyzed by considering its Nash equilibria. One possible Nash equilibrium coincides with the competitive solution. Given that all other individuals put in $\overline{I}$ units of leisure, it is optimal for an agent to supply $\overline{I}$ units of leisure himself. Supplying more than $\overline{I}$ units does not lead to more consumption because of the Leontiev technology, but is costly as it provides less leisure. Supplying less than $\overline{I}$ units leads to less consumption and more leisure, but involves a utility loss if the utility function is strictly quasi-concave.

Surprisingly, the reasoning above remains true for any effort level I less than $\overline{I}$. If all other individuals put in $I < \overline{I}$ units of leisure, an agent does not want to supply less than I units himself, since the marginal rate of substitution of leisure for consumption is more than one at effort level $I < \overline{I}$. An agent does not want to supply more than I units, since that does not lead to more consumption because of the Leontiev technology. The strategy tuples where all individuals supply $I \leq \overline{I}$ units of labor characterize the set of symmetric Nash equilibria. There is a continuum of equilibria, ranging from a no-trade equilibrium to a competitive equilibrium. Almost all equilibria display coordination failures. Even though the competitive equilibrium is an outcome, the arguments of Section 2 imply that it is not obvious that agents are able to coordinate on it.

Indeed, the structure of Bryant's game is similar to the one of the stag hunt game. Suppose that there are only two sites, there is only one individual per site and individuals have only two effort levels; either an individual remains inactive or an individual provides the competitive effort level. In the former case, the individual knows for sure that there will be no consumption of the final output. In the latter case, there will only be consumption of the final output if the other individual provides the competitive effort level, otherwise the individual will not consume any final output, while still making all the effort. It might well be that remaining inactive corresponds to the risk dominant Nash equilibrium.

Bryant's model has been generalized in Cooper and John (1988), who consider a special class of symmetric normal form games with I agents. Each agent i has to choose an effort level $e_i \in [0, E]$. Given effort levels e_{-i} chosen by the opponents of player i, a payoff of $\sigma(e_i, e_{-i})$ results for player i. Typical assumptions are that payoffs increase when all effort levels are increased simultaneously, and that the marginal payoff of increasing one's effort level diminishes in effort.

Since the normal form games considered are symmetric, it is useful to introduce the notation $V(e_i, \overline{e})$ for the payoff to agent i when he chooses effort level e_i, whereas all others take action $\overline{e}$. The utility maximizing effort level for agent i, when all other agents choose effort level $\overline{e}$, is denoted by $e_i^*(\overline{e})$.

The model is analyzed by considering its symmetric Nash equilibria. Although there might in general be Nash equilibria with different agents providing different effort levels, the imposition of an anonymity requirement implies a restriction to symmetric Nash equilibria. It is easily verified that effort levels chosen in a symmetric Nash equilibrium belong to the set S given by

$$S = \{e \in [0, E] \mid V_1(e, e) = 0\},$$

where V_1 is the partial derivative of V with respect to its first argument.

An obvious question to ask is under what conditions the effort levels chosen in the model are Pareto optimal. The symmetric Pareto optimal effort levels are found by maximizing $V(e, e)$ over e. The effort levels which satisfy the necessary first order conditions are given by

$$\widetilde{S} = \{e \in [0, E] \mid V_1(e, e) + V_2(e, e) = 0\},$$

where V_2 is the partial derivative of V with respect to its second argument.

The effort levels chosen are inefficient whenever $V_2(e, e) \neq 0$. In plain economic terms, the condition that $V_2(e, e)$ be zero is called the absence of externalities. Both positive and negative externalities lead to suboptimal effort levels, with positive externalities leading to too high effort levels and negative externalities to too low effort levels. Since it seems plausible that higher effort levels by others increase the value of V, it is natural to assume positive externalities. As a consequence, effort levels chosen in a decentralized economy are lower than first best effort levels.

It is here that a first weakness of the model is exhibited. Higher effort levels by other agents do not imply higher values of V in an economic analysis that is carried out in the traditional way. Higher effort levels lead to higher output, which, at a given price per unit, implies higher revenues for the agent that made the effort. Externalities caused by the effort levels chosen by others are absent as soon as trade in commodities is guided by prices. When prices are around to coordinate economic decisions, the underinvestment in effort may well disappear.

Under what conditions may coordination failures arise? To answer that question, it is analyzed first under what conditions multiple equilibria may arise. Since an effort level $\overline{e}$ that corresponds to a symmetric Nash equilibrium satisfies $\overline{e} = e_i^*(\overline{e})$, multiple equilibria can only exist if the best response function e_i^* is upward sloping. The best response function is characterized by solutions to $V_1(e_i, \overline{e}) = 0$. By an application of the implicit function theorem, it follows that the slope of e_i^* at $\overline{e}$ equals $-V_{12}(e_i^*(\overline{e}), \overline{e})/V_{11}(e_i^*(\overline{e}), \overline{e})$.

Under the standard assumption that V is concave, i.e. the marginal benefits from effort are decreasing, we find that $V_{12} > 0$ is a necessary condition for the existence of multiple equilibria. The latter condition is well-known in economics, and is called the case of strategic complementarities in Bulow, Geanakoplos and Klemperer (1985). It states that an increase in the effort level of others increases the marginal benefits of an increase in the individual's effort level.

If there are multiple equilibria, and positive externalities, then multiple equilibria give rise to coordination failures. Evaluating the utility of an agent along the best response function yields

$$\begin{aligned}\frac{dV(e_i^*(\overline{e}), \overline{e})}{d\overline{e}} &= V_1(e_i^*(\overline{e}), \overline{e})\frac{de_i^*(\overline{e})}{d\overline{e}} + V_2(e_i^*(\overline{e}), \overline{e}) \\ &= V_2(e_i^*(\overline{e}), \overline{e}) > 0,\end{aligned}$$

where the second equality follows from the fact that e_i^* is characterized by the condition $V_1(e_i^*(\overline{e}), \overline{e}) = 0$. It follows that the symmetric Nash equilibria are Pareto ranked by the equilibrium action. Moreover, higher action equilibria are preferred to lower action equilibria.

4 Strategic Models of Coordination Failures with a Coordinating Role for Prices

Coordination failures may result when there are strategic complementaries and positive externalities. The latter is a natural assumption in an abstract model as in Section 3, but less so when trade takes place on markets and is guided by prices. It is of crucial importance to give more economic flesh to the skeleton set up in the previous section. To achieve this goal, one needs a closed and complete economic model with a role for prices and wages to coordinate economic activity.

In a model with an explicit role for prices, it is possible to analyze a number of interesting issues. It would be possible to study whether involuntary unemployment may occur, where involuntary unemployment is said to exist if there are workers who are unable to sell as much of their labor as they desire at the given wage, or to study whether unemployment is possible with competitive prices.

Roberts (1987) presents a model that meets our criteria. It is the simplest model of a closed economy that allows for coordination failures in a strategic setting. There are five commodities, X, Y, R, S, and M, four types of agents, A, B, J, and K, and $n \geq 2$ agents of each type. The commodities X and Y are outputs, R and S are inputs, and M is a commodity that serves as a store of value. Agents of type A and B are producers, and agents of type J and K are workers.

Producers of type A and B have endowments μ_A and μ_B of commodity M. This is the only commodity that enters their utility functions. Producer A has the knowledge to transform input R into output X. Producer B knows how to transform input S into output Y. Production takes place by means of a constant returns to scale technology; one unit of input yields one unit of output.

A worker of type J has an endowment (ρ, μ_J) of R, M, and a worker of type K has an endowment (σ, μ_K) of S, M. Worker J derives utility from consumption of Y, R, and M, and worker K from X, S, and M. Notice that no pair of agents has a mutually advantageous trade. Producer A needs inputs from worker J, but sells his output to worker K. Producer B needs inputs from worker K, but sells his output to worker J.

The separation of a firm's customers and workers models consumers that specialize in supplying labor, but generalize in consuming outputs. Not only is this a characteristic of the way most economic activity takes place. It also limits strategic effects. A firm cannot increase the demand for its output by raising its wages, and the supply of inputs to a firm does not directly depend on its price and output levels.

The model consists of three stages. In the first stage each firm states a price for its output and a wage for its input. Next each worker announces the amount of input they want to sell to each firm and of output they want to buy. Finally, each firm decides how much of these offers and orders to accept. Since there are n agents of each type, the first stage results in an n-tuple of prices and wages for output X and input R and an n-tuple of prices and wages for output Y and input S. When announcing the amount of inputs to be sold to each firm, workers are required to take into account the feasibility constraint that they cannot sell more inputs than they initially own. When accepting orders and offers, firms are subject to the technological constraints that the total of accepted output orders cannot exceed the total of accepted input offers.

The model described above specifies explicitly how all prices and quantities result from individual choices. It also leads to consistent modeling of out-of-equilibrium behavior. Any sequence of actions, rational or not, leads to clearly defined and feasible outcomes. The typical way to analyze this model is by determining its subgame perfect Nash equilibria. In a subgame perfect Nash equilibrium, each agent correctly forecasts the strategies the others are using and these strategies are best responses to each other at every decision point, including those that would not be reached if behavior actually is generated by the specified strategies.

The model can also be seen as a special case of a general equilibrium model, and can be analyzed by means of its Walrasian solutions. A Walrasian solution is a price-wage vector and an allocation that satisfies a number of conditions. Prices and wages should be identical for all agents of the same type. Given prices and wages, workers maximize their utility subject to their budget constraint and the condition that the most they can supply of the input is their initial endowment. Producers maximize profits, that is consumption of good M, subject to the available technology. Finally, prices and wages should be such that all markets clear.

Roberts (1987) shows that, given any Walrasian solution, there is a subgame perfect Nash equilibrium such that all firms announce the Walrasian prices and wages, and the resulting allocation is the Walrasian one. To be more precise, the prices and wages specified by all producers in the first stage equal Walrasian prices and wages. Because of the constant returns to scale technologies, this implies that the price of output X is less than or equal to the wage for input R, and the price of output Y is less than or equal to the wage for input S, with equality holding if there is a positive production level of an output. The total input supplied by a worker in the second stage and the total output delivered by a producer in the third stage, following the quote of Walrasian prices in the first stage, should be equal to the Walrasian amounts. This still allows for many possible equilibrium strategies, since the Walrasian solution does not specify who delivers to whom. Any equilibrium strategy that is consistent with the total amounts specified by the Walrasian solution will do.

A subgame perfect Nash equilibrium should also specify the behavior off the equilibrium path. If, given Walrasian prices and wages, consumer choices differ from Walrasian ones, then let firms maximize profits given the actual orders and offers. No firm gains by cutting its price while raising its wage. If any firm deviates from the Walrasian price and wage in a different way, then specify all orders to and offers from it as zero. Workers dealing with the deviating firm go to other firms and have their Walrasian orders and offers filled. Notice that if all workers act in this way, then this kind of behavior is optimal, even when a deviating firm has raised both its price and wage or lowered both. At any price-wage vector at which more than one firm is deviating from the Walrasian levels, specify any subgame equilibrium, for instance one with all orders and offers being zero. It can be verified that the strategies specified above constitute a subgame perfect Nash equilibrium that leads to the Walrasian outcome.

The Walrasian outcome is not the only one that can be sustained as a subgame perfect Nash equilibrium. Consider some Walrasian solution, and any integer k, $0 \leq k \leq n$. There exists an equilibrium in which all firms announce the Walrasian prices and wages, and in the equilibrium k of the consumers of each type get their Walrasian allocations and the $n-k$ remaining consumers of each type receive their initial endowments, buying and selling zero amounts. The arguments needed to show this result are similar to the ones used in the previous paragraph. Note in particular that an inactive firm sees no demand for the output that hiring a worker yields, and will therefore decide not to hire such a worker in the first place. This phenomenon is caused by the strict separation of workers and consumers.

There exist at least n equilibria in the model, most of which are characterized by inefficiently low levels of output and involuntary unemployment, everything at Walrasian prices. Since the framework used is rather specific, it is important to consider how robust these results are to the exact specification of the game.

Jones and Manuelli (1992) consider a slightly different game. Up to now it is assumed that goods are produced on a "made-to-order" basis. A precommitment on the part of the buyers is needed before any production can take place. Jones and Manuelli (1992) consider the case where goods are manufactured for inventory. They assume exactly the same primitives as before (five goods, four types of agents, utility functions, endowments and technology), but a slightly different institutional set-up.

Suppose now that the second stage is disentangled into two stages, separating the announcement of input offers and the placement of output orders. The latter is now supposed to occur at the very end. The complete institutional set-up of the game is therefore as follows. In the first stage firms set wages and prices. In the second stage workers announce the amount of input they want to sell. In the third stage firms decide how much of these offers to accept, and production takes place. In the fourth stage workers announce how much output they want to buy.

Production on a made-to-order basis and production for inventory lead to quite distinct conclusions. In particular, it is not difficult to show that the no-trade equilibrium is no longer an equilibrium in the latter specification, unless the endowments are already Pareto optimal. Suppose, on the contrary, that the no-trade equilibrium can be sustained as a subgame perfect Nash equilibrium. Then, at the beginning of the fourth stage, workers have supplied none of their inputs. They will express a demand for output, to be paid out of their holdings of commodity M. As a consequence, firms facing some demand for sure in the fourth stage are willing to hire labor in the third stage, and workers are willing to offer labor in the second stage. The no-trade solution cannot be sustained as a subgame perfect equilibrium.

Jones and Manuelli (1992) show that for any choice of prices and wages in the first stage, there is a unique pure strategy equilibrium continuation of the game. The vast multiplicity of equilibria obtained in the made-to-order model is therefore ruled out. For separable utility functions, it can also be shown that workers are neither rationed in the labor market, nor in the goods market. The intuition is that a producer would otherwise either lower its wage or increase its price. In particular, there is no involuntary unemployment in this case.

5 General Equilibrium Models of Coordination Failures

Strategic models of coordination failures lead to quite distinct conclusions, dependent on seemingly minor issues in the precise modeling of an economy. Strategic models of coordination failures lack the generality that one would need to be more confident about the importance of their results. It is therefore natural to search for general equilibrium models of coordination failures. Conclusions in a general equilibrium setting tend to be robust to both minor modeling details and the particularities of the economic primitives.

A robust multiplicity of equilibria was shown to occur in the traditional fix-price model in Herings (1996), thereby extending earlier work of van der Laan (1982). Moreover, in Herings (1998) it was shown that the set of equilibria possesses a particular connectedness property. Although the fix-price model seems to explain underutilization of resources as a consequence of trade at "wrong" prices, there is no need to rely on such prices for the multiplicity of equilibria to appear, as was also observed in observed in Citanna, Crès and Villanacci (1997).

The interpretation of the multiplicity of equilibria as coordination failures and the link to the macro-economic literature on coordination failures surveyed in the previous two sections was made by Jacques Drèze, resulting in the papers Drèze (1997) and Herings and Drèze (1998). The exposition in this section is a simplified version of the latter paper.[1]

Consider the classical general equilibrium model with H households and F firms. A firm f is described by its production possibility set Y^f. A household h is characterized by its consumption set X^h, utility function u^h, initial endowment e^h, and shares θ^{fh} in the profits of firm f. Suppose that trade takes place against Walrasian prices p^*, so unconstrained total supply equals unconstrained total demand at prices p^*.

A deviation from the standard framework occurs by assuming that it is *not* common knowledge that the prices are Walrasian. Even when all agents know that prices are Walrasian, it is not necessarily the case that all agents know that all other agents know that prices are Walrasian, and it is even less likely that all agents know that all other agents know that all other agents know that prices are Walrasian, and so on. Common knowledge that prices are Walrasian requires structural knowledge about the economy, very much at odds with the standard general equilibrium paradigm that in a decentralized economy agents only have to maximize utility given the prices that are quoted in the market place.

When it is not common knowledge that prices are Walrasian, it no longer makes sense for agents to express their unconstrained demands and supplies. In particular, they should form expectations on supply and demand possi-

[1] A merger of that paper with Citanna, Crès and Villanacci (1997) resulted in the one of Citanna, Crès, Drèze, Herings and Villanacci (2001).

bilities of all commodities. One particular instance of such expectations are the very optimistic expectations that supply and demand possibilities always exceed supply and demand desirabilities. If all agents hold such optimistic expectations, they will find out that their expectations are justified, and they will consume the Walrasian allocation.

There can be other expectations apart from the very optimistic ones. The intriguing question is whether such expectations can be self-confirming. An underemployment equilibrium of the economy consists of expectations on supply possibilities that are self-confirming.[2] Those expectations are said to be self-confirming if the utility maximizing demand of households given these expectations equals the profit maximizing supply of firms given these expectations, all optimization taking place at Walrasian prices.

The set of underemployment equilibria is reduced further by imposing an additional rationality constraint on the expectations. It makes sense to model explicitly that in case of excess supply, there is some rationing system that specifies how much each agent may supply. The simplest case of such a rationing system is the uniform rationing system, where for each commodity there is a maximal amount to be supplied that is common to all agents. The set of underemployment equilibria is reduced by requiring that the expectations of the agents are compatible with the prevailing rationing system.

It is straightforward that apart from the very optimistic Walrasian expectations, there is also an equilibrium with very pessimistic no-trade expectations. When all agents expect that no supply of any commodity is possible such expectations become self-confirming. Firms maximize profits by remaining inactive. Indeed, they expect no demand for their output, and avoid any cost of production. As a consequence, households have no way to earn income, resulting in the absence of any purchasing power, and the consumption of their initial endowments. Total constrained demand matches total constrained supply and is equal to zero for all commodities.

There are many more underemployment equilibria. The formal result states that there is a connected set of underemployment equilibria that ranges from a no-trade equilibrium to an underemployment equilibrium with very optimistic expectations in at least one market. Examples show that the latter equilibrium may or may not be the Walrasian equilibrium. Typically, the set of underemployment equilibria is a one-dimensional continuum. Intuitively, this can be explained as follows. There are as many instruments, in this case expectations levels, to achieve market clearing as there are commodities. But by virtue of Walras' law there is no need to clear the last market when all but one markets are in equilibrium. One degree of freedom is left over, leading to the one-dimensional continuum. There are not only multiple equilibria, but also coordination failures as the Walrasian equilibrium is typically strictly

[2] Because of the interest to analyze unemployment, the focus is here on expectations on supply possibilities. The extension to expectations on demand possibilities is straightforward.

Pareto superior to the no-trade equilibrium, or any underemployment equilibrium in the neighborhood of a no-trade equilibrium.

A highly stylized example with one production sector and one representative household can be used to illustrate the result mentioned above. The production sector produces output from labor. The representative household consumes output and supplies labor. The production sector and the household have to form expectations about the maximal supply of output and the maximal supply of labor possible. If the production sector expects that the total demand for output is low, they will hire only a limited amount of labor. This has a negative impact on the income of the representative household, and thereby leads to a low demand for outputs, exactly confirming the expectations of the production sector. The result implies that infinitely many expectation levels, varying from very pessimistic to very optimistic, are consistent with equilibrium.

The result makes clear that the observation of excess supply is not sufficient to infer the existence of price and wage distortions. Downwards pressure on prices or wages is not very helpful to get out of this situation. When prices or wages are not at Walrasian values, similar coordination failures exist, and will even be compounded by the traditional consequences of "wrong" prices. Pareto improvements can only be made by a simultaneous recoordination of expectations of all agents, by no means an easy task to accomplish.

6 Conclusion

The paper exhibits coordination failures as an interesting source of unemployment, which complements the traditional neo-classical and Keynesian views. Economic models with coordination failures can be classified in three categories; two types of game-theoretic models and one class of general equilibrium models. In all cases, coordination failures arise as a consequence of self-confirming pessimistic expectations.

An issue not touched upon in Section 5 is the distinction between nominal prices and real prices. Section 5 formulates the model in real terms. The nominal specification is probably more desirable, since in reality commodities are not exchanged against commodities, but against money. A model in nominal terms necessitates an explicit modeling of money demand and supply. Although there seem to be no major obstacles to extend the model by a monetary sector as in Drèze and Polemarchakis (1999), it remains to be verified that its structural properties will not change.

For very simple game-theoretic models, theories of equilibrium selection exist. These theories amount to theories of expectation formation in the setting with coordination failures. The available theories of expectation formation make clear that there is no reason that coordination on the Pareto dominant equilibrium takes place. Considerations related to risk dominance turn out to be of crucial importance. The extension of the ideas of equilibrium selection in game theory to equilibrium selection in general equilibrium theory seems a natural next step to take.

A particular problem for equilibrium selection in general equilibrium theory is that this theory has not made much progress in modeling out of equilibrium behavior. With infinitely many equilibria around to coordinate on, coordination on one equilibrium seems a task too heroic to impose on agents. A model with a full-fledged dynamic specification of expectation formation, in a general equilibrium setting, and including the specification of trading processes out of equilibrium, remains highly desirable.

References

1. Aumann, R.J. (1990), "Nash Equilibria Are Not Self-Enforcing," in J.J. Gabszewicz, J.F. Richard, and L.A. Wolsey (eds.), *Economic Decision-Making: Games, Econometrics, and Optimization,* 201-206.
2. Bénassy, J.-P. (1975), "Neo-Keynesian Disequilibrium Theory in a Monetary Economy," *Review of Economic Studies, 42,* 503-523.
3. Blad, M.C. (1978), "On the Speed of Adjustment in the Classical Tatonnement Process: A Limit Result," *Journal of Economic Theory, 19,* 186-191
4. Bryant, J. (1983), "A Simple Rational Expectations Keynes-Type Model," *Quarterly Journal of Economics, 98,* 525-528.
5. Bulow, J., J. Geanakoplos, and P. Klemperer (1985), "Multimarket Oligopoly: Strategic Substitutes and Complements," *Journal of Political Economy, 93,* 488-511.
6. Carlsson, H., and E. van Damme (1993), "Global Games and Equilibrium Selection," *Econometrica, 61,* 989-1018.
7. Citanna, A., H. Crès, J.H. Drèze, P.J.J. Herings, and A. Villanacci (2001), "Continue of Underemployment Equilibria Reflecting Coordination Failures, Also at Walrasian Prices," forthcoming in *Journal of Mathematical Economics.*
8. Citanna, A., H. Crès, and A. Villanacci (1997), Underemployment of Resources and Self-confirming Beliefs : Nonwalrasian Allocations at Walrasian Prices, Working Paper, pp. 1-31.
9. Cooper, R., and A. John (1988), "Coordinating Coordination Failures in Keynesian Models," *Quarterly Journal of Economics, 103,* 441-463.
10. Drèze, J.H. (1975), "Existence of an Exchange Equilibrium under Price Rigidities," *International Economic Review, 16,* 301-320.
11. Drèze, J.H. (1997), Walras-Keynes Equilibria-Coordination and Macroeconomics, *European Economic Review, 41,* 1735-1762.
12. Drèze, J.H. (2001), "On the Macroeconomics of Uncertainty and Incomplete Markets," *Recherches Economiques de Louvain, 67,* 5-30.
13. Drèze, J.H., and C. Gollier (1993), "Risk Sharing on the Labour Market," *European Economic Review, 37,* 1457-1482.
14. Drèze, J.H., and H.M. Polemarchakis (1999), "Money and Monetary Policy in General Equilibrium," in A.P. Kirman and L.A. Gérard-Varet (eds.), *Economics Beyond the Millennium,* Oxford University Press, Oxford, pp. 55-77.
15. Ellison, G. (1993), "Learning, Local Interaction, and Coordination," *Econometrica, 61,* 1047-1071.

16. Friedman, M. (1968), "The Role of Monetary Policy," *American Economic Review, 58,* 1-17.
17. Grandmont, J.-M. (1977), "Temporary General Equilibrium Theory," *Econometrica, 45,* 535-572.
18. Harsanyi, J.C., and R. Selten (1988), *A General Theory of Equilibrium Selection in Games,* MIT Press, Cambridge.
19. Herings, P.J.J. (1996), "Equilibrium Existence Results for Economies with Price Rigidities," *Economic Theory, 7,* 63-80.
20. Herings, P.J.J. (1997), "Endogenously Determined Price Rigidities," *Economic Theory, 9,* 471-498.
21. Herings, P.J.J. (1998), "On the Existence of a Continuum of Constrained Equilibria," *Journal of Mathematical Economics, 30,* 257-273.
22. Herings, P.J.J., and J.H. Drèze (1998), "Continua of Underemployment Equilibria," CentER Discussion Paper 9805, CentER, Tilburg University, Tilburg, pp. 1-32.
23. Herings, P.J.J., A. Mauleon, and V. Vannetelbosch (2001), "Fuzzy Play, Matching Devices and Coordination Failures," METEOR Research Memorandum 01/00, Universiteit Maastricht, pp. 1-11.
24. Jones, L.E., and R.E. Manuelli (1992), "The Coordination Problem and Equilibrium Theories of Recessions," *American Economic Review, 82,* 451-471.
25. Kalai, E., and D. Samet (1985), "Unanimity Games and Pareto Optimality," *International Journal of Game Theory, 14,* 41-50.
26. Kandori, M., G. Mailath, and R. Rob (1993), "Learning, Mutation and Long Run Equilibria in Games," *Econometrica, 61,* 29-56.
27. Laan, G. van der (1982), "Simplicial Approximation of Unemployment Equilibria," *Journal of Mathematical Economics, 9,* 83-97.
28. Roberts, J. (1987), "An Equilibrium Model with Involuntary Unemployment at Flexible, Competitive Prices and Wages," *American Economic Review, 77,* 856-874.
29. Tuinstra, J. (2000), "The Emergence of Political Business Cycles in a Two-sector General Equilibrium Model," *European Journal of Political Economy, 16,* 509-534.
30. Weddepohl, C. (1997), "Simulating Tatonnement in a Production Economy," *Economic Theory, 10,* 551-558.
31. Younès, Y. (1975), "On the Role of Money in the Process of Exchange and the Existence of a Non-Walrasian Equilibrium," *Review of Economic Studies, 42,* 489-501.
32. Young, H.P. (1993), "The Evolution of Conventions," *Econometrica, 61,* 57-84.

Dynamic Implementation of Competitive Equilibria in Renegotiation Proof Equilibrium

Hans Keiding

University of Copenhagen, Copenhagen

Abstract. In the design of mechanisms for the purpose of allocation in economies over time, there is a natural restriction on the choice of mechanism, which should be dynamic in the sense of respecting the dynamic nature of the problem. With dynamic mechanisms, a version of the celebrated folk theorem can be proved, showing that the class of allocations which are implementable in subgame perfect Nash equilibrium is very large indeed. Therefore, in order to implement competitive allocations in a dynamic context, the solution concept must be stronger. We show that the concept of renegotiation proof equilibrium can be used: Competitive allocations may be dynamically implemented in renegotiation proof equilibrium.

Keywords: Implementation, dynamic mechanisms, renegotiation proof equilibrium.

1 Introduction

The theory of implementation is concerned with rules for the decision making of individuals or groups such that the equilibrium results of these decisions respect certain preassigned properties or coincides with given social choice or allocation rules. The literature on implementation has been rapidly expanding, and the possibilities of implementing particular types of social choice or allocation rules have been studied in great detail. However, if implementation of competitive allocations in ordinary Arrow-Debreu economies seems to be a reasonably well understood phenomenon, problems of another type arise when this phenomenon is studied over time.

From a formal point of view, economies over time differ from the standard economies mainly in the fact that the commodity spaces involved are not finite-dimensional. However, for the interpretation of allocation in the economy as a process taking place over time the mechanism should let the outcome evolve over time, so that the commodity bundles obtained by individuals at time t are determined by the messages sent up to this point of time.

Restricting attention to mechanisms with such a dynamic structure, so that allocation over time is the result of a sequence of short-run outcomes, one runs into a problem related to the celebrated Folk Theorems: Almost every allocation can be obtained in Nash equilibrium, even in subgame perfect Nash equilibrium, a result which seems rather devastating to the prospects of building a meaningful theory of implementation in economies over time. A way of solving this problem can be found by narrowing down the game-theoretical solution concept used in the implementation. In the present case, where we consider mechanisms for implementing competitive allocations, a solution concept almost suggests itself, namely that of renegotiation proof equilibrium.

The paper is structured as follows: In Section 2, we introduce the main concepts of this paper, namely economies over time, dynamic mechanisms, and dynamic implementation. In the following Sections 3 and 4, we then consider the particular version of the Folk Theorem which applies in our setting; in Section 5 we introduce the concept of renegotiation proof equilibrium and show that competitive allocations over time may be implemented in renegotiation proof equilibrium.

2 Economies over Time and Dynamic Mechanisms

In the present section, we introduce the main concepts to be used in the following: We define economies over time as the set of possible environments of the mechanisms considered, and we introduce the notion of a dynamic mechanism.

Economies over time. We shall study exchange economies in discrete time and, at least from the following section and onwards, infinite horizon; for notational convenience, we introduce a general time set T of possible dates of delivery of commodities, which at present may be a finite set or the set $\mathbf{N} = \{1, 2, \dots\}$. At each $t \in T$ there is a finite set $L = \{1, \dots, l\}$ of goods available, and there is a finite set M of consumers (assumed to be active at all the dates in T), with $|M| \geq 3$. For each $i \in M$ and $t \in T$, x_t^i denotes the consumption of consumer i at time t. The set X of consumption bundles of consumer i consists of arrays $x^i = (x_t^i)_{t \in T}$, where $x_t^i \in \mathbf{R}_+^L$ is the consumption of the l commodities at time t. For $y = (y_\tau)_{\tau \in T}$ and $t \in T$, we use notation $y_{\geq t}$ for $(y_\tau)_{\tau \in T, \tau \geq t}$.

Each consumer $i \in M$ has a utility function $u_i : X \to \mathbf{R}$; if $(x_1, \dots, x_t)$ is an initial segment of some consumption bundle, define

$$u_i(\tilde{x}|x_1, \dots, x_t) = u_i(x_1, \dots, x_t, \tilde{x}_1, \tilde{x}_2, \dots),$$

as the utility of $\tilde{x} = (\tilde{x}_1, \tilde{x}_2, \dots) \in X$ if appended to the initial segment. Finally, the consumer i has an initial endowment $\omega^i = (\omega_t^i)_{t \in T}$ belonging to X, that is an endowment (possibly 0) of each commodity at each date. The collection $\mathcal{E} = (u_i, \omega^i)_{i \in M}$ defines an infinite horizon economy.

A consumption bundle x^i in the economy $\mathcal{E}$ gives rise to a net trade $z^i = (z_t^i)_{t \in T} \in (\mathbf{R}^L)^T$ for consumer i, where the time t component of the net trade, $z_t^i \in \mathbf{R}^L$, is defined by $z_t^i = x_t^i - \omega_t^i$, $t \in T$. An allocation in $\mathcal{E}$ is an array $z = (z^1, \dots, z^m)$ assigning to each $i \in M$ a net trade z^i and thereby a vector $x^i = z^i + \omega^i \in (\mathbf{R}^L)^T$. It is said to be feasible if x_t^i belongs to $\mathbf{R}_+^L$ for each t (that is if x^i is a consumption bundle), and balanced if $\sum_{i \in M} z_t^i \in \mathbf{R}_-^L$ for each t. Finally, an allocation $z = (z^1, \dots, z^m)$ is said to be voluntary if each consumer is as least as well off with the net trade prescribed than with the zero net trade, that is if

$$u_i(\omega^i + z^i) \geq u^i(\omega^i).$$

The set of balanced allocations is denoted by Z; it does not depend on the utility functions and the endowments of the consumers but only on the structural parameters L, M, and T. Thus, classes of economies where only utility functions and endowments differ have the same set of balanced allocations. Note that by slight abuse of language, we consider allocations as consisting of individual *net trade* vectors rather than of individual consumption bundles, since this has the advantage of simplifying notation and terminology.

A price system p in $\mathcal{E}$ is an array $p = (p_t)_{t \in T} \in (\mathbf{R}_+^L)^T$ with $p_t \neq 0$, all t; For each consumer i, the net trade z^i is said to satisfy the budget constraint at p, written $z^i \in H(p)$, if either

$$\sum_{\tau=1}^{\bar{t}} p_\tau \cdot z_\tau^i \leq 0$$

in the case where T is a finite set $T = \{1, \dots, \bar{t}\}$, or

$$\operatorname{limsup}_t \sum_{\tau=1}^{t} p_\tau \cdot z_\tau^i \leq 0$$

for $T = \mathbf{N}$.

For later use, we define the family of *linear economies* $\mathbf{E}_L$ consisting of all economies $\mathcal{E}(p)$ such that all consumers have (identical) utility functions $x^i \mapsto \operatorname{limsup}_t \sum_{\tau=1}^{t} \theta_\tau \cdot x_\tau^i$, for $\theta = (\theta_\tau)_{\tau \in T}$ a family of nonnegative linear forms on $\mathbf{R}^L$. Also, for $\mathcal{E} = (u_i, \omega^i)_{i=1}^m$ an economy, $z \in Z$ a balanced allocation, and $t \in T$ a date, we let

$$\mathcal{E}[z,t] = (u_i(\cdot|(z_1^i + \omega^i, \dots, z_{t-1}^i + \omega_{t-1}^i), z_{\geq t}^i + \omega_{\geq t}^i)_{i=1}^m$$

be the economy where the initial endowments before t have been modified by the net trades specified by z, and starting at time t.

With the terminology introduced, the following notion of a competitive equilibrium is quite standard: The pair (z,p) consisting of a feasible and balanced allocation z and a price p is a *competitive equilibrium* if for each i,

(i) z^i satisfies the budget constraint at p,
(ii) if $\tilde{z}^i$ satisfies the budget constraint at p, then $u_i(z^i + \omega^i) \geq u_i(\tilde{z}^i + \omega^i)$.

An allocation z belonging to a competitive equilibrium (z,p) is called competitive.

Game forms, implementation. In the present paper, we shall be concerned with obtaining competitive allocations through the strategic behavior of individuals in suitably defined mechanisms or game forms. The possibility of implementing competitive allocations has been investigated in the literature, cf., e.g., Hurwicz [4], Maskin [8], and even if the results are obtained in the framework of economies with a finite horizon, they often generalize to economies with infinite horizon, cf. e.g. Keiding [6]. As we shall see in the following, the dynamic structure of the economies considered poses some additional problems.

We start with the standard notion of a game form or a *mechanism*, which is an array $G = (\Sigma^1, \dots, \Sigma^m; f; Z)$, where for each $i \in M = \{1, \dots, m\}$, Σ^i is a set of strategies available to individual i, where Z is the set of balanced allocations introduced above, and where $f : \Sigma = \prod_{i \in M} \Sigma^i \to Z$ is an outcome function which to each strategy array $\sigma = (\sigma^1, \dots, \sigma^n) \in \Sigma$ assigns an allocation $f(\sigma) \in Z$. If for each player $i \in M$, there is given a utility function v_i on net trades, then the game form G together with the array $v = (v_1, \dots, v_m)$ of utility assignments defines a *game* (in strategic form) $\Gamma = (G, v)$. A strategy array σ is a Nash equilibrium in Γ if

$$v_i(\sigma^i, \sigma^{-i}) \geq v_i(\tau^i, \sigma^{-i}), \text{ all } \tau^i \in \Sigma^i,$$

and the outcome $f(\sigma)$ is called a Nash equilibrium outcome.

The game form G is said to *implement* the competitive equilibria in a given class $\mathbf{E}$ of economies if for each $\mathcal{E} \in \mathbf{E}$, the set of competitive allocations coincides with the set of Nash equilibrium outcomes of the game $G(\mathcal{E}) = (\Sigma^1, \dots, \Sigma^m, u_1(\omega^1 + f_1(\cdot)), \dots, u_m(\omega^m + f_m(\cdot)))$, that is the game in which the payoff for individual i resulting from the strategy array σ is the utility level obtained by the consumption bundle corresponding to the net trade $f_i(\sigma)$, all $i \in M$.

Regular mechanisms. Before proceeding to dynamic mechanisms, it will be useful to consider a particular class of game forms, which is a modification of the class of strategic outcome functions introduced by Schmeidler [13]: The mechanism $G = (\Sigma^1, \dots, \Sigma^n; f, Z)$ is said to be *regular* if

(a) there is a strategy array $\hat{\sigma}$ such that $f^i(\sigma^i, \hat{\sigma}^{-i}) = 0$ for each individual i and each σ^i (a no-action strategy array, where no individual can achieve a nonzero net trade),

(b) for each price p and allocation z with $z^i \in H(p)$, $i = 1, \dots, m$, there is a strategy array $\sigma \in \Sigma$ such that
  (i) $f(\sigma) = z$,
  (ii) $f(\Sigma^i, \sigma^{-i}) \subset H(p)$ for each individual i (at the strategy array σ, no individual can change the outcome to something outsice $H(p)$).

The strategy σ is called a *commitment* strategy (w.r.t. p and z) in the following.

An example of a regular mechanism is the following (adapted from Schmeidler [13] to the context of infinite horizon economies): Consumers choose triples $(\tilde{p}^i, \tilde{z}^i, \nu^i)$, where p^i is a price, $\tilde{z}^i$ a net trade satisfying $\tilde{z}^i \in H(\tilde{p}^i)$, and $\nu^i \in \{0,1\}$; given these choices, the consumers are grouped into coalitions G consisting of all the singleton coalitions with consumers choosing $\nu^i = 0$, and all coalitions $G(p) = \{i \mid \tilde{p}^i = p, \nu^i = 1\}$; the resulting net trade for consumer $i \in G$ is

$$z^i = \tilde{z}^i - \frac{\sum_{j \in G} \tilde{z}_j}{|G|},$$

where $|G(p)|$ is the cardinality of $G(p)$. It is easily seen that this mechanism satisfies (a) (the net trade component of the strategies can be chosen as 0 and the third component is 0) and (b) (the strategy choices $(p, z^i, 1)$ will do as commitment strategy).

It is easily seen that competitive allocations in $\mathcal{E}$ may be obtained as Nash equilibrium outcomes of a regular mechanism. Actually only part (b) of the definition is used; we shall need part (a) at a later point of our analysis.

Lemma 1. *Let (z,p) be a competitive equilibrium in $\mathcal{E}$, and let $G = (\Sigma^1, \dots, \Sigma^n; f; Z)$ be a regular mechanism. Then there is a Nash equilibrium σ in $G(\mathcal{E})$ such that $z = f(\sigma)$.*

Proof. By regularity, there is a commitment strategy σ such that (i) and (ii) of part (b) above are satisfied w.r.t. the pair (z,p). Thus $f(\sigma) = z$ by (i), and for any $i \in M$ and $\tau^i \in \Sigma^i$, the outcome $f(\tau_i, \sigma^{-i})$ belongs to $H(p)$ by (ii). Since (z,p) is a competitive equilibrium, we have that

$$u_i(\omega^i + f_i(\tau^i, \sigma^{-i})) \leq u_i(\omega^i + z^i) = u_i(\omega^i + f_i(\sigma)),$$

which shows that σ is indeed a Nash equilibrium in $G(\mathcal{E})$. □

Dynamic mechanisms. While implementation of competitive equilibria in the general context has been treated at length in the literature, less attention has been given to problems of implementation where the strategic choices of individuals are adapted to the dynamic structure of the problem. In this situation, the choices of each individual at time t may depend on information about the choices of the others at any time $\tau < t$. This is formalized by the notion to be introduced now. For a more detailed treatment of game forms with a dynamic structure, the reader is referred to e.g. Rosenmüller [9].

A *dynamic mechanism* (for the family $\mathbf{E}$) is an array $\Gamma = (Q, S^1, \dots, S^m; \delta, \pi, q_0)$, where Q is a set of *states,* and for each i, S^i is a set of actions (assumed to be the same at every date t). The map $\delta : Q \times S^1 \times \cdots \times S^m \to Q$ selects a new state of the dynamic mechanism, and the outcome map $\pi : Q \times S^1 \times \cdots \times S^m \to (\mathbf{R}^L)^M$ assigns a time t net trade

$$z_t = (z_t^1, \dots, z_t^m) = \pi(q, s^1, \dots, s^m)$$

in $(\mathbf{R}^L)^M$ to any array $(s^1, \dots, s^m)$ of moves chosen at time t with the mechanism in state q. Finally, there is a given initial state q_0 of the dynamic mechanism.

The dynamic mechanism gives rise to a game form $G = (\Sigma^1, \dots, \Sigma^m; \Pi; Z)$ in the following way: For each i, the strategy space Σ^i consists of all sequences

$$\sigma^i = (\sigma^i_t)_{t \in T}$$

where $\sigma^i_1 \in S^i$ is an action in the first period, whereas $\sigma^i_t : Q \times H_{t-1} \to S^i$, $H_{t-1} = (S^1 \times \dots \times S^m)^{t-1}$, for $t > 1$ is a map assigning to any state q at time $t-1$ and any history of previous moves h_{t-1} a move $\sigma^i_t(q, h_{t-1}) \in S^i$. If the players have chosen strategies $\sigma^1, \dots, \sigma^m$ with $\sigma^i = (\sigma^i_t)_{t \in T}$, the outcome $\Pi(\sigma^1, \dots, \sigma^m; q_0)$ is the sequence of net trades $(z_t)_{t \in T}$ determined by

$$z_1 = \pi(q_0, \sigma^1_1, \dots, \sigma^m_1),\ q_1 = \delta(q_0, \sigma^1_1, \dots, \sigma^m_1), h_1 = (\sigma^1_1, \dots, \sigma^m_1)$$

in the first period, and given that z_{t-1}, q_{t-1} and h_{t-1} have been defined, we get

$$\begin{aligned} z_t &= \pi(q_{t-1}, \sigma^1_t(q_{t-1}, h_{t-1}), \dots, \sigma^m_t(q_{t-1}, h_{t-1})), \\ q_t &= \delta(q_{t-1}, \sigma^1_t(q_{t-1}, h_{t-1}), \dots, \sigma^m_t(q_{t-1}, h_{t-1})), \\ h_t &= (h_{t-1}, \sigma^1_t(q_{t-1}, h_{t-1}), \dots, \sigma^m_t(q_{t-1}, h_{t-1})). \end{aligned}$$

The sequence $(q_t)_{t \in T}$ is said to be induced by the strategy choices $(\sigma^1, \dots, \sigma^m)$. We notice that the outcome functions and thus the game form $G(q_0)$ depends on the choice of initial state q_0.

For $\mathcal{E}$ an exchange economy, an outcome $z = (z_t)_{t \in T}$ of the dynamic mechanism $G(q_0)$ defines an allocation z in $\mathcal{E}$; this allocation is balanced but not necessarily feasible, meaning that strictly speaking payoffs is defined only for some strategy arrays. Assigning payoffs of $-\infty$ to outcomes for which $z^i + \omega^i \notin (\mathbf{R}^L_+)^T$ we may still consider the array $(G(q_0), v_1, \dots, v_m)$ as a game where it makes sense to discuss Nash equilibria and Nash implementation of Walras equilibria by dynamic mechanisms. This is the theme of the following section.

3 Nash Implementation and Folk Theorems

As it is seen from the definition, the problem of Nash implementation of competitive equilibria consists of two parts, namely (1) finding for each equilibrium allocation an array of strategies which is Nash and yields exactly this allocation, and (2) asserting that only competitive equilibrium allocations are Nash outcomes of the mechanism. While part (1) is easily taken care of, an attempt to prove (2) runs into difficulties of the type known from the theory of Folk Theorems for repeated games. Indeed, the result stated below can be seen as an extension of the this theory (cf. Sorin [12]) from the context of repeated games to the present one.

From now on, we assume infinite horizon, that is $T = \mathbf{N}$. Let Γ be a dynamic mechanism, and let $\mathcal{E}$ be an economy. We need to introduce the notion of a minmax utility for individual i as the worst outcome that the other players inflict on i. Formally, let $q \in Q$ be any state of the mechanism and let $u : X \to \mathbf{R}$ be a utility function. We then define the minmax at state q for a individual i with utility u and given actions $s = (s^1, \ldots, s^m)$ as

$$\operatorname{minmax}(i, q, u_i, s) = \inf_{(\sigma^j)_{j \neq i} : \sigma^j_1 = s^j} \sup_{\sigma^i} u_i(\Pi_i(\sigma^i, (\sigma^j)_{j \neq i}; q)).$$

The reason why the minmax has to be defined relative to a state and a utility function is that we work not with repeated games but with games depending on a state variable, so that the parameters of the game and the assessment of the future change over time. Also the general nature of our utility functions makes this somewhat complicated definition necessary.

Theorem 1. *Let $\mathcal{E} \in \mathbf{E}$, and assume that z is a feasible allocation which is obtainable in the dynamic mechanism Γ from initial state q_0 by a strategy choice $\sigma = (\sigma^1, \ldots, \sigma^m)$, let $x = \omega + z$ and let $(q_t)_{t \in T}$ be the sequence of states induced by σ. If for each $t \in T$,*

$$u_i((x^i_\tau)_{\tau > t} | x^i_1, \ldots, x^i_t) > \operatorname{minmax}(i, q, u_i(\cdot | x^i_1, \ldots, \tilde{x}^i_t)),$$

where $\tilde{x}^i_t = \pi(q_{t-1}, s^i, (\sigma^j_t)_{j \neq i}) + \omega^i_t$, some $s^i \in S^i$, then z is a Nash equilibrium of the mechanism Γ for $\mathcal{E}$.

We shall restrict ourselves to an outline of the proof of Theorem 1, which follows the lines of the standard proofs of the Folk Theorem for Nash equilibrium; the result will not be used directly in the sequel, rather it shows the need for a more selective equilibrium concept.

Proof of Theorem 1 (outline): Let $(\bar{s}^1_t, \ldots, \bar{s}^m_t)_{t \in T}$ be the sequence of moves associated with a play of the strategies σ from q_0, and similarly, and define strategies $\bar{\sigma}^i$ for each i such that at time t, individual i chooses $\bar{s}^i_t$ if everyone has chosen in accordance with $(\bar{\sigma}^1_t, \ldots, \bar{\sigma}^m_t)_{t \in T}$ for $\tau \leq t - 1$. If some player i has deviated from this choice, say at $\tau = t$, let all j choose actions in accordance with a strategy which yields a payoff u'_i for i with

$$u_i((x^i_\tau)_{\tau > t} | x^i_1, \ldots, x^i_t) > u'_i \geq \operatorname{minmax}(i, q, u_i(\cdot | x^i_1, \ldots, \tilde{x}^i_t));$$

then the final outcome for i will be inferior to that of no deviation. Consequently, the strategies $\bar{\sigma}^i$, $i = 1, \ldots, m$, define a Nash equilibrium. □

The proof of Theorem 1 hinges on the fact that all the allocations, which are obtained in such a way that deviators may be subsequently punished by the remaining individuals, are Nash equilibrium outcomes. This argumentation is of course subject to the objection that it neglects the question of individual rationality of the punishments: It is not clear that the punishments would actually be carried them out, in particular if punishers are themselves hurt by doing so. To follow up on these objections, we shall have to restrict our solution concept, in a first step to subgame perfect Nash equilibrium, and this is indeed what we shall do in the following section.

Before doing so, however, we follow up on the special case of regular mechanisms introduced in the previuous section, since these mechanisms will turn out to be useful as we proceed. First of all, we say that a dynamic mechanism is *adapted* to the class $\mathbf{E}$ of economies if the state space Q equals Z, the set of all balanced allocations common to all economies in $\mathbf{E}$. A dynamic mechanism adapted to $\mathbf{E}$ is said to be *regular* if it is regular at each state q, in the sense that there is a regular (static) mechanism $(\Sigma^1, \dots, \Sigma^n; f, Z)$ with $\pi(z, \sigma) = \mathrm{pr}_1(z + f(\sigma))$ and $\delta(z, \sigma) = \mathrm{pr}_{t>1}(z + f(\sigma))$, where $\mathrm{pr}_1(\cdot)$ is the projection $(\hat{z}_t^1, \dots, \hat{z}_t^m)_{t=1}^{\infty} \mapsto (z_1^1, \dots, z_1^m)$, and $\mathrm{pr}_{t>1}(\cdot)$ is given by $\hat{z} = (\mathrm{pr}_1(\hat{z}), \mathrm{pr}_{t>1}(\hat{z})))$.

Thus, in a regular dynamic mechanism, at date t and in state (= allocation) z there is a (state-dependent) regular mechanism the outcome of which, again an allocation, determines both current outcome (namely the part pertaining to the first period) and future state (the remaining part). Intuitively, what happens is that current outcome is realized according to previous commitments as recorded in the current state of the mechanism, whereas future commitments incurred in the current period may be stored as a change in the state.

We can now extend the result of Lemma 1 to the present context of dynamic mechanisms.

Lemma 2. *Let $\Gamma = (Q, S^1, \dots, S^n, \delta, \pi, q_0)$ be a regular dynamic mechanism adapted to a class $\mathbf{E}$ of economies, let $\mathcal{E} \in \mathbf{E}$, and let (z,p) be a Walras equilibrium in $\mathcal{E}$. Then there is a Nash equilibrium σ such that $z = \Pi(\sigma, q_0)$.*

Proof: Define equilibrium strategies σ in the regular dynamic mechanism as follows: The actions of σ at $t = 1$ are defined as a commitment strategy σ_1 in the static mechanism at time 1 which realizes the competitive allocation (z,p) in Nash equilibrium (such strategies exist by Lemma 1), and for $t > 1$, as a no-action strategy array $\hat{\sigma}$ in the static mechanism satisfying properties (b.i) and (b.ii) in the definition of regularity with respect to the pair $(0,p)$ (corresponding to the no-trade equilibrium at z with price p). For the static mechanism corresponding to any other (z', p') and $t > 1$, the strategies may be defined arbitrarily.

It is easily checked that playing these strategies will give the allocation z. We check that the strategies constitute a Nash equilibrium: Suppose that individual i changes moves in some or all periods, while the remaining players keep their choices. Since the other individuals have chosen their no-action strategy for $t > 1$, we have that the change of outcome can occur only at $t = 1$. But then it follows easily from Lemma 1 that individual i can obtain only net trades in $H(p)$, and since (z,p) is a competitive equilibrium, this cannot give i a better result than z. We conclude that (z, p) is indeed obtained as a Nash equilibrium outcome. □

It may be noticed that the particular strategies used in the proof of Lemma 2 consisted of a commitment strategy at $t = 1$ followed by no-action strategies. We shall make repeated use of this type of strategies in the sequel.

4 Implementation in Subgame Perfect Nash Equilibrium

In order to exclude strategies based on threats of punishment which will not be carried out by the individuals called on to do so, we consider strategies which are not only Nash equilibrium but also subgame perfect (cf. Selten [11]). In our context, a subgame perfect Nash equilibrium of Γ with respect to the economy $\mathcal{E}$ is a strategy array $\sigma = (\sigma^1, \dots, \sigma^m)$ such that for any $t \in T$ and $q \in Q$, if $(q_\tau, s_\tau^1, \dots, s_\tau^m)_{\tau<t}$ is history of moves in Γ with

$$z_{\tau+1} = \pi(q_\tau, s_\tau^1, \dots, s_\tau^m)$$

$$q_{\tau+1} = \delta(q_\tau, s_\tau^1, \dots, s_\tau^m)$$

then the restriction of σ to all $\tau \geq t$ is a Nash equilibrium in $\Gamma(q_t)$ for the economy $\mathcal{E}[(x^\tau)_{\tau<t}])$, where $x_\tau = z_\tau + \omega_\tau$, $\tau < t$. are given by the history.

The result of Theorem 1 extends to subgame perfect equilibria under suitable (and more restrictive) assumptions. for subgame perfect Nash equilibria; the only change is that the inequality in the minmax condition has to be proper; the proof again follows the standard folk theorem approach.

Theorem 2. *Let $\mathcal{E} \in \mathbf{E}$, let z be a feasible allocation which is obtainable in $G(q_0)$ from initial state q_0 by a strategy choice $\sigma = (\sigma^1, \dots, \sigma^m)$, $x = z + \omega$, and $(q_t)_{t\in T}$ be the sequence of states induced by σ. Assume that there is $(v_t^1, \dots, v_t^m) \in \mathbf{R}^M$ for each $t \in T$ such that*

$$u_i((x_\tau^i)_{\tau>t}|x_1^i, \dots, x_t^i) > v_i^t$$

and for each i, strategies $\sigma_{(i)}^j \in G(q_t)$ for $j \neq i$ such that

$$u_i(\Pi^i(\sigma_{(i)}^j, \tau^i; q_t) + (\omega_\tau^i)_{\tau>t}|x_1^i, \dots, x_t^i) \leq v_t^i$$
$$u_j(\Pi^j(\sigma_{(i)}^j, \tau^i; q_t) + (\omega_\tau^j)_{\tau>t}|x_1^i, \dots, x_t^i) > v_t^j$$

for all τ^i and $j \neq i$. Then z is a subgame perfect Nash equilibrium of the mechanism Γ for $\mathcal{E}$.

As was the case with Theorem 1 above, we restrict ourselves to an outline of the proof of Theorem 2, which anyway is not directly used in our main results to follow. The assumptions are such that the proof follows that of the Folk Theorem for subgame perfect Nash equilibrium (cf. Sorin [12]).

Proof of Theorem 2 (outline): As in the proof of Theorem 1, define a sequence of actions for each player yielding the allocation z as outcome, and define strategies at each t as this action if all players have chosen it previously. If player i is the first player to deviate at time t, the strategies prescribe that the remaining players $j \neq i$ switch at $t+1$ to the strategies $\sigma^j_{(i)}$ in the game $G(q_t)$. This will reduce the future outcome of individual i to at most v^i_t, and if in addition the equilibrium strategy prescribes that failure to carry out this punishment will in its turn be punished, then the individuals $j \neq i$ choose the prefer to adhere to the equilibrium strategies. □

Since a version of the Folk Theorem still holds even for subgame perfect Nash equilibria, the perspective of dynamically implementing competitive equilibria in subgame perfect Nash equilibria does not look too promising; too many equilibrium outcomes will not be Walras equilibria. Therefore, in order to reduce the set of possible outcomes, we must refine the equilibrium concept still further, and this is indeed what will be done in the following section.

As usual we close this section with a further consideration of the special class of regular (dynamic) mechanisms: We show that the result of Lemma 2 can be extended to subgame perfect Nash equilibria, a result which will be used in the following section.

Lemma 3. *Let $\Gamma = (Q, S^1, \dots, S^n, \delta, \pi, q_0)$ be a regular dynamic mechanism adapted to $\mathbf{E}$, let $\mathcal{E} \in \mathbf{E}$ be an economy, and let (z,p) be a competitive equilibrium in $\mathcal{E}$. Then there is a subgame perfect Nash equilibrium σ such that $z = \Pi(\sigma, z_0)$.*

Proof: Let $\tilde{z}$ be any state of the mechanism (since the mechanism is adapted to $\mathbf{E}$ containing $\mathcal{E}$, its states are the balanced allocations); we define strategies for each player by specifying actions depending on the state z of the game at time t (and thus not on the history of previous actions). For $t = 1$ and state $\tilde{z} = q_0$, let σ_1 be a strategy of the static mechanism at q_0 which satisfies (b.i) and (b.ii) of the definition of a regular mechanism w.r.t. (z,p). For $t > 0$ and $\tilde{z}$ an arbitrary state, the action at t is determined as follows:

If $(\tilde{z}, p')$ is a competitive equilibrium in the economy $\mathcal{E}[\tilde{z}, t]$, then the individuals choose the no-action strategies $(\hat{\sigma}^1_t, \dots, \hat{\sigma}^m_t)$ in the mechanism at state $\tilde{z}$ and time t. If $\tilde{z}$ is not a competitive allocation in $\mathcal{E}[\tilde{z}, t]$, then the individuals choose actions $(\sigma^1_t(\tilde{z}), \dots, \sigma^m_t(\tilde{z}))$ in the mechanism at $\tilde{z}$ which gives an allocation z' belonging to a competitive equilibrium (z', p') in $\mathcal{E}[\tilde{z}, t]$ and which satisfies (b.i) and (b.ii) w.r.t. (z', p').

Since the strategies defined above give rise to the same play as those used in the proof of Lemma 2 (they are identical on the Nash equilibrium path), we have that $\Pi(\sigma) = z$. Using Lemma 2 we get that the strategies defined above are Nash equilibrium strategies of the mechanism Γ with initial state $\tilde{z}$ for the economy $\mathcal{E}[\tilde{z}, t]$, for each $t \in T$ and each state (= net trade allocation) z. But this means that the strategies define a subgame perfect Nash equilibrium. □

As in the proof of Lemma 2, we have made use of strategies for which the actions have the form of either a commitment strategy or an array of no-action strategies of the relevant static mechanisms, no action being chosen whenever the state of the mechanism defined a no-trade equilibrium of the economy considered. We shall use the term *normal* for such strategies.

With this terminology, Lemma 3 states that competitive allocations can be obtained as subgame perfect Nash equilibrium in normal strategies. The converse is not true, but rather little remains, as will be seen in the next section.

5 Implementation in Renegotiation Proof Equilibrium

We have seen that competitive allocations are implementable in subgame perfect Nash equilibrium. However, many other allocations may also have this property, depending on the economy and the mechanism. Thus, the implementation property does not characterize the competitive allocations, at least as long as we consider implementation in subgame perfect Nash equilibrium.

This picture may change drastically if we consider other solution concepts. Among those which may be eligible for a closer consideration, it seems natural to choose the so-called renegotiation proof equilibria, (cf. e.g. Farrell and Maskin [3]) which are equilibria with the additional property that the subgame perfect equilibrium induced in any subgames must not be Pareto dominated (in this subgame) by another equilibrium. This property of renegotiation proofness introduces an aspect of collective rationality into our considerations of non-cooperative implementation, and as such it may seem out of place. However, a similar combination of non-cooperative implementation and collective rationality lies behind the fundamental result by Hurwicz [4] on static implementation, so that it may be considered as a reasonable additional property of a solution concept used in implementation theory.

Let $(\sigma^1, \dots, \sigma^m)$ be a subgame perfect Nash equilibrium in the mechanism $\Gamma = (Q, S^1, \dots, S^m; \delta, \pi, q_0)$ for the economy $\mathcal{E}$. For each subgame $\Gamma[t, q, h_{t-1}]$, defined by a date $t \in \mathbf{N}_0$, a state q and a history of previous actions h_{t-1}, the restriction of the strategies to this subgame defines a Nash (actually, a subgame perfect Nash) equilibrium of $\Gamma[t, q, h_{t-1}]$. If there is another subgame perfect Nash equilibria of $\Gamma[t, q, h_{t-1}]$ such that its outcome (as allocation in $\mathcal{E}[\Pi(\sigma, q_0), t]$) is better for all i than the restriction of $\Pi(\sigma, q_0))$ to all $\tau \geq t$, then σ is said to be Pareto dominated in $\Gamma[t, q, h_{t-1}]$ (for $\mathcal{E}$) A *renegotiation proof equilibrium* in $\Gamma = (Q, S^1, \dots, S^m; \delta, \pi, q_0)$ is a strategy array σ which is not Pareto dominated in any subgame.

There is a small detail still to be added to our model: Since in models with infinite horizon, competitive allocations need not be Pareto optimal (cf. e.g. Malinvaud [7], Samuelson [10]), we need to single out in the following those economies as *well-behaved* where competitive allocations *are* Pareto optimal. There are general conditions on preferences (specifying that the level surfaces of utility functions restricted to consumption at any two consecutive dates should be sufficiently curved) which guarantees well-behavedness (see e.g. Balasko and Shell[1], Borglin and Keiding [2]).

We are now ready to state our main result on dynamic implementation of competitive allocations, giving a characterization of competitive allocations in a dynamic model in terms of implementation properties, Pareto optimality, and individual rationality.

Theorem 3. *Let* $\mathbf{E}$ *be a class of economies containing* $\mathbf{E}_L$*, and let* $\Gamma = (Q, S^1, \ldots, S^n, \delta, \pi, q_0)$ *be a regular dynamic mechanism. Assume that all renegotiation proof equilibria are voluntary. Then* Γ *implements the competitive allocations for well-behaved economies* $\mathcal{E} \in \mathbf{E}' \subset \mathbf{E}$ *in normal renegotiation proof Nash equilibrium strategies.*

Proof. Let $\mathcal{E} \in \mathbf{E}'$ be a well-behaved economy, and let (z, p) be a competitive equilibrium in $\mathcal{E}'$. Let σ be the (normal) strategies defined in the proof of Lemma 3; then for each subgame defined by time t, state $\tilde{z}$ and history h_{t-1}, the actions defined by σ result in a competitive allocation in $\mathcal{E}[\tilde{z}, t] \in \mathbf{E}'$, and by well-behavedness this allocation is Pareto optimal; consequently no other subgame perfect equililbrium in this subgame can yield a better allocation for all the players. We conclude that σ is renegotiation proof.

For the converse, let z be the allocation associated with an arbitrary renegotiation proof equilibrium σ of Γ for $\mathcal{E} \in \mathbf{E}'$. We claim that z is a Pareto optimal allocation in the economy $\mathcal{E}$. Suppose not; choose an allocation $\bar{z}$ in $\mathcal{E}$ which Pareto dominates z. Now any competitive allocation in the economy obtained from $\mathcal{E}$ by replacing the original endowments ω^i with $\bar{z}^i + \omega^i$, all $i \in M$, is Pareto better than z. By Lemma 3, this competitive allocation can be obtained as subgame perfect Nash equilibria Γ, contradicting renegotiation proofness.

Since z is Pareto optimal, it is price supported; let $p = (p_t)_{t \in T}$ be a supporting price. Consider now the linear economy $\mathcal{E}(p)$ where all consumers have the linear preferences given by p. We claim that the equilibrium strategies σ are regotiation proof also in $\mathcal{E}(p)$. Indeed, for each subgame $\Gamma[t, q, h_{t-1}]$, σ prescribes a commitment strategy followed by no-action, which is subgame perfect also in $\mathcal{E}(p)$. The no-domination part is trivial, so σ is indeed renegotiation proof also in $\mathcal{E}(p)$.

Now we use the assumption that negotiation proof equilibria are voluntary; it follows that ω^i is not better than $\omega^i + z^i$ for each individual i in the linear economy $\mathcal{E}(p)$, from which it follows that $z^i \in H(p)$ for each i. But then (z, p) is a competitive equilibrium in $\mathcal{E}(p)$, and a fortiori in $\mathcal{E}$. □

The result of Theorem 3 shows that competitive allocations may indeed be implemented by suitable mechanisms when we add the equilibrium condition of negotiation proofness. Our use of adapted mechanisms, however, shows that the mechanism itself may have to be rather complex; at each stage not only a current outcome is determined but also a prospect for the whole future of the economy is set up, which is either kept by subsequent no-action strategies or modified. Thus, while the mechanism is dynamic it is nevertheless concerned with the whole allocation over time at each of its stages. This seems unavoidable and to some extent fits with intuition: To determine allocation over time future must enter, at least as conjectures or – as here – commitments.

References

1. Balasko,Y. and K.Shell (1980), The overlapping-generations model, I: The case of pure exchange without money, Journal of Economic Theory **23**, 281 - 306.
2. Borglin,A. and H.Keiding (1986), Optimality in infinite horizon economies, Springer-Verlag, Berlin.
3. Farrell, J., and E. Maskin, Renegotiation in repeated games, Games and Economic Behavior **1**, 327–360, 1991.
4. Hurwicz, L. (1979), On allocations attainable through Nash equilibria, in J.-J. Laffont (ed.), Aggregation and Revelation of Preferences, North Holland.
5. Keiding, H. (1977), A note on the game theoretic interpretation of Walras equilibria, Journal of Economic Theory, 518-520.
6. Keiding, H. (1995), Infinite horizon economies, a survey. in: J.Guddat (ed.) Approximation and Optimization in the Carribean 1993, Proceedings, European Science Publishers, 468 - 493.
7. Malinvaud, E. (1953), Capital accumulation and efficient allocation of resources, Econometrica **21**, 233 - 268.
8. Maskin, E.S. (1977), Nash equilibrium and welfare optimality, mimeo, M.I.T.
9. Rosenmüller, J. (2000) Game theory: Stochastics, information, strategies and cooperation, Kluwer Academic Publishers, Boston.
10. Samuelson, P.A. (1958), An exact consumption-loan model of interest with or without the social contrivance of money, Journal of Political Economy **66**, 467 - 482.
11. Selten, R. (1965), Spieltheoretische Behandlung eines Oligopolmodelles mit Nachfrageträgheit, Zeitschrift für die gesamte Staatswissenschaft **12**, 301 - 324.
12. Sorin, S. (1992), Repeated Games With Complete Information, ch.4 in: R.J.Aumann and S.Hart (eds.), Handbook of Game Theory, Elsevier Science Publishers.
13. Schmeidler, D. (1980), Walrasian Analysis via Strategic Outcome Functions, Econometrica **18**, 281–290.

On the Economics of Managing and Returns to Communications Technology

Michael Keren and David Levhari

Department of Economics, Hebrew University of Jerusalem

Abstract. The last couple of decades have witnessed revolutionary changes in information and communication technology, which have surely left their mark on the structure of firms and other organizations. In this note we are trying to examine how a greater efficiency in the conveyance of information between superiors and subordinates in a hierarchic organization would influence the span of control of each level and the total number of levels in the hierarchy. We do this by modifying the communication time function in a model of a hierarchical organization that we have used in several papers. The result of the exercise is that as marginal communication time with each additional subordinate gets shorter, the span of control in the hierarchy grows and the depth of the hierarchy, i.e., the total number of levels, shrinks. This result reflects changes which have been reported in corporate structures during recent phases of 'downsizing'.

Keywords: firms, hierarchies, span-of-control, downsizing, informational revolution.

1 Introduction

Technical change has caused in the last years a sharp increase in the efficiency of information technology and a sharp reduction in the costs of processing information. As an example, *Barron's Financial Weekly* of Sept. 3, 2001 has a lead article on "The Disappearing Secretary," in which it is reported, "The ranks of secretaries in America have fallen from a record high of 5.2 million in 1987 to 3.4 million today. The U.S. now has one secretary for every 12 executives and professionals, compared with one for every five in 1987. Chances are that by 2015 there will be one for every 25. Blame it on e-mail, voice mail and Palm Pilots." These changes are also reflected in the labor force and its organization. The proportion of employees involved in direct production ("Blue Collar") in most modern corporations has been falling, while that of employees involved in controlling the operations and processing of data, the bureaucratic part of the corporation ("White Collar"), has been increasing.

The drastic changes that have occurred over the years in the technology of information, in our ability to communicate and to process information, surely affected the way firms are organized. One way of theorizing on these matters is to divide the labor inputs of the corporation into blue collar workers, whose task is to transform physical inputs into goods and services, and white collar workers or bureaucrats, whose task is to process information and make decision that guide the operations of the blue collar workers. Yet most research focuses on the production tasks of the blue-collar workers, and little theorizing has been carried out to try to understand the contribution of these bureaucrats to production and the manner of their organization. It is our aim in this note to calculate some of the effects of these changes in IT on the firm's organization. As we shall see in our simple model, this would imply flattening of the hierarchy and increasing the number of members in each tier.

We are using the basic structure of a model that we developed in our previous papers on the structure of hierarchies (1979, 1983, and 1989). Other contributions have been made by Radner, van Zandt, and Meagher (various years)[1]. They are in some manner isomorphic to literature in computer science, in particular to the problem of the optimal set up of parallel computation by setting up "hierarchy" of computers or microprocessors. R. Radner (1993) derives somewhat analogous results by following a similar structure. Our papers were based on a technology of supervision and communication that assumed time-costs linear in the span of control of the bureaucrat, i.e., in the number of direct supervisee's of each member of the hierarchy. This satisfied a first stab at the problem. It is however clear that the marginal costs of supervision must be increasing, and we do therefore replace the linear costs with exponential costs, changes in which indicate changes in the efficiency of informational technology.

The paper proceeds as follows: section 2 outlines a discrete model of the hierarchy and section 3 characterizes its mathematical solution. Section 4 shows the economic rationale of the solution, and section 5 concludes with a brief discussion. Appendix A contains a continuous model, analogous to the model of section 2, which is required for certain results.

2 The Model

Assume that the firm has N "productive units" to control. We assume that communication is done solely along hierarchical lines: one communicates either with one's subordinates or with one's superior. No communication is performed between individuals at the same level or tier–we use both terms interchangeably–of the hierarchy.

We denote by $t \in T$ the number of levels or tiers, with 1 the top and $T+1$ the lowest tier. We refer to T as the length of the hierarchy. Let s_t be the span of control of level t, i.e., the number of level $t+1$ members supervised by a member of tier t. Let x_t be the number of individuals needed to staff

[1] See also Tarng Ming-Yueh and Chen Miao-Sheng (1986, 1987). Very interesting contributions using, or arriving at, a similar framework are van den Brink and Ruys (2001) and Ruys (2002).

level t. Thus,

$$s_t = \frac{x_{t+1}}{x_t},\ t = 1, ..., T, \tag{1}$$

and

$$x_{t+1} = x_t \cdot s_t = s_1 s_2 ... s_t, \tag{2}$$

where $x_{T+1} = N$, and $x_l = 1$.

The time of communication at level t, is assumed to be $z_t = s_t^\alpha + b$. If we denote by y_t the total communication or planning time taken by tiers 1 to t, then total communication time is

$$y_T = \sum_{t=1}^{T} z_t = a \sum_{t=1}^{T} s_t^\alpha + bT. \tag{3}$$

$\alpha > 0$ is the new parameter that we are adding in the present study. The higher α, the greater the burdens that each additional subordinate adds to his superior's data-processing time. When $\alpha < 1$, marginal communication time is decreasing as the number of subordinates grows. When $\alpha > 1$ it is increasing. α is the variable that we assume to be declining with the changes in information technology.

We assume that the productive units of the firm operate under a sub-optimal work program until the hierarchy completes its planning tasks. The task of the hierarchy therefore is to minimize this planning time, and the task of the architect who constructs the hierarchy is to minimize the sum of the costs created by slow planning and the cost of maintaining the hierarchy. We assume that planning costs, $C(y_T)$, are a convex function of the planning time with no loss for instantaneous planning, i.e., $C(0) = 0, C' > 0, C" > 0$. As for the costs of maintaining the hierarchy, we assume as a first simplifying approximation, that all "bureaucrats" receive the same wage w. The total costs, to be minimized, are

$$C(y_T) + w \sum_{t=1}^{T} x_t. \tag{4}$$

3 The Path of s_t

Let us form the Lagrangian

$$L(y_T, x_1, ..., x_T) = C(y_T) + w \sum_{t=1}^{T} x_t$$
$$-\lambda \left[y_T - \left(bT + a \left(\frac{x_2}{x_1}\right)^\alpha + a \left(\frac{x_3}{x_2}\right)^\alpha + \cdots + a \left(\frac{N}{x_T}\right)^\alpha \right) \right]$$

The first order conditions are:

$$\lambda = C'(y_T),$$

and

$$-\frac{x_t^{\alpha-1}}{x_{t-1}^{\alpha}} + \frac{x_{t+1}^{\alpha}}{x_t^{\alpha+1}} = \frac{w}{a\alpha\lambda} = \frac{w}{a\alpha C'(y_T)} \equiv \theta \geq 0, \quad t = 1, \ldots\ldots T, \tag{5}$$

and $x_{T+1} = N$. θ will be shown to be an important shadow price that determines the shape of the hierarchy. Eq. (5) can be rewritten,

$$-\frac{x_t^{\alpha}}{x_{t-1}^{\alpha}} + \frac{x_{t+1}^{\alpha}}{x_t^{\alpha}} = s_t^{\alpha} - s_{t-1}^{\alpha} = \theta x_t, t = 1, \ldots\ldots T, \tag{6}$$

This leads us to Proposition 1:

Proposition 1: *The path of the span-of-control s_t depends on θ.*

1. *When $\theta = 0$ the span-of-control is uniform throughout the hierarchy.*
2. *When $\theta > 0$ the span-of control grows as t grows, and the rate of growth increases as we descend the tiers of the hierarchy.*

Observe that θ, $\theta = \frac{w}{a\alpha C'(y_T)}$, vanishes when the wage, w, vanishes. A more interesting case is where C' is very large relatively to w, that is, where time is of paramount importance, as in matters of defense. This explains why armies, whose θ is surely very small, maintain fixed spans-of-command throughout their ranks. Increases in α and a also reduce θ, but for the former the total effect on the path of s deserves a demonstration, since α appears also in the left-hand-side of Eq. (6). Let us rewrite the latter as

$$\psi = s_t^{\alpha} - s_{t-1}^{\alpha} - \frac{w x_t}{a\alpha C'(y_T)}, \quad t = 1, \cdots, T,$$

and $\frac{ds_t}{d\alpha} = -\frac{\psi_\alpha}{\psi_{s_t}}$.

$$\psi_\alpha = s_t^{\alpha} \log s_t^{\alpha} - s_{t-1}^{\alpha} \log s_{t-1}^{\alpha} + \frac{w x_t}{a\alpha^2 C'(y_T)} > 0, \quad \psi_{s_t} = \alpha s_t^{\alpha-1} > 0,$$

where we are using the fact that $s_t \geq s_{t-1} > 1$.

We shall first solve the path s_t for the simplest case, namely the case where time only counts, i.e., $\theta = 0$, and of linear time-costs, where $\alpha = 1$. In this case, by Proposition 1, $s_t = s$ for all t, and $s^T = N$, and $T = \frac{\log N}{\log s}$. Substituting this in Eq. (4.2),

$$y_T = T(b+s) = \frac{\log N}{\log s}(b + as), \tag{7}$$

and the minimand becomes $\frac{b+as}{\log s}$. The condition for minimization is

$$s(\log s - 1) = \frac{b}{a}.$$

The sign of the second derivative at this point shows this to be a minimum. Thus $\frac{ds}{db} = \frac{1}{a \log s} > 0$since $s > 1$, and the span increases with the fixed costs per tier, b, and declines with$a$. For the simple case of b=0, i.e., no fixed costs per tier, $s = e$, and given that s is an integer, one finds that the minimizer is $s = 3$, because of the two neighbors of e,

$$\frac{3}{\log 3} < \frac{2}{\log 2}.$$

To solve the more general case of increasing costs of communication where $\alpha \neq 1$, we retain for the first shot the simplifying assumption that time is "all that matters". The time of communication, Eq. (4.2), now becomes, as in Eq. (7),

$$y_T = \frac{N}{\log s}(as^\alpha + b). \tag{8}$$

We have to minimize $\frac{as^\alpha+b}{\log s}$. The first order condition for minimization is:

$$\frac{\alpha a s^{\alpha-1} \log s - \frac{1}{s}(as^\alpha + b)}{(\log s)^2} = 0, \tag{9}$$

or

$$\psi(s, \alpha) = \alpha a s^{\alpha-1} \log s - \frac{1}{s}(as^\alpha + b) = 0. \tag{10}$$

The second order condition, evaluated at the solution to the first order condition is:

$$\psi_s = \alpha(\alpha - 1) a s^{\alpha-2} \log s + \alpha a s^{\alpha-2} - \left[a(\alpha - 1) s^{\alpha-2} - \frac{b}{s^2}\right] =$$

$$= \frac{\alpha - 1}{s}\left[\alpha s^{\alpha-1} \log s - as^{\alpha-1}\right] + \frac{b}{s^2} + \alpha a s^{\alpha-2}.$$

Using the first order condition we find that

$$\psi_s = \alpha \frac{b}{s^2} + \alpha a s^{\alpha-2} > 0 \quad , \tag{11}$$

i.e., the second order condition indicates that the first order condition (9) defines a minimum. To compute the effect of changes in α on the optimum

s, observe $ds/d\alpha$. Since $\frac{ds}{d\alpha} = -\frac{\psi_\alpha}{\psi_s}$ and $\psi_s > 0$ by Eq. (11), $\frac{ds}{d\alpha}$ has the sign of $-\psi_\alpha$. But

$$\psi_\alpha = as^{\alpha-1}\log s + \alpha as^{\alpha-1}(\log s)^2 - as^{\alpha-1}\log s = \alpha as^{\alpha-1}(\log s)^2 > 0.$$

Hence, the lower α, the larger the span-of-control and the flatter or shorter is the hierarchy.

In the special case where b = 0, (8) becomes $y_T = \frac{Nas^\alpha}{\log s}$ and we minimize $\frac{s^\alpha}{\log s}$, with the result that $s = e^{\frac{1}{\alpha}}$. Thus, if α=1/2, s=e^2, and where α=2, s=$e^{1/2}$, and so on. We should however remember that the logic of the problem requires that s≥2.

Proposition 2: *For the "time is all that matters" case, when θ=0, s_t has to obey $\alpha as^{\alpha-1}\log s - \frac{1}{s}(as^\alpha + b) = 0$. For α=1, this reduces to $s(\log s - 1) = \frac{b}{a}$. In all cases $\frac{ds}{db} > 0$ and $\frac{ds}{d\alpha}, \frac{ds}{d\alpha} < 0$.*

We are now ready to characterize the path of s_t in the general case. Proposition 3 builds upon a result which we import from Appendix A, the solution to the continuous case equation (A-2), namely that the span of the top tier obeys

$$s_1^\alpha\left(\log s_1 - \frac{1}{\alpha}\right) = \frac{b}{\alpha a} + \theta.$$

Hence $\frac{ds_1}{d\alpha} < 0$.

Proposition 3: *For any given N, the span of control at each tier grows (declines) and the length of the hierarchy T shortens (lengthens) when:*

- θ grows (declines).
- α declines (grows).

Figure 1 illustrates Proposition 3. It shows the span, t, on the horizontal axis and the logarithm of the number workers in tier t, $\log x_t$, on the vertical axis. Four paths of x_t are shown, labeled $x(\theta,\alpha)$, whose end points at $T(\theta,\alpha)$ are also labeled. All run from the top (which is shown at the origin), x_1=1, to x_{T+1}=N. The rightmost path $x(0,\alpha)$, ending at $T(0,\alpha)$, and the third from the right, $T(0,1)$, are straight lines, since, given that for θ=0 the span-of-control is constant, $\log x_t = t\log s$. The leftmost curve, $x(\theta,1)$, ending in $T(\theta,1)$, and the second from the right, $x(\theta,\alpha)$, ending in $T(\theta,\alpha)$, where $\theta > 0$ and $\alpha > 1$, show increasing slopes due to the increase of s_t in t. In the figure $\alpha > 0$: paths for $\alpha < 1$ would lie further to the left, because the spans at each tier would be larger. What the figure illustrates is that higher θ curves the path upwards, and higher α shifts the curves to the left. The steeper (i.e., the lower α) and the more curved (the higher θ) the path, the closer is the end point, T, hence the squatter the hierarchy.

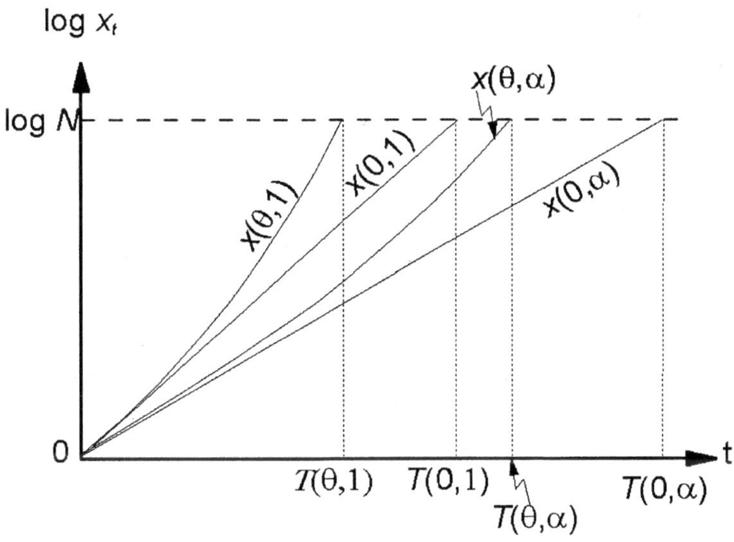

Fig. 1. The path of x_t

4 The Characteristics of the Organization

The previous section has characterized the path of $s_t$from the boss at t=1 to N. The present section will translate those results to the level of the organization at large. This is done in Proposition 4:

Proposition 4: *An increase in the wage costs, w, or the fixed time costs of each tier, b, increases the optimum number of members of the hierarchy at any level but reduces the total number of workers in the organization for any given number of productive workers, and the height of the hierarchy. An increase in the marginal time costs, C', or in the rate of increase of marginal time costs with the span of control, α, reduces the number of members in each level x_t, increases the total number of members of the hierarchy and the height of the hierarchy.*

Figure 1 illustrates Proposition 4. The ranking of the number of members in each level can be seen in the height of the graphs, which meet only at the

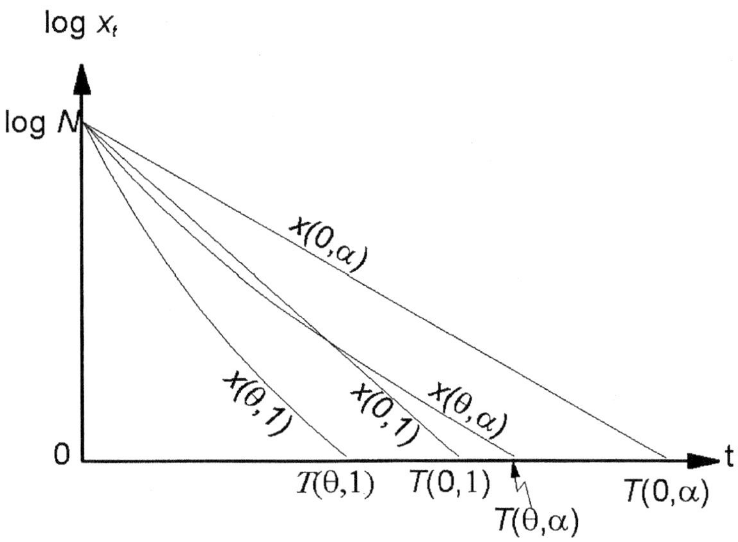

Fig. 2. Comparing total employment

origin t=1: $\theta = 0$ is the lowest curve for any given α, and $\alpha = 1$ is the highest curve for any given θ. The length of the hierarchy is given by $T(\theta,\alpha)$: here the lowest curve provides the longest hierarchy, and vice versa for the highest. As for the total number of organizational employees, these are shown by the integral of the area under the relevant curve. It can easily be seen that the shorter the hierarchy, the smaller the number of employees in the hierarchy.

Proposition 4 and Figure 1 serve to show the economic sense of the paths characterized in section 3. Figure 2 has been added to facilitate the message of Figure 1. The graphs of the latter have been turned around, so that for each x_{T+1} now runs flush with the vertical axis. The area under the $x(\theta,\alpha)$ curve is the sum of the memberships of all tiers of the hierarchy for the relevant path. It is easy to see that paths with higher θ lie outside those with a lower θ, i.e., the number of white collar workers in them is larger. In other words, "time only matters" hierarchies, where $\theta \to 0$, are organizations that are ready to waste decision making manpower to save time, and they do so by having small spans of control and long hierarchies. When we compare high and low α hierarchies, we see that higher α paths lie outside those with a lower α, i.e., their total administrative employment is larger. The effect of marginal coordination time that is increasing, i.e., a high α, is to shorten spans of control and lengthen hierarchies, with the consequence that the total number employed in the hierarchy increases. A *reduction* in α, on the other hand, has the opposite effect: it permits higher spans and a saving in hierarchical employment, and may explain some aspects of 'downsizing', the pruning of middle-management through the elimination of some hierarchical levels, that has been documented in recent years. Two paths here cannot be compared: $x(\theta,\alpha)$ and $x(0,1)$ intersect because of the conflicting effects of the two parameters.

5 Discussion

The assumptions we made on the "hierarchy" and its roles are obviously simplistic and ignore many realistic aspects of the components of the bureaucratic structures. Thus, we ignore the ability to communicate across pure hierarchical lines and we assume that decision-making is totally centralized and disregard the delegation of decision-making to lower levels of the hierarchy. In reality decision-making is per force quite dispersed, with looser control from the center. It is not simple to add some of these factors to our model in order to make it more realistic and in order to study in its framework the effects of some recent changes in IT and organizational practices. We have in the past built a model in which some of the earlier changes could be examined, where we added a "language" variable to account for the aggregation of information in the dialog between top and bottom of the hierarchy, but we do not deem it relevant to the problem at hand.

We have nonetheless managed to touch in this paper upon one of the important phenomena of recent decades, the phenomenon of "downsizing": see the end of section 4. We claim that some part of downsizing, possibly only a small one, can be explained by our model. Downsizing was accompanied by a reduction in the number of mid-rank management, i.e., in the height of the hierarchy as well as in the total number of managers, the white-collar employees.

Research on the control, management and processing of information and its effects on structure of corporations has not received the prominence it deserves, but there is no question that in the coming years it will occupy more of the profession's attention. Questions such as why firms spin off, or why they form conglomerates are just beginning to be asked, and we hope that research on control and hierarchy may shed some light on them.

Appendix A: The Continuous Case

Models of hierarchical structure are models of paths, leading from the top of the hierarchy, occupied by a single 'boss', to its bottom, the productive workers or units. We have in the past used both continuous and discrete models to characterize these paths. Discrete models are more intuitive, because the phenomena we discuss are naturally discrete. They do however suffer from a serious weakness: since the solution is a second-degree difference equation, it is impossible to tie the path to the end points. In this paper we are therefore compromising: we are expounding in full the discrete model, but adding this appendix for a concise presentation of the continuous model, whose results regarding the initial qualities of the path are imported into Proposition 3.

Only the bare outlines of this case are given here, and the reader is requested to look at our 1983 paper for more detail.

The general form of the function we have to minimize is

$$\min F_0 + \int_1^T F_t dt.$$

Our state variables are x_t and y_t, and the instrument is s_t. $F_0 = C(y_t)$ and $F_t = wx_t$, subject to the 'equations of motion' that define the evolution of the state variables y_t and $x_t$over the path. The latter are $\dot{x} = x_t \log s_t$ and $\dot{y}_t = z_t = b + as_t^\alpha$. The full minimand is therefore

$$F_0 + \int_1^T F_t dt = C(y_T) + \int_1^T (wx_t + \mu_t x_t \log s_t + \lambda_t (b + as_t^\alpha))\, dt.$$

We construct the Hamiltonian

$$H_t = wx_t + \mu_t x_t \log s_t + \lambda_t (b + as_t^\alpha).$$

Since $y_t$does not appear in the Hamiltonian, $\frac{dH_t}{dy_t} = 0$, and $\dot{\lambda}_t = 0,\ t \in [1, T]$.

Hence λ_t is constant along the whole path. By the transversality conditions, $\lambda_T = \frac{dF_0}{dy_T} = \frac{dC(y_T)}{dy_T} = C'(y_T)$. Thus λ_t=λ=C'.

The partial derivative of the Hamiltonian with respect to the instrument, s_t,

$$\frac{dH_t}{ds_t} = \frac{\mu_t x_t}{s_t} + \lambda \alpha a s_t^{\alpha - 1} = 0,$$

and

$$\mu_t = -\frac{C' \alpha a s_t^\alpha}{x_t}.$$

In a t-independent, autonomous problem like the present one, H_t=0.This leads to $wx_t - C' \alpha a s_t^\alpha \log s_t + C'(b + as_t^\alpha) = 0$.

Rearrangement leads to

$$s_t^\alpha \left(log s_t - \frac{s_t^\alpha}{\alpha} \right) = \frac{b}{\alpha a} + \theta x_t. \qquad (A-1)$$

For t=1,

$$s_1^\alpha \left(log s_1 - \frac{1}{\alpha} \right) = \frac{b}{\alpha a} + \theta. \qquad (A-2)$$

For the effect of α on s_1(remembering that θ is also a function of α),

$$\psi = s_1^\alpha \left(log s_1 - \frac{1}{\alpha} \right) - \frac{b}{\alpha a} - \theta$$

$$\frac{ds_1}{d\alpha} = -\frac{\psi_\alpha}{\psi_{s_1}} = -\frac{+\frac{b}{\alpha^2 a}}{s_1^\alpha \log s \left(\log s_1 - \frac{1}{\alpha} \right) + \frac{s_1^\alpha}{\alpha^2} + \frac{b}{\alpha^2 a} - \frac{d\theta}{d\alpha}} < 0. \qquad (A-3)$$

References

1. van den Brink, Ren, and Pieter H. M. Ruys (2001), "On the Depth of a Hierarchically Structered Firm", CentER Discussion Paper, Tilburg University
2. Keren, Michael, and David Levhari (1979), "The Optimum Span of Control in a Pure Hierarchy." *Management Science*, 25,11, 1162-1172.
3. Keren, Michael, and David Levhari (1983), "The Internal Organization of the Firm and the Shape of Average Costs." *Bell Journal of Economics*, Autumn, 474-486.
4. Keren, Michael, and David Levhari (1989), "Decentralization, Aggregation, Control Loss, and Costs in a Hierarchical Model of the Firm." *Journal of Economic Behavior and Organization*, 11,2, 213-236.
5. Meagher, Kieron (2001), "The impact of hierarchies on wages." *Journal of Economic Behavior & Organization*, 45, 441-458.
6. Meagher, Kieron, and Timothy Van Zandt (1998) "Managerial Costs for One-Shot Decentralized Information Processing." *Review of Economic Design* 3, 329-345.
7. Meagher, Kieron, Hakan Orbay and Timothy Van Zandt (2001), "Hierarchy, Size and Environmental Uncertainty." Mimeo.
8. Radner, Roy (1992), "Hierarchy: The Economics of Managing." *Journal of Economic Literature*, 30, 1382-1415.
9. Radner, Roy (1993), "The Organization of Decentralized Information Processing." *Econometrica*, 61, 1109-1146.
10. Radner, Roy, and Timothy Van Zandt (1992), "Information Processing in Firms and Returns to Scale." *Annales d'Economie et Statistique* No. 25/26, 265-298. (Reprinted in *The Economics of Informational Decentralization: Complexity, Efficiency and Stability*, Norwell, MA: Kluwer Academic Press, 1995.)
11. Radner, Roy, and Timothy Van Zandt (2001), "Real-Time Decentralized Information Processing and Returns to Scale." *Economic Theory* 17, 497-544.
12. Ruys, Pieter H. M., (2002), "A General Equilibrium of Managed Services", this volume.
13. Tarng Ming-Yueh and Chen Miao-Sheng (1986), "Note on the Optimum Span of Control in a Pure Hierarchy", mimeograph.
14. Tarng Ming-Yueh and Chen Miao-Sheng (1987), "On the Optimum Structure of the Hierarchy in an Organisation", *Mathematical Social Sciences* **14**,3, 239-50.
15. Van Zandt, Timothy (1995), "Continuous Approximations in the Study of Hierarchies." *Rand Journal of Economics* 26, 575-590.
16. Van Zandt, Timothy (1995), "Hierarchical Computation of the Resource Allocation Problem." *European Economic Review* 39, 700-708.
17. Van Zandt, Timothy (1999), "Real-Time Decentralized Information Processing as a Model of Organizations with Boundedly Rational Agents." *Review of Economic Studies* 66, 633-658.

Dynamic Adjustment of Supply Constrained Disequilibria to Walrasian Equilibrium

Gerard van der Laan[1] and Dolf Talman[2]

[1] Department of Econometrics and Tinbergen Institute, Vrije Universiteit, Amsterdam
[2] Department of Econometrics & Operations Research and CentER, Tilburg University, Tilburg

Abstract. It is well-known that the classical Walrasian tatonnement process of adjusting prices does not need to converge. In several papers Weddepohl has shown cyclic or even chaotic behavior in discrete time tatonnement processes. Starting from a trivial non-trade disequilibrium with complete rationing on the supplies at a given vector of initial prices, a quantity and price adjustment process is proposed, converging always to a Walrasian equilibrium. At any price and rationing system generated by the process all markets clear, only supply rationing occurs, in the short run only rationing schemes are adjusted and in the long run both prices and rationings are adjusted. Furthermore, prices can only take values equal or above their initial values, while supply rationing only may occur when the corresponding price is on its initial value.

Keywords: disequilibrium, supply rationing, price and quantity adjustments

1 Introduction

It is well-known that the classical Walrasian tatonnement process of adjusting prices in an economy has a number of drawbacks. First, it is not guaranteed that the price adjustment process converges to Walrasian equilibrium prices. In Scarf (1960) examples of economies have been given for which the process fails to converge. Other studies have shown that the discrete time process does not need to converge, even if the economy satisfies conditions that guarantee convergency of the continuous time adjustment, see e.g. Bela and Majumdar (1992) and Weddepohl (1994, 1995). In fact, Saari (1985) has shown that any process based on a finite amount of local information may fail to converge. This lack of convergence has been solved by Smale (1976), van der Laan and Talman (1987a,b) and Kamiya (1990), where price adjustment processes have been presented in which the adjustment is based on local information about the excess demands combined with global information on the starting price system.

Economists since Walras have been focussing on the convergency properties of price adjustment processes. However, stimulated by the computer making simulations easily executable, more recently a shift of interests to the economic interpretation of the adjustment process has taken place. This brings us to the second drawback of the tatonnement process, namely that demand and supply are not in equilibrium as long as the process has not achieved Walrasian equilibrium prices. So, trade must be excluded until an equilibrium is reached. Although the process can be seen as taking place in a sequence economy, where in each period the excesses or shortages determine prices for the next period and trade occurs at disequilibrium prices, see e.g. Weddepohl (1997), also such an interpretation is unsatisfactory because it leaves the way in which the trade is determined unspecified. A possible way out has been provided by Veendorp (1975), who noticed that the relevant market signals for an adjustment process are based on the effective demands associated with an equilibrium under rationing constraints, and who specified a process generating a sequence of rationed equilibria in which prices are adjusted according to effective excess demands and supplies. However, also such a process does not necessarily converge to a Walrasian equilibrium price system and even chaotic behavior may be expected, see Day and Pianigiani (1991). Using computer simulations the possibility of chaotic behavior has been confirmed in several studies, see e.g. Böhm (1993) and Weddepohl (1997).

Weddepohl (1999) argues that both the classical Walrasian tatonnement process and price adjustments based on effective demands with in each period trade under some rationing scheme at disequilibrium prices are not very satisfactory; the Walrasian tatonnement because of the drawbacks already mentioned above, and the latter one because the determination of the rationings is at least as complicated as solving for an equilibrium price. Therefore he takes a more realistic approach by assuming that markets are open consecutively, not simultaneously. Applying such a process to a simple model with two goods, labor and a produced good, and money as numeraire commodity, shows that again cyclical movements and even chaos may appear, albeit that for moderate step sizes long cyclic movements similar to business cycles in reality can occur. This illustrates nicely that not only the convergency properties, but also the non-convergent behavior of adjustment processes is a very interesting topic of research, because it may help us to obtain a better understanding in real phenomena.

In this contribution in honor to Weddepohl's work on 'Equilibrium, Disequilibrium and Dynamics', we focus on convergency and consider an alternative approach to solve the disequilibrium problem as long as the price adjustment process has not reached yet a Walrasian equilibrium price system. This approach has been introduced in Herings et al. (1997), in which an adjustment process is proposed satisfying that at any price system generated by the process all markets clear endogenously by rationing constraints. In this process the problem of finding the market clearing rationings does not appear. Instead, any state generated by the process gives a feasible price system and a rationing scheme, i.e. at these prices and rationings the corresponding total constrained excess demand is equal to zero on each market. Moreover, convergency of the process to an unconstrained Walrasian equilibrium state is guaranteed under standard assumptions with respect to the economy.

For an economy with restrictions on the set of feasible prices, an equilibrium concept has been introduced by Drèze (1975). A Drèze equilibrium is given by a feasible price vector and a rationing scheme on net demands and net supplies, such that (i) the constrained demands of the agents are feasible, i.e. all markets clear, (ii) on each market rationing may affect either supply or demand, but not both sides of the market simultaneously, and (iii) rationing of a commodity only occurs when its price restriction is binding. In the price and quantity adjustment process considered in Herings et al. (1997) the existence of a connected set of Drèze equilibria leading to a Walrasian equilibrium is shown, such that at any point in time rationing on the net demands may occur, while the net supplies are never rationed. That process was defined to incorporate features of Eastern European economies, where rationing on demand used to be the rule rather than the exception. On the other hand, in van der Laan (1980) it is argued that 'constraints on the supply side can often more easily be realized than constraints on the demand side', while Kurz (1982) argues that 'demand rationings rarely occur in market economies while resource unemployments are very common'. This motivated these authors to consider Drèze equilibria with only rationing on the net supplies. In van der Laan (1980, 1982) and Kurz (1982) the existence of a Drèze equilibrium, satisfying that only rationing on the net supplies occurs and at least one commodity is not rationed at all, has been proven. Such an equilibrium is also called an unemployment equilibrium. In Dehez and Drèze (1984) and van der Laan (1984) it has been shown that in economies with money as numeraire commodity, there exists a supply constrained equilibrium without rationing on money, when the set of feasible prices allows for enough flexibility of the price level of the non-numeraire commodities. Weddepohl (1987) discusses extensively linkages between prices through index functions and shows that a supply constrained equilibrium in which at least one commodity is not rationed exists if the system of indexes is non-circular, i.e. no commodity is directly or indirectly indexed by itself. In these papers it is also argued that the equilibrium condition that rationing of a commodity is not allowed unless its price restriction is binding, can not be maintained when prices are tied to each other through index functions.

In this paper we propose an adjustment process satisfying that at any price system generated by the process all markets clear endogenously by rationing on the net supplies only. The starting point of the process is a historically given price system. At this system of fixed prices, the process starts with a trivial non-trade disequilibrium in which all markets clear by imposing complete rationing on the supplies of all commodities, and then subsequently adjusts rationings and prices in such a way that at any moment in time all markets clear by rationing on the net supplies only, while there is no rationing on the demand side of the markets. In the beginning of the process only the rationing schemes are simultaneously adjusted until one of the commodities is no longer rationed and thus an unemployment equilibrium is reached. This part of the process in which the rationings are reduced, can be seen as short term adjustment of the rationings at a given fixed initial price system. Continuing from this unemployment equilibrium, the supply rationings on the other commodities can not be reduced further without either imposing demand rationing on the unrationed commodity or allowing that its price adjusts upwards. So, from this short term market clearing disequilibrium at given fixed prices, the prices of the unrationed commodities are allowed to adjust upwards. Keeping all markets in equilibrium, the process therefore continues with simultaneously adjusting the prices of the unrationed commodities above their initial level and adjusting the rationings on the supplies of the other commodities. This long term process continues until none of the commodities is rationed anymore and thus a Walrasian equilibrium has been obtained. Summarizing, a process is proposed satisfying that at any price and rationing system generated by the process (i) all markets clear, so that trade can take always place, (ii) only supply rationing occurs, (iii) in the short run only rationing schemes are adjusted, (iv) in the long run both prices and rationings are adjusted, (v) prices are never below their initial value, (vi) supply rationing only occurs when the corresponding price is on its initial value, and (vii) eventually a Walrasian equilibrium is reached.

In Herings et al. (1997) the existence of a path of approximate market clearing states with only demand rationing connecting the initial state with a Walrasian equilibrium has been shown by using simplicial techniques. Using the existence of such a path, they show that there is a connected set of market clearing disequilibria with demand rationings containing both the trivial starting state and the Walrasian equilibrium state. In this paper we will show the existence of a connected set of supply constrained market clearing disequilibria directly by using Browder's fixed point theorem. Again simplicial path following techniques can be used to obtain an implementable algorithm for finding a path of approximate market clearing states.

The paper is organized as follows. In Section 2 we introduce the model and define the concept of a supply constrained disequilibrium. In Section 3 we discuss the equilibrating mechanism by constructing a reduced total excess demand function satisfying that any zero point of this function yields a supply constrained disequilibrium. In Section 4 we prove the existence of a connected set of supply constrained disequilibria. The dynamics of the adjustment process is discussed in Section 5.

2 The Model

We consider a pure exchange economy $\mathcal{E} = (\{X^i, \succeq^i, w^i\}_{i=1}^m, P)$. In this economy there are m consumers, indexed $i = 1, \ldots, m$, and n commodities, indexed $j = 1, \ldots, n$. For k a positive integer, we denote $I_k = \{1, \ldots, k\}$. Each consumer $i \in I_m$ is characterized by a consumption set X^i, a preference preordering $\succeq^i$ on X^i, and a vector of initial endowments w^i. The vector w is defined by $w = \sum_{i \in I_m} w^i$. We consider the situation that the economy $\mathcal{E}$ is faced with a fixed price $\underline{p}$ in the short run. In the long run prices are allowed to increase and are therefore restricted to the set $P = \{p \in \mathbb{R}^n \mid p \geq \underline{p}\}$. Since prices are regulated and restricted, quantity constraints have to accompany the price mechanism to equilibrate the markets. The following standard assumptions X, U and W with respect to the economy $\mathcal{E}$ are made.

Assumption X
For every consumer $i \in I_m$, the consumption set X^i is a closed and convex subset of $\mathbb{R}^n_+$ and $X^i + \mathbb{R}^n_+ \subset X^i$.

Assumption U
For every consumer $i \in I_m$, the preference preordering $\succeq^i$ on X^i is complete, transitive, reflexive, continuous, strongly monotonic, and strictly convex.[1]

Assumption W
For every consumer $i \in I_m$, the vector of initial endowments w^i belongs to the interior of X^i.

Notice that the assumption of strict convexity allows us to work with demand functions instead of demand correspondences. We further assume that the short run price vector $\underline{p}$ is a strictly positive vector. Following van der Laan (1980) and Kurz (1982) we only allow for quantity rationing on the supply side of the market. Given an arbitrary price system $p \in \mathbb{R}^n_{++}$ and rationing scheme on supply $\ell \in \mathbb{R}^n_+$, the constrained budget set of consumer $i \in I_m$ is given by

$$B^i(p, \ell) = \{x^i \in X^i \mid p \cdot x^i \leq p \cdot w^i \text{ and } w^i_k - x^i_k \leq \ell_k w_k, \ \forall k \in I_n\}.$$

Observe that the number ℓ_k is the fraction of the total endowment of commodity k each consumer is allowed to supply maximally. The corresponding constrained demand $d^i(p, \ell)$ of consumer i is defined as the best element for $\succeq^i$ in $B^i(p, \ell)$. Because of the strict convexity and strong monotonicity assumptions, this element is unique and lies on the budget hyperplane, i.e. $p \cdot d^i(p, \ell) = p \cdot w^i$. A supply constrained disequilibrium is defined as follows.

[1] A preference preordering $\succeq^i$ is said to be strongly monotonic if $\overline{x}^i, \widehat{x}^i \in X^i$, $\overline{x}^i \leq \widehat{x}^i$, and $\overline{x}^i \neq \widehat{x}^i$ implies $\widehat{x}^i \succ^i \overline{x}^i$. A preference preordering $\succeq^i$ is said to be strictly convex when for any pair $\overline{x}^i$, $\widehat{x}^i \in X^i$, such that $\overline{x}^i \neq \widehat{x}^i$, $\overline{x}^i \sim^i \widehat{x}^i$, holds $\lambda \overline{x}^i + (1-\lambda)\widehat{x}^i \succ^i \overline{x}^i$ for $\lambda \in (0,1)$.

Definition 2.1 (Supply Constrained Disequilibrium)
A Supply Constrained Disequilibrium for the economy $\mathcal{E} = (\{X^i, \succeq^i, w^i\}_{i=1}^m, P)$ is a price system $p^* \in P$, a rationing scheme $\ell^* \in \mathbb{R}_+^n$, and, for every consumer $i \in I_m$, a consumption bundle $x^{*i} \in X^i$ such that

(i) for all $i \in I_m$, $x^{*i} = d^i(p^*, \ell^*)$;
(ii) $\sum_{i=1}^m x^{*i} = w$;
(iii) for all $k \in I_n$, if $\exists i \in I_m$ such that $w_k^i - x_k^{*i} = \ell_k^* w_k$, then $p_k^* = \underline{p}_k$.

In the definition, the rationing schemes on supply are assumed to be uniform, i.e. the same for each consumer. This assumption can be easily relaxed. Condition (i) requires that the consumption of each consumer equals his constrained demand, while condition (ii) is the market clearing condition. Condition (iii) precludes supply rationing on the market of some commodity as long as its price is not on its lower bound. Observe that in a supply constrained disequilibrium all markets clear and so trade is possible. However, trade takes place against disequilibrium prices as long as at least one of the rationings is binding. In case for every commodity $k \in I_n$ it holds that $w_k^i - x_k^{*i} < \ell_k^* w_k$ for all $i \in I_m$, no supply constraint is binding and p^* and x^{*i}, $i \in I_m$, form a Walrasian equilibrium. Further, the definition implies the existence of a trivial disequilibrium with allocation $x^{*i} = w^i$, for all $i \in I_m$, by setting $p = \underline{p}$ and $\ell = 0^n$. Notice that $B^i(\underline{p}, 0^n) = \{w^i\}$ for all $i \in I_m$. We will show that there exists a connected set of supply constrained disequilibria containing both the trivial disequilibrium with complete supply rationing and a Walrasian equilibrium without any rationing.

3 The Equilibrating Mechanism

To prove the existence of a connected set of SCDs, we first focus on the equilibrating mechanism to find price systems and rationing schemes satisfying the equilibrium conditions of Definition 2.1. To formalize the equilibrating mechanism, we introduce a set $Q \subset \mathbb{R}^n$ containing the set P in its interior and define for every $q \in Q$ a price $p(q) \in P$ and a rationing scheme $\ell(q) \in \mathbb{R}_+^n$. The set Q is taken to be equal to

$$Q = \{q \in \mathbb{R}^n \mid q \geq \underline{p} - e\},$$

where e is the n-vector of ones. For $q \in Q$, the price $p(q) \in P$ is defined to be the orthogonal projection of q on P, i.e.

$$p_k(q) = \max\{q_k, \underline{p}_k\}, \ k \in I_n,$$

and the rationing scheme $\ell(q) \in \mathbb{R}_+^n$ is defined by

$$\ell_k(q) = 1 - \underline{p}_k + q_k, \ k \in I_n.$$

Observe that $\ell_k(q) = 0$ and $p_k(q) = \underline{p}_k$ if $q_k = \underline{p}_k - 1$ and that $\ell_k(q) \geq 1$ and $p_k(q) = q_k$ if $q_k \geq \underline{p}_k$. Clearly, both $p(q)$ and $\ell(q)$ are continuous in q. For any $q \in Q$ we now define for every consumer $i \in I_m$ his reduced budget set $B^i(q)$ by

$$B^i(q) = B^i(p(q), \ell(q)).$$

Finally, for any consumer $i \in I_m$ we define his reduced excess demand correspondence $d^i : Q \to \mathbb{R}^n$ by

$$d^i(q) = \{x^i \in B^i(q) | x^i \succeq^i y^i, \text{ for all } y^i \in B^i(q)\}.$$

Because $\underline{p}_k > 0$ for all k, we have that $p(q)$ is strictly positive for all $q \in Q$. According to Herings (1996) it follows that B^i is a continuous correspondence at any $q \in Q$. With Assumption U it then follows that d^i is a continuous function and so is the reduced excess demand function $z : Q \to \mathbb{R}^n$ defined by

$$z(q) = \sum_{i \in I_m} d^i(q) - w.$$

Since for any $i \in I_m$ the budget constraint $p(q) \cdot d^i(q) \leq p(q) \cdot w^i$ is always satisfied with equality, Walras' law holds, i.e. $p(q) \cdot z(q) = 0$ for all $q \in Q$. Moreover, for any $j \in I_n$, we have that $q_j = \underline{p}_j - 1$ implies $\ell_j(q) = 0$ and therefore $z_j(q) \geq 0$. So, we have the following lemma.

Lemma 3.1 Under Assumptions X, U and W, the reduced excess demand function $z : Q \to \mathbb{R}^n$ satifies the following properties:
i) z is continuous on Q;
ii) for all $q \in Q$ it holds that $p(q) \cdot z(q) = 0$ (Walras' law);
iii) $q_j = \underline{p}_j - 1$ implies $z_j(q) \geq 0$.

The next theorem shows that any zero point of z induces an SCD.

Theorem 3.2 Let q^* be a zero point of z in Q, i.e. $z(q^*) = 0^n$. Then the price system $p^* = p(q^*) \in P$, the rationing scheme $\ell^* = \ell(q^*) \in \mathbb{R}^n_+$, and for each $i \in I_m$, the bundle $x^{*i} \in X^i$ given by $x^{*i} = d^i(q^*)$, form a supply constrained disequilibrium.

Proof.
We have to show that the three conditions of Definition 2.1 are satisfied. Clearly the conditions (i) and (ii) hold by construction of the reduced demand functions. For any $i \in I_m$ and $k \in I_n$ it holds that $x_k^{*i} \geq 0$ and $w_k^i < w_k$ and thus $w_k^i - x_k^{*i} < w_k$. So, if for some $i \in I_n$ it holds that $w_k^i - x_k^{*i} = \ell_k(q^*) w_k$, we must have $\ell_k(q^*) < 1$, and therefore $q_k^* < p_k(q^*)$, which implies $p_k(q^*) = \underline{p}_k$. This proves condition (iii) of Definition 2.1. Q.E.D.

When q^* is a zero point of z satisfying $q^* \geq \underline{p}$, i.e. $q^* \in P$, then $p(q^*) = q^*$ and $\ell(q^*) \geq e$ and hence for every $k \in I_n$ it holds that $w_k^i - x_k^{*i} < \ell_k(q^*)w_k$ for all $i \in I_m$, which implies that q^* induces a Walrasian equilibrium. On the other hand, the point $q^0 = \underline{p} - e$ induces the trivial non-trade disequilibrium with complete rationing on all supplies.

In the next section we show that there exists a connected set of zero points of the reduced excess demand function z containing both q^0, inducing the trivial disequilibrium, and some $q^* \in P$, inducing a Walrasian equilibrium. This implies that there exists a connected set of supply constrained disequilibria leading from the trivial disequilibrium with complete supply rationing to a Walrasian equilibrium without rationing.

4 Connected Set of Disequilibria

To show the existence of a connected set of zeros of the reduced excess demand function z containing both $q^0 = \underline{p} - e$ and a vector q^* in P, let $s_0 = \sum_{k=1}^n q_k^0$ and define for $s \geq s_0$ the set $Q(s)$ by

$$Q(s) = \{q \in Q | \sum_{k=1}^{n} q_k \leq s\}.$$

For any $s > s_0$, let $\overline{Q}(s) = \{q \in Q(s) | \sum_{k=1}^n q_k = s\}$ be the upper boundary of $Q(s)$ and let $\underline{Q}(s) = \{q \in Q(s) | q_k = \underline{p}_k - 1$ for some $k \in I_n\}$ be the lower boundary of $Q(s)$. The next theorem says that for any $s > s_0$ there exists a connected set of zero points of z in $Q(s)$ containing both q^0 and a point in $\overline{Q}(s)$.

Theorem 4.1 Let $\mathcal{E} = (\{X^i, \succeq^i, w^i\}_{i=1}^m, P)$ be an economy satisfying Assumptions X, U and W. Then for any $s > s_0$ there exists a connected set $C(s)$ of zero points of the reduced excess demand function z in $Q(s)$ such that $q^0 \in C(s)$ and $C(s) \cap \overline{Q}(s) \neq \emptyset$.

Proof.
Take any $s > s_0$. For some $M > s_0$, define X_0 by

$$X_0 = \{y \in \mathbb{R}^n | \sum_{k=1}^{n} y_k = s_0, \ \max_{k \in I_n} y_k \leq M\}$$

and, for $0 < \alpha \leq 1$, define X_α by

$$X_\alpha - \{x \in \mathbb{R}^n | x - y + \frac{\alpha}{n}(s - s_0)e, \ y \in X_0\}.$$

Clearly, for $x \in X_\alpha$, it holds that $\sum_{k=1}^n x_k = \sum_{k=1}^n y_k + \alpha(s - s_0) = s_0 + \alpha(s - s_0) = (1 - \alpha)s_0 + \alpha s$. Further, define $X = \cup_{\alpha \in [0,1]} X_\alpha$ and take M

so large that $Q(s) \subset X$ and X_1 contains $\overline{Q}(s)$ in its relative interior. Notice that $Q(s) \cap X_0 = \{q^0\}$ and $Q(s) \cap X_1 = \overline{Q}(s)$. Let $Y = \text{aff}(X_0)$ be the affine hull of X_0, i.e. $Y = \{y \in \mathbb{R}^n | \sum_{k=1}^n y_k = s_0\}$, and for $x \in \mathbb{R}^n$ let $\tau(x)$ be the projection of x on Y , i.e. $\tau(x)$ is the unique point in Y satisfying $\tau(x) = x - \lambda e$ for some $\lambda \in \mathbb{R}$. Next, let the set Y_0 be defined by

$$Y_0 = \{y \in Y | y = \tau(q + z(q)),\ q \in Q(s)\} \cup \{y \in Y | y = \tau(p(q)),\ q \in \underline{Q}(s)\}.$$

Since the set $Q(s)$ is compact and the functions z, p and τ are continuous, it follows that Y_0 is a bounded subset of Y, and thus M can be taken so large that X_0 contains Y_0 in its relative interior.

For any $y \in X_0$ and $\alpha \in [0,1]$, let $y^\alpha \in X_\alpha$ be given by $y^\alpha = y + \frac{\alpha}{n}(s - s_0)e$. Then the point-to-set mapping $\varphi\colon X_0 \times [0,1] \to X_0$ is defined by

$$\varphi(y,\alpha) = \begin{cases} \{\tau(y^\alpha + z(y^\alpha))\} \text{ if } y^\alpha \in Q(s) \setminus \underline{Q}(s), \\ \text{Conv}(\{\tau(y^\alpha + z(y^\alpha))\} \cup \{\tau(p(y^\alpha))\}) \text{ if } y^\alpha \in \underline{Q}(s), \\ \{\tau(p(q(y^\alpha)))\} \text{ if } y^\alpha \in X \setminus Q(s), \end{cases}$$

where Conv(.) denotes the convex hull of a set and $q(x)$ is the orthogonal projection of a point $x \in \mathbb{R}^n$ on $Q(s)$. Clearly, X_0 is nonempty, convex and compact, φ is upper semi-continuous, and for every $(y,\alpha) \in X_0 \times [0,1]$ it holds that $\varphi(y,\alpha)$ is compact, convex and nonempty. According to Browder's theorem, see Browder (1960), there exists a connected set $\overline{C}(s)$ of fixed points of φ in $X_0 \times [0,1]$ such that $\overline{C}(s) \cap (X_0 \times \{0\}) \neq \emptyset$ and $\overline{C}(s) \cap (X_0 \times \{1\}) \neq \emptyset$, i.e. there exists a connected set $\overline{C}(s)$ in $X_0 \times [0,1]$ satisfying

$$y \in \varphi(y,\alpha),\ \text{for all } (y,\alpha) \in \overline{C}(s)$$

and containing a point $(\widetilde{y}, 0) \in X_0 \times \{0\}$ and a point $(\widehat{y}, 1) \in X_0 \times \{1\}$. Hence, the set $C(s)$ in X defined by

$$C(s) = \{x \in X | x = y^\alpha,\ (y,\alpha) \in \overline{C}(s)\}$$

is a connected set in X satisfying $C(s) \cap X_0 \neq \emptyset$ and $C(s) \cap X_1 \neq \emptyset$. It remains to show that every element of $C(s)$ lies in $Q(s)$ and is a zero point of z. Since $Q(s) \cap X_0 = \{q^0\}$ and $Q(s) \cap X_1 = \overline{Q}(s)$, it then follows that $\widetilde{y}^0 = q^0$ and $\widehat{y}^1 \in \overline{Q}(s)$, so that $C(s)$ is a connected set in $Q(s)$ of zero points of z satisfying $q^0 \in C(s)$ and $C(s) \cap \overline{Q}(s) \neq \emptyset$.

Take any $y^\alpha \in C(s)$. First, suppose $y^\alpha \in X \setminus Q(s)$. By definition of φ it then follows that

$$y = \tau(p(q(y^\alpha))) = p(q(y^\alpha)) - \lambda e$$

for some $\lambda \in \mathbb{R}$. On the other hand, $y = y^\alpha - \frac{\alpha}{n}(s - s_0)e$, implying that

$$y^\alpha - \frac{\alpha}{n}(s - s_0)e = p(q(y^\alpha)) - \lambda e$$

so that $y^\alpha - p(q(y^\alpha)) = \beta e$ with $\beta = \frac{\alpha}{n}(s - s_0)e - \lambda$. However, since $y^\alpha \notin Q(s)$, there must be a component k with $y_k^\alpha < \underline{p}_k - 1$ and a component h with $y_h^\alpha \geq$

$\underline{p}_h - 1$. Taking first the projection $q(y^\alpha)$ of y^α on Q and then the projection $p(q(y^\alpha))$ of $q(y^\alpha)$ on P implies that $y_k^\alpha - p_k(q(y^\alpha)) < -1 \le y_h^\alpha - p_h(q(y^\alpha))$, yielding a contradiction. Consequently, $y^\alpha \in Q(s)$ for every $y^\alpha \in C(s)$.

Now, suppose $y^\alpha \in Q(s) \setminus \underline{Q}(s)$. Then

$$y^\alpha - \frac{\alpha}{n}(s - s_0)e = y = \tau(y^\alpha + z(y^\alpha)).$$

Since $\tau(y^\alpha + z(y^\alpha)) = y^\alpha + z(y^\alpha) - \lambda e$ for some $\lambda \in \mathbb{R}$ it follows that

$$z(y^\alpha) = (\lambda - \frac{\alpha}{n}(s - s_0))e.$$

Because of Walras' law, we have

$$p(y^\alpha) \cdot z(y^\alpha) = (\lambda - \frac{\alpha}{n}(s - s_0))p(y^\alpha) \cdot e = 0.$$

Hence, since $p(y^\alpha) \cdot e > 0$, we obtain that $\lambda = \frac{\alpha}{n}(s - s_0)$ and thus $z(y^\alpha) = 0^n$, showing that y^α is a zero point of z.

Finally, suppose $y^\alpha \in \underline{Q}(s)$. Then there exists $\beta_1 \ge 0$ and $\beta_2 \ge 0$ with $\beta_1 + \beta_2 = 1$ such that

$$\beta_1\tau(y^\alpha + z(y^\alpha)) + \beta_2\tau(p(y^\alpha)) = y = y^\alpha - \frac{\alpha}{n}(s - s_0)e.$$

Using that $\tau(x)$ is the projection of x on Y, it follows that this can be rewritten as

$$\beta_1 y^\alpha + \beta_1 z(y^\alpha) + \beta_2 p(y^\alpha) = y^\alpha - \mu e$$

for some $\mu \in \mathbb{R}$. Since $\beta_1 + \beta_2 = 1$ we obtain

$$\beta_1 z(y^\alpha) = \beta_2(y^\alpha - p(y^\alpha)) - \mu e.$$

When $\beta_1 = 0$, we have $p(y^\alpha) = y^\alpha - \mu e$, which implies that $y^\alpha = q^0$, yielding the trivial equilibrium. When $\beta_1 > 0$, we obtain

$$z(y^\alpha) = \frac{\beta_2}{\beta_1}(y^\alpha - p(y^\alpha)) - \frac{\mu}{\beta_1}e.$$

Since $y^\alpha \in \underline{Q}(s)$, we have that $y_k^\alpha - p_k(y^\alpha) = -1$ for some $k \in I_n$ and $y_h^\alpha - p_h(y^\alpha) \ge -1$ for all $h \neq k$. Hence,

$$z_k(y^\alpha) \le z_h(y^\alpha) \text{ for all } h \in I_n.$$

On the other hand, $\ell_k(y^\alpha) = 0$ and thus $z_k(y^\alpha) \ge 0$. Therefore, $z_h(y^\alpha) \ge 0$ for all $h \in I_n$, and so $z(y^\alpha) = 0^n$ due to Walras' law. Q.E.D.

The theorem guarantees that for any $s > s_0$ there is a connected set of supply constrained disequilibria connecting the trivial disequilibrium with a supply constrained disequilibrium induced by a point $\overline{q}(s)$ in $\overline{Q}(s)$. The next theorem shows that when s is large enough $\overline{q}(s)$ induces a Walrasian equilibrium.

Theorem 4.2 Let $\mathcal{E} = (\{X^i, \succeq^i, w^i\}_{i=1}^m, P)$ be an economy satisfying Assumptions X, U, W. Then there exists an $\overline{s} > s_0$ such that for all $q \in Q$ with $\sum_{k=1}^n q_k \geq \overline{s}$ it holds that $z(q) \neq 0^n$ when $q \notin P$.

Proof.
Suppose, without loss of generality, that there exists a commodity $j \in I_n$ and a sequence $(q^r)_{r \in \mathbb{N}}$ in Q, satisfying $\sum_{h=1}^n q_h^r \geq r$, $z(q^r) = 0^n$ and $q_j^r < \underline{p}_j$ for all $r \in \mathbb{N}$. For any $r \in \mathbb{N}$, consider the vector $p^r = p(q^r)/\sum_h p_h(q^r)$. Clearly, we have

$$d^i(p^r, \ell(q^r)) = d^i(q^r),\ i \in I_m.$$

Since $z(q^r) = 0^n$ and thus $0^n \leq d^i(q^r) \leq w$, we may restrict the budget set of any consumer i to the set of consumption bundles with upper bound $w^i + w$. Now, denote $\ell^r = \ell(q^r)$ and $d^{i,r} = d^i(q^r)$ for all $i \in I_m$, $r \in \mathbb{N}$. Then $(p^r, \ell^r, d^{1,r}, \ldots, d^{m,r})_{r \in N}$ is a sequence in a compact set. Hence there is a subsequence converging to some $(\overline{p}, \overline{\ell}, \overline{d}^1, \ldots, \overline{d}^m)$ satisfying $\sum_{i=1}^m \overline{d}^i = w$. Moreover, since for all $r \in \mathbb{N}$, $\sum_{k=1}^n p_k^r = 1$ and $p_j(q^r) = \underline{p}_j$, we also have that $\overline{p}_j = 0$ and there exists some $h \neq j$ with $\overline{p}_h > 0$. This latter property implies that $q_h^r > \underline{p}_h$ for sufficiently large r and hence $\overline{\ell}_h \geq 1$. Following Drèze (1975), this implies that the demand functions are continuous at $(\overline{p}, \overline{\ell})$. Consequently, $\overline{d}^i = d^i(\overline{p}, \overline{\ell})$ for all $i \in I_m$. Since $\overline{p}_j = 0$ and $\overline{\ell}_k = 1$ for some $k \neq j$, it follows from the monotonicity of the preferences that $\overline{d}_j^i = w_j^i + w_j$ for all $i \in I_m$, and thus $\sum_{i=1}^m \overline{d}_j^i > w_j$, which contradicts $z_j(q^r) = 0$ for all $r \in \mathbb{N}$. Q.E.D.

The theorem implies that for s sufficiently large any point $\overline{q}(s)$ of the connected set $C(s)$ in $\overline{Q}(s)$ lies in P and hence induces a Walrasian equilibrium price system $p^* = \overline{q}(s)$. From this it follows that there exists a connected set C of zero points of z in Q containing both q^0 and a q^* inducing a Walrasian equilibrium. Notice that due to Walras' law $z(q) = 0^n$ is generically a system of $n-1$ independent equations in n unknowns, so that C is typically a 1-dimensional set, e.g. see Herings (1996). Hence, there exists a connected set of market clearing supply constrained disequilibria connecting the trivial non-trade disequilibrium to a Walrasian equilibrium.

5 The Dynamics of the Adjustment Process

In this section we consider the adjustment process induced by the connected set of supply constrained disequilibria. Under suitable differentiability conditions it can be shown that the connected set C contains a piecewise smooth path of zero points of z in Q connecting $q^0 = \underline{p} - e$ inducing the trivial non-trade market clearing disequilibrium and a $q^* \in P$ inducing a Walrasian equilibrium. The path starts in q^0 with full supply rationing and price vector $\underline{p}$ and in the short run adjusts rationings keeping all markets in equilibrium and all prices fixed until a point q^u is reached with $q_j^u = \underline{p}_j$ for at least one $j \in I_n$. At this point an SCD is reached with commodity j not being rationed, inducing a so-called unemployment equilibrium. Then the process continues with long run adjustment of prices and rationings. Keeping all markets in equilibrium, prices of commodities are allowed to increase above their initial value when they are unrationed, while simultaneously the supply rationings of the other commodities are adjusted with their prices still on the initial values. This means that as soon as a commodity becomes unrationed its price is allowed to increase. However, it may happen that in the long run the price of an unrationed commodity becomes equal again to its initial value. In that case the price of that commodity is kept equal to its initial value and the commodity is allowed to become rationed in its supply again. In case in the long run all prices return to their initial values, all commodities are being rationed in their supply and the process switches back to the short run. As soon as one of the commodities becomes unrationed, the process switches back to the long run. Eventually, the process reaches a point q^* in P and thus no commodity is rationed anymore. This q^* induces a Walrasian equilibrium without rationing.

To follow a path of zero points several mathematical path following techniques are available, for instance the well-known predictor-corrector continuation methods and techniques based on piecewise linear approximation, see Allgower and Georg (1980). Predictor-corrector methods can not be applied directly because they require differentiability of the reduced excess demand function z, which is not the case due to the switches in regimes between price and quantity adjustments and between rationing and nonrationing. Therefore, to use such a method a transformation of all variables as proposed in Garcia and Zangwill (1981) is needed. The piecewise linear approximation technique is based on taking a simplicial subdivision of the underlying set Q and linearizing the function z on each subsimplex, resulting in a piecewise linear function Z from Q to $\mathbb{R}^n$. Then there exists a piecewise linear path in Q from q^0 to a point q^* in P satisfying that for any point q on the path it holds that

$$Z(q) = \beta(q)e$$

for some $\beta(q) \in \mathbb{R}$. For every $\epsilon > 0$, it can be guaranteed that $|\beta(q)| < \epsilon$ for any q on the path by taking the mesh size of the simplicial subdivision small enough. Hence, any point on the piecewise linear path gives an approximating supply constrained disequilibrium, i.e. it induces a price system and rationing scheme with the absolute values of the reduced excess demands less than ϵ. Each linear piece of the path lies in one of the subsimplices of the subdivision and can be followed by a linear programming pivot step in a system of $n+1$ equations and variables. The pivot step determines endogenously which vector q has to be considered in the next step. This makes that the path can be seen as an adjustment process in discrete time. At any point in time the linear programming step determines a new point based on the vectors of excess demands in at most $n+1$ of the points generated before. This new point induces a new system of prices and rationings yielding an approximating SCD. This solves the problem of finding the next price and rationing scheme, as was raised by Weddepohl (1999) as an argument against adjustment of prices through a sequence of market clearing disequilibria.

References

1. E.L. Allgower and K. Georg (1980), *Numerical Continuation Methods, An Introduction* (Springer-Verlag, Berlin).
2. V. Bela and M. Majumdar (1992), Chaotic tatonnement, *Economic Theory* 2, 437-445.
3. V. Böhm (1993), Recurrence in Keynesian macroeconomic models, in: F. Gori, L. Geronazzo and M. Galeotti, eds., *Nonlinear Dynamics in Economics and the Social Sciences* (Springer-Verlag, Berlin), pp. 69-94.
4. F.E. Browder (1960), On continuity of fixed points under deformation of continuous mappings, *Summa Brasilia Mathematica* 4, 183-191.
5. R.H. Day and G. Pianigiani (1991), Statistical dynamics and economics, *Journal of Economic Behaviour and Organization* 16, 37-83.
6. P. Dehez and J.H. Drèze (1984), On supply-constrained equilibria, *Journal of Economic Theory* 33, 172-182.
7. J.H. Drèze (1975), Existence of an exchange economy under price rigidities, *International Economic Review* 16, 310-320.
8. C.B. Garcia and W.I. Zangwill (1981), *Pathways to Solutions, Fixed Points, and Equilibria* (Prentice-Hall, Englwood Cliffs).
9. P.J.J. Herings (1996), *Static and Dynamic Aspects of General Equilibrium Theory* (Kluwer, Dordrecht).
10. P.J.J. Herings, G. van der Laan, A.J.J. Talman and R.J.G. Venniker (1997), Equilibrium adjustment of disequilibrium prices, *Journal of Mathematical Economics* 27, 53-77.
11. K. Kamiya (1990), A globally stable price adjustment process, *Econometrica* 58, 1481-1485.
12. M. Kurz (1982), Unemployment equilibria in an economy with linked prices, *Journal of Economic Theory* 26, 110-123.
13. G. van der Laan (1980), Equilibrium under rigid prices with compensation for the consumers, *International Economic Review* 21, 53-73.
14. G. van der Laan (1982), Simplicial approximation of unemployment equilibria, *Journal of Mathematical Economics* 9, 83-97.
15. G. van der Laan (1984), Supply-constrained fixed price equilibria in monetary economies, *Journal of Mathematical Economics* 13, 171-187.
16. G. van der Laan and A.J.J. Talman (1987a), Adjustment processes for finding economic equilibria, in: A.J.J. Talman and G. van der Laan, eds., *Computation*

and Modelling of Economic Equilibria (North-Holland, Amsterdam), pp. 85-123.
17. G. van der Laan and A.J.J. Talman (1987b), A convergent price adjustment process, *Economics Letters* 23, 119-123.
18. D.G. Saari, Iterative price mechanisms (1985), *Econometrica* 53, 1117-1131.
19. H. Scarf (1960), Some examples of global instability of the competitive equilibrium, *International Economic Review* 1, 157-172.
20. S. Smale (1976), A convergent process of price adjustment and global Newton methods, *Journal of Mathematical Economics* 3, 107-120.
21. E.C.H. Veendorp (1975), Stable spillovers among substitutes, *Review of Economic Studies* 42, 445-456.
22. H.N. Weddepohl (1987), Supply-constrained equilibria in economies with indexed prices, *Journal of Economic Theory* 43, 203-222.
23. H.N. Weddepohl (1994), Erratic dynamics in a restricted tatonnement process, in: J. Grasman and G. van Straten, eds., *Predictability and Nonlinear Modelling in Natural Science and Economics* (Kluwer, Dordrecht).
24. H.N. Weddepohl (1995), A cautious price adjustment mechanism: chaotic behavior, *Journal of Economic Behavior and Organisation* 27, 293-300.
25. H.N. Weddepohl (1997), Simulating tatonnement in a production economy, *Economic Theory* 10, 551-558.
26. H.N. Weddepohl (1999), Price adjustment models in a production economy, *Annals of Operations Research* 89, 149-164.

A General Equilibrium of Managed Services

Pieter H.M. Ruys*

CentER, Tilburg University

Abstract. A service is a value-creating interaction between individuals, each in the role of either a performer or a receiver. A service is represented by a network on the set of individuals. There are many service networks possible, which all belong to the service network possibility set. The service structure consists of this set and a value function on the service networks. Specialization of individuals leads to separating the performer interaction from the receiver interaction in the service structure. The service concept is thus extended to a stratified service with interactive groups on each level, represented by two hierarchical graphs. The main idea behind this paper is that services cannot be rendered unless they are managed. So a governance structure is introduced with agents who obey specific rules of management. There are two types of agents, managers - both for a performers organization and for a receivers organization - and consumers. An equilibrium for an economy with managed stratified services is defined and shown to exist for a market economy with rather strong conditions on both the service structure and the governance structure. It is shown to be a generalization of a neoclassical general equilibrium.

Keywords: Service economy, management of cooperation, general equilibrium, corporate governance, organization theory, and network industries.

1 Introduction

The service economy as introduced here involves some hard problems. Some are too hard to solve for the moment. A service is rendered between individuals, but economic services are closely tied to and identified by the institutions rendering those services. The term 'institution' has, however, two meanings: (i) that of an entity rendering a service or (ii) a set of institutional arrangements governing a variety of social interactions, which meaning is called here a governance. As Hurwicz (1994, p.11) remarked, "institutional arrangements are what one may call 'categorical', in the sense that they apply to a category of situations and/or agents, rather than to specific cases and places. This amounts to regarding institutional arrangements as particular kinds of game forms, rather than arbitrary ones." The approach introduced here focuses on roles people and institutions play in relation to other people, and on

* This paper is part of a long-term research project *Competition and Cooperation* launched in 1980 by the MathEcon groups of the UvA and the VU at Amsterdam, the UM in Maastricht and the KUB in Tilburg. I want to thank René van den Brink and Rob Gilles, my closest companions in developing the ideas introduced here. I also thank Dave Furth and the referee of this paper for valuable comments.

the relational characteristics of services, rather than on the input and output characteristics. This is completely in line with Hurwicz's observation that "as for the property of categoricity, it is characteristic of legal or customary rules that they are formulated for persons in certain roles, say buyer, seller, parent, spouse, etc." (1994, p.11). Following the lead of Williamson (1967) and Keren and Levhari (1979), van den Brink and Ruys (1996) have applied the role approach in a partial equilibrium framework. They achieved determining the (hierarchical) structure of an organization endogenously rather than it being imposed exogenously. Later, they introduced cultural traits in organizations and framed the partial model in a market equilibrium (Ruys e.a., 2000).

Many authors have written about the service economy[1], but the service concept has not been defined formally in a general equilibrium framework This paper introduces such a framework. A service is introduced as a value-generating interaction between agents. A network describes the interactions between the agents. The value-creating properties of this interaction are carried by the relation or by the network of the service. That generalizes the concept of a commodity, where the attractive properties are embodied in an object, represented by a point in the Euclidean space. It follows, however, that the aggregation of services will be much harder to define than the aggregation of commodities. Only similar networks can be added, so networks have to be made similar. That is achieved by the role approach and by distinguishing receivers from performers of a service. An economic service is characterized by the possibility of substitution between services and between individuals. If an individual is involved in a service, he or she is playing an economic role as a performer or as a receiver of that service. Performers are specialized and cooperate, which features are present in the given service network. The service opportunity structure provides a large but manageable variety of service networks. Since there are more service networks possible than can be realized, a selection and organization process has to be designed. That is the task of agents and managers, whose behavior is described by the governance structure of the society. Both structures together constitute a *service society.*

The management of a service society is based on the capability of individuals to engage in transactions. The individual becomes an agent. Some agents will become manager of a performers organization, called a provider; some will become manager of a receivers organization, called a procurer. All agents are consumers who demand a managed service and supply their labor service to some organization. An *economic service* is defined as a monetary transaction between two agents concerning a value-generating interaction be-

[1] Consider, for example, the description given by Giarini and Stahel (1993). They observe that in the Service Economy people are buying functioning systems, not tools or products. Services mean performances, in real periods of time, which means that values must be based on probabilities. They describe the relevance of these notions, but don't present them in a coherent model.

tween receivers and performers of that service. The service society becomes a *service economy*. All agents are maximizing specific behavioral rules, which allows for defining a Nash-equilibrium. The governance aims at coordinating decisions of agents such that expected values are managed and realized. In equilibrium the agents meet again as performer or as receiver. So the equilibrium concept manages the risks due to interdependencies via institutional cooperation. This view is in line with the findings of sociologists like Coleman (1988, 1990) and Granovetter (1985), who maintain a simple behavioral theory of rational choice, and focus attention on the impact of social conditions, such as institutional cooperation or embeddedness of interactions, on individual behavior.

In order to arrive at operational results, rather strong conditions are imposed on the service opportunity structure and the governance structure. One is that the size of an organization can be controlled by a single parameter: the depth or length of an organization. The equilibrium concept will eventually be restricted to a market economy, which governance is well known. This excludes institutions that describe cooperation among receivers, imply social choices or procure public services and therefore require another type of governance.

Organization theory focuses on the interaction between roles that agents play. It comes in several forms. The currently dominating paradigm sees organizational forms as responding to a variety of incentive problems in which agents act strategically, see, e.g., Calvo and Wellisz (1979). An alternative view is to consider the organization as solving an information processing problem, see, e.g., Marschak and Radner (1972), Keren and Levhari (1979), or more recently, Bolton and Dewatripont (1994) and van Zandt (1997). A third view is to consider the organization as solving a coordination problem by assigning tasks to members of the organization, see, e.g., Williamson (1967, 1979), Alchian and Demsetz (1972), and Rosen (1982), or in terms of delegation of authority, see, e.g., Hart and Moore (1999). This paper presents a model of governance along the latter lines. It contributes to the question of boundaries, as Holmstrom (1999) has named it: the boundary between the internal organization of the firm and the operation of markets, which is part of the external organization of the firm. Ichiishi (1993) has explored cooperation in a general equilibtium context. Aoki (2001) identifies an institution as a self-sustaining system of collectively shared beliefs about non-technical, self-enforcing rules of the game that governs the strategic interactions of the agents. His approach applies to some concepts used in this paper. The network approach, as introduced by a.o. Jackson and Wolinsky (1996), is fundamental for the design developed here.

The paper is organized as follows. The next section presents the conceptual framework of the service opportunity structure. The management of services and the equilibrium concept is introduced in Section 3. Section 4 concludes.

2 The Service Opportunity Structure

2.1 The Set of Possible Value-generating Networks

A *service* is defined as a value generating interaction between one or more performers and one or more receivers. A set $M = \{1, \ , m\}$ of individuals are connected in a reciprocal network relationship, modeled as a nondirected graph. Individuals are the nodes in the graph and links indicate bilateral relationships between individuals. Thus, a *network* z is simply a list of which pairs of agents are linked to each other. Consider a pair of individuals i and j, then $\{i, j\} \in z$ indicates that i and j are linked under the network z. The relation is bilateral and requires consent of both parties. More formally, let z^M denote the set of all subsets of M of size 2. Then $Z = \{z \subset z^M\}$ denotes the set of all possible networks or graphs on M, with z^M being the full or complete network.

Different network configurations lead to different values of overall production or overall utility to a society. These various possible valuations are represented via a *value function* $w : Z \to \mathbb{R}$, with $w(\emptyset) = 0$. Let $\mathcal{W}$ denote the set of all possible value functions. Jackson (2001), from whom this section draws, notes that different networks connecting the same individuals may lead to different values. This makes the value function a much richer object then a characteristic function used in cooperative game theory. A value function keeps track of how total societal welfare varies across different networks. The allocation or distribution of this value among the individuals forming the network will eventually be defined by an *allocation rule,* $d : Z \times \mathcal{W} \to \mathbb{R}^M$, which can be defined for any game $(M, v_{z,w})$ with the characteristic function $v_{z,w}(S) = w(z \mid S)$, for $S \subset M$. This function depends both on the network z and the value function w, where $z \mid S$ means that only those parts of the network z are considered in which the individuals belong to the coalition S. The allocation rule may assign, for example, a Shapley-value with $d(z, w) = \varphi^S(M, v_{z,w})$. The distribution of a value can only take place within a governance framework. That concept is specified below in Section 3.

The triple (M, Z, w) is the primitive of the model and describes the set of possible value-generating networks. So a non-directed network represents here the value-generating interaction between two or more individuals. This interaction between individuals is therefore symmetric and both sides of the relation assign the same value to that interaction. These properties erode when specialization occurs and relations become more anonymous. The step that transforms a service into an economic service is set next.

2.2 Stratification of Services by Performers and by Receivers

The interaction between individuals is the cornerstone of value-generation. Individuals in an economy are building on this cornerstone to increase the value generated by specialization. Three developments in social behavior describe this specialization. Firstly, certain types of interaction are aggregated into a category of interactions that have approximately the same content, which category is called a service or a service sector. One may think of the financial services, care, or agricultural services. Secondly, an individual may adopt one of two roles, the role of performer in some service or the role of receiver of some service. This distinction is comparable to an individual being a producer or a consumer. The set of possible service networks is so rich that any individual in the economy can serve as a performer and as a receiver for any service. Value is always generated between some pair of a performer and a receiver, but the height of value will vary greatly among these pairs, of course. A selection from these networks will eventually be made by means of the rules of management.

So the individuals in the set of possible value-generating networks (M, Z, w) are being identified as a performer of a service and as a receiver of a service. A finite set N of services is given, indexed by $j = 1, \dots, n$. Each individual $i \in M$ may assume the role of performer of a service, denoted by $i \in M_j^P$, and the role of receiver of a service, denoted by $i \in M_j^R$. So a possible network for some service, z_j, is now a subset of $M_j^P \times M_j^R$ instead of the set $M \times M$. If $(f, g) \in z_j$, then $f \in M_j^P$ and $g \in M_j^R$. The set $Z = Z_1 \times Z_2 \times \cdots \times Z_n$ is the set of possible service networks. The value function is still defined on Z, with $w(\emptyset) = 0$. Since $z \subset M_j^P \times M_j^R$ is a bipartite graph, only the interaction between one or more performers and one ore more receivers generates value. A bipartite service network is represented in Figure 1.

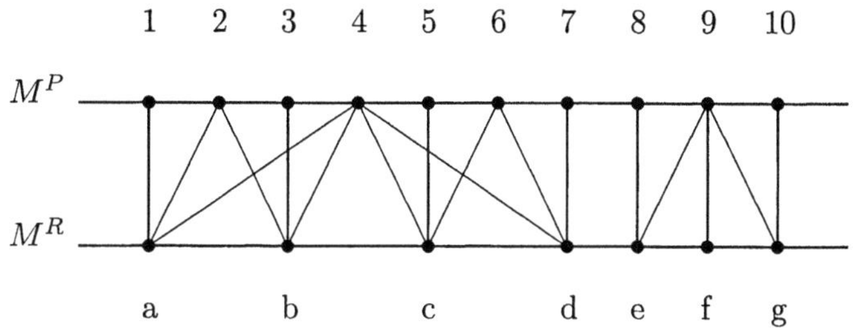

Figure 1: A bipartite service network with 2 connected sub-networks

Thirdly, people specialize again in the performance or in the reception of a service. Consider first the case of performance. A performer may receive support from other performers to render the service more efficiently: the case of vertical interaction between performers. Or performers may cooperate between each other to increase the value generated by the group: the case of horizontal interaction. For example, the interaction between a teacher and her students –the service-relation between a performer and a set of receivers–

is part of a service-network rendered by a team of school teachers, which team is in turn assisted by a school organization. This last service-network is part again of the whole service-network that is called the 'education-service'. So a hierarchy emerges in the set of roles of a service network, which can be described by a directed graph. Some roles are reserved for performers who directly interact with receivers: those are the *front roles* of a service-network. Consider the smallest set of horizontally interacting front roles. Some performer is coordinating these front-performers on a vertically higher level, and so on. These covering or succeeding performers have indirect performers roles. So the roles in a performers' service-network are vertically composed of various layers or *strata* from the front roles up, each of which is specialized in rendering a more abstract part of the composite service. Each stratum is horizontally partitioned in groups of interacting performers. A performer on a higher level of the service network coordinates each such group.

For each service, such a hierarchical interaction structure of performers is described by the graph-theoretical concept of a forest. That concept is introduced as follows. Given a network $z \in Z$ a *path* between i and j is a sequence $i_1, \dots, i_K$ of individuals such that $(i_k, i_{k+1}) \in z$ for each $k \in \{1, \dots, K-1\}$, with $i_1 = i$ and $i_K = j$. A *cycle* is a path between i and j with $i = j$ and containing at least one link. A *tree* is a connected graph $z' \subset z$ that contains no cycles. A set of trees is a *forest*. Any two individuals in a tree are connected by exactly one path in the tree, and there is no path connecting any two individuals in different trees. Both conditions define a component of a network. So a tree is component of a stratified performers' network. A tree is also *partially ordered*, denoted by $(M^T, \preceq^T)$, if the relation $\preceq^T$ on M^T is reflexive, antisymmetric and transitive. Individual j *succeeds* or *coordinates* individual i in the tree T if $j \succ^T i$. If j coordinates i and there is no h in M^T such that $j \succ^T h \succ^T i$, then j coordinates i directly. A partially ordered tree has a unique top individual coordinating all other individuals. The set of individuals coordinating nobody in the tree is called the set of pendants or *front individuals* of the tree and is denoted by $M(T)$.

Definition 2.1: Given a set of possible service networks (M, Z, w), a finite set of services N, and, for each service $j \in N$ with service network z_j, a set of performers $M_j^P \subset M$ and a set of receivers $M_j^R \subset M$. Then a *stratified performers' network* (M_j^P, z_j^P) is a forest in which each tree is partially ordered by a coordination relation that is derived from the service network. Some individual performer is said to *coordinate* a set of directly preceding performers if these performers interact or cooperate horizontally according to a criterion in which the succeeding performer plays a distinctive role. Similarly, a *stratified receivers' network* (M_j^R, z_j^R) is a forest in which each tree is partially ordered by a coordination relation.

So both performers and receivers in a stratified service network assume specialized roles that are horizontally comparable and vertically ordered. Horizontal interactions within a stratum arise when performers cooperate to serve a receiver. Such a set of interdependent performers with an internal horizontal interaction is called a *group*. Think of a medical team assisting a surgeon who performs a surgery on a patient. Vertical externalities emerge when a service to be rendered requires a complementary infrastructure service. So a service is decomposed into sub-services, which are integrated into a service. The medical team needs the support of a hospital, and the hospital is embedded in a health system. The supply of electricity at home requires an ordered set of complementary sub-services that are vertically integrated. Similarly, from the point of view of the receivers of a service, children receiving education belong to some family, which family cares for a school and itself belongs to a community that establishes an education system, etceteras. Examples of vertical complementarities are abundant. So each stratified service can be decomposed into partially ordered sets or groups of interacting performers, with groups on the same level rendering the same (sub) service.

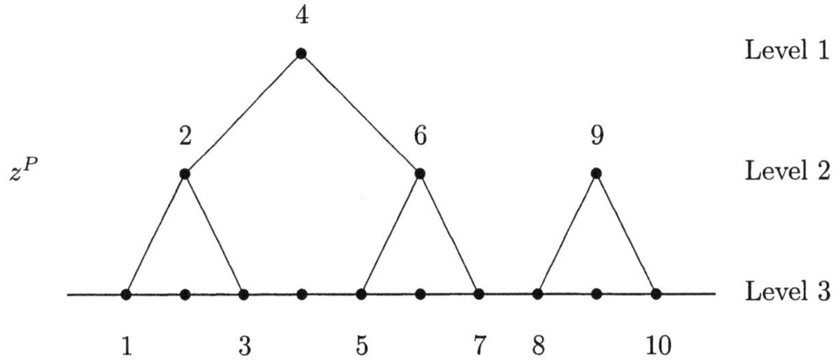

Figure 2: A stratified performers' network

The stratified performers' network in Figure 2 corresponds with an underlying bipartite service network in Figure 1. This can be seen as follows. The performers who serve only one receiver can only be front-performers (the performers 1, 3, 5, and 7). The front performers 1 and 3 interact in serving receivers a and b because a coordinating performer 2 is also serves a and b. The front-performers 5 and 7 also interact and are coordinated by performer 6. Finally, the performers 2 and 6 interact because the coordinator 4 covers their receivers. It is understood that the performer who covers front-performers serves their receivers indirectly. The value of this covering relation between performers is derived from the basic value-generating interaction between the front-performer and the receiver of the service.

The stratified services in the service structure and their value are potential services. They need to be realized by managers who themselves are ruled by a governance to be specified in the next Section. In this section the service structure will be specified further to obtain the concept of a manageable stratified service.

2.3 Valuation of Stratified Services

The value of interaction within each type of service network is defined as follows. A performer forest interacts with some receiver forest if the individuals on the front positions of both forests interact and generate welfare for those involved according to the social welfare function. Since all individuals involved are playing either the role of performer or the role of receiver, one individual may substitute another in some role. This assumption is a kind of anonymity condition, which is fundamental for the institutional approach developed in the next section. It is in contrast to the case of reciprocal exchange, where the individual cannot be substituted by another (see Kranton, 1996). So the social welfare generating capacity of interacting *individuals* in networks may be replaced by the interaction between *roles* in a performer forest and roles in a receiver forest. Then, given a receiver forest, the value of some performer forest can be derived from the social welfare function, as well as the value of a receiver forest follows from the social welfare function, given some performer forest. This is done as follows.

A value function w is given and defined on the set of possible (bipartite) networks for all services, Z. The function $w : Z \to \mathbb{R}$ assigns a value $w(z)$ to each interaction structure between performers and receivers as described by the network z in Z. This value is generated by the interaction between one or more performers and one ore more receivers, including those in the front roles, of the stratified services represented by z.

Consider a service $j \in N$ with a stratified performer' network (M_j^P, z_j^P) and a stratified receivers' network (M_j^R, z_j^R). So the service network $z_j \in Z_j$, from (M, Z, w), the set of possible value-generating networks, is decomposed into two forests z_j^P and z_j^R. The value function w, however, is defined on the whole set $z \in Z$. Both decomposed sets are 'glued' together by the following operation.

A forest z_j^P in Z_j^P is decomposed by a set of connected components, called trees, $\{z_j^{P,1}, z_j^{P,2}, \ldots, z_j^{P,t}\}$. Each tree $z_j^{P,k} \subset M_j^{P,k} \times M_j^{P,k}$, is a directed graph for $k = 1, 2, \ldots, t$. Let $[M_j^P]$ denote the partition of the performer set M_j^P into subsets of performers within some tree, $\{M_j^{P,1}, M_j^{P,2}, \ldots, M_j^{P,t}\}$. Similarly, a receivers' forest z_j^R can be decomposed on the set M_j^R of receivers. The set $[M_j^R]$ represents the partition of the receiver set into subsets of receivers within a tree. Both the value function on the set of performer forests and the value function on the set of receiver forests can now be derived from a well-chosen social welfare function $w \in \mathcal{W}$.

Let $\zeta_j = [M_j^P] \times [M_j^R]$ denote the complete[2] bipartite network between the tree populations. The service network in Z_j that is generated by some performer tree z_j^P and some receiver tree z_j^R, denoted by $z_j^P \otimes z_j^R \subset M_j^P \times M_j^R$, is defined as follows. For each $(g,h) \in \zeta_c$, if $p \in M_j^{P,g}$ and $r \in M_j^{R,h}$, then $(p,r) \in z_j^P \otimes z_j^R$. Let $z = (z_j, z_{-}j)$. The value of the performer forest z_j^P is a function $v_j : Z_j^P \times Z_j^R \times Z_{-j} \to \mathbb{R}$, defined by:

$$v_j(z_j^P | z_j^R, z_{-j}) = \max\nolimits_{z_j \in Z_j} \{w(z_j, z_{-j}) | z_j \subset z_j^P \otimes z_j^R\}.$$

The value of a performer forest is thus contingent on the realization of a corresponding receiver forest for the same service and on the realization of specific networks of the other services. Although this realization is crucial for the valuation of the forest, it falls outside the scope of the individuals within the forest. That is expressed by saying that the value v_j implies that the individuals concerned *believe* for some reason that the networks (z_j^R, z_{-j}) will be realized. Later, when selected networks are managed, this belief is transformed into an expectation, based on the probability of realization as determined by the governance structure. Similarly, it is assumed that the value-enhancing interaction between receivers, which is also implicitly given in the social welfare function, can be made explicit. The utility or user value of some receiver forest is defined by:

$$u_j(z_j^R | z_j^P, z_{-j}) = \max{}_{z_j \in Z_j} \{w(z_j, z_{-j}) | z_j \subset z_j^P \otimes z_j^R\}.$$

If beliefs fully match, that is, if z_j^P in the performers function equals the conditional z_j^P in the receivers function, and z_j^R in the receivers function equals the conditional z_j^R in the performers function, then $v_j(z_j^P | z_j^R) = u_j(z_j^R | z_j^P)$. The value of the performers' forest for all services is conditional on the receivers' forest for all services:

$$v(z^P | z^R) = \max\nolimits_{z \subset\, z^P \otimes z^R} w(z).$$

Similarly,

$$u(z^R | z^P) = \max\nolimits_{z \subset\, z^P \otimes z^R} w(z).$$

So in equilibrium, $v(z^P | z^R) = u(z^R | z^P) = w(z)$, which shows consistency between these concepts. They have an independent meaning in the planning procedure, when beliefs are not (yet) realized and an imperfect match may be planned. This implies that somewhere services become scarce due to a managerial misfit, which scarcity is eventually balanced by the price mechanism in the equilibrium concept.

[2] Completeness implies that all performer trees are substitutes for a receiver tree, and vice versa. Reciprocal exchange between individuals would require additional restrictions on this network.

So the value of cooperation between performers (or the value of integration within a performer forest) can be derived from the value of interaction between performers and receivers. That last type of interaction is fundamental in the model, the first type, the interaction between performers themselves, is instrumental and its value is derived. The same is true for cooperation between receivers.

2.4 Manageable Stratified Services

The stratification of performers in a service network is characteristically determined by the content of the service. A health service, for example, has other chains of support than a transport service. The same observation can be made about the stratification of receivers in a receiver' network. A service that is a public service (a public good), for example, has at the receivers' side a coordinator at the top who coordinates, possibly indirectly, all members of the society. So two dimensions characterize a service: its content and its form, both on the receivers' side and on the performers' side. When the demand for a service varies, that will affect its size, but not its content and its form. So a concept or a rule has to be developed that allows for changing the size of a forest and the set of receivers, but that leaves characteristic elements of its form invariant. A general rule for contracting or expanding networks leaving its interaction or cooperation characteristics (represented by dependent sets) intact is given by the concept of a partially ordered antimatroid[3]. A simpler rule is proposed below.

Definition 2.2: The set $\mathcal{P}_j$ of *manageable performers' networks for service* j contains all and only forests $P_j = (M_j^P, z_j^P)$ that are *congruent*, that is, have similar interaction or cooperative properties that are characteristic for the jth service. Similarly, the set $\mathcal{R}_j$ of *manageable receivers' networks for service* j contains all and only congruent forests $R_j = (M_j^R, z_j^R)$.

For simplicity, it is assumed here that removing a layer of front-positions in a manageable component of a network, which makes the succeeding layer a layer of front-positions in that component, is an admissible expansion or contraction rule for a manageable service network. A forest P_j in $\mathcal{P}_j$ is decomposed by a set of connected components, called trees, $\{P_j^1, P_j^2, \ldots, P_j^t\}$. Each tree P_j^k is a directed graph, represented by a set of nodes in M_j^P and a set of arcs in z_j^P, that is, $P_j^k = (M_j^{P,k}, z_j^{P,k})$, with $z_j^{P,k} \subset M_j^{P,k} \times M_j^{P,k}$, for

[3] The concept of a poset-antimatroid (see Bilbao, 2000) describes the desired structure in a more general way. The purpose is to describe a structure that allows for contraction or expansion of a service network leaving its interaction or cooperative characteristics (represented by dependent sets) intact.

$k = 1, 2, \ldots, t$. Since there is a unique path from each front-position to the top-position in a tree it can be assigned length $\ell(p, P_j^k)$.

In order to simplify the analysis further, the trees in each forest P_j or R_j are assumed to be isomorphic. So for each forest the length of all its trees is equal and $\ell(p, P_j^k) = \ell(P_j)$, for all k. A similar assumption is made for the receivers' side of a manageable service network, so $\ell(r, R_j^k) = \ell(R_j)$, for all k. It follows that both networks are indexed by their length or depth. The number of performers' trees in the forest is equal to $|P_j, (\ell_i^P)|$. Similarly, the receivers forest with depth ℓ_j^R is given by $R_j(\ell_j^R)$, where ℓ_j^R is a real number denoting the number of levels.

Since for each service, the variable ℓ_j parameterizes both the performers' forest and the receivers' forest, the value functions mentioned above can be redefined in terms of this parameter: $u_j(\ell_j^R|P_j, z_{-j}) = u_j(R(\ell_j)|P_j, z_{-j})$. Let $R(\ell^R) = (\mathrm{R}_1(\ell_1^R),\ R_1(\ell_2^R),\ \ldots,\ R_1(\ell_n^R))$ and $\ell^R = (\ell_1^R, \ell_2^R, \ldots, \ell_n^R)$; a similar convention applies for $P(\ell^P)$ and ℓ^P. Then the performers value function for service j is:

$$v_j(\ell_j^P|\ell_j^R, \ell_{-j}) = v_j(P_j(\ell_j^P)|P_{-j}(\ell_{-j}^P), R(\ell^R)) \tag{2.1}$$

and the receivers value function for service j is:

$$u_j\left(\ell_j^R|\ell_j^P, \ell_{ij}\right) = u_j\left(R_j\left(\ell_j^R\right)|R_{-j}\left(\ell_{-j}^R\right), P\left(\ell^P\right)\right) \tag{2.2}$$

Both functions are derived from the underlying welfare function w, so their properties can in principle be derived from this function. Both functions are supposed to be continuous and monotonously increasing in ℓ. This assumption is needed to prove existence, but implies that the number of levels is also a continuous variable. This methodological assumption is not unusual for network models, see Keren and Levhari (2002). It is not so strong an assumption if the variety of networks for some stratified service is large. The receivers value function u is quasi-convex, the performers value function v is quasi-concave. The variable ℓ refers to the depth of a service and therefore to its quality of a service rather than to the quantity of a commodity, which is the usual domain of a utility function. This quality is a potential and the optimum quality will be determined by the tradeoff between its benefits and costs, included its costs of governance.

Finally, assume that only the front-performers of a performer tree interact with only the front-receivers of a receiver tree and generate value from interaction. This implies that all the value added by increasing the length of the path is discounted in the front-subservice.

Definition 2.3: The series $S = (M, N, (\omega_0^i)_{i\in M}, \{\mathcal{P}_j(\ell_j^P),\ v_j(\ell^P|\ell^R\}_{j\in N}, \mathcal{R}_j(\ell_j^R), u_j(\ell^R|\ell^P)\}_{j\in N}, w)$, consisting of a set of individuals M, a set of services $N = \{0, 1, \ldots, n\}$, with a vector $(\omega_0^i)_{i\in M}$ of labor-service resources, a set of manageable performers' networks with corresponding service-value functions $\{\mathcal{P}_j(\ell_j^P), v_j(\ell^P|\ell^R)\}_{j\in N}$, a set of manageable receivers' networks with corresponding service-value functions $\{\mathcal{P}_j(\ell_j^R), u_j(\ell^R|\ell^P)\}_{j\in N}$, and a social welfare function w on Z, describes a *society with a manageable service opportunity structure.*

This structure still offers a large variety of value-generating service networks. The agents and managers of the society will select some of these, empowered by the governance structure of the society.

3 The Governance Structure

The basic idea in this paper is expressed by the assumption that a service cannot be rendered unless it is managed. People can choose from a large variety of potential services, much larger than can be realized. A choice from these alternatives has to be made. Any individual has a variety of potential services at her disposal, from which she has to select and to realize an optimal one. Realization of a service usually means the management of transactions to arrive at a reciprocal exchange. This feature is modeled in the service economy. An *agent* in the service society is an individual capable of making a choice and of entering into a transaction with other agents. An *economic service* is defined as a monetary transaction embodying a value-generating interaction between receivers and performers of that service.

In a service economy, adequate institutions have been developed to choose and to manage potential services. Institutions are based on roles or positions and on relations between these positions. Individuals assume certain roles. Principal-agent contracts govern the relations. This methodological device does not imply that role characteristics are more important than individual characteristics, but it focuses attention on the systematic aspects of the interaction between the roles in organizations. An adequate and well-defined governance structure is required to facilitate and coordinate management. An *organization* is a set of *positions* connected by contractual *relations* describing task and competence of each pair of positions, aimed at realizing some stratified service. It is represented by a directed graph. The governance structure distinguishes between the internal organization of decision units buying and selling services and the external organization between decision units coordinating their decisions. This section introduces two internal organizations. The only external organization in this paper is the market, introduced in Section 4.

3.1 The Provider of a Service

An agent in the service economy becomes a *manager of a performers' service network* if, in the process of transforming (values from) interactions into (values from) transactions, he establishes an organization that supports a performers' service network, which is called a *provider* of the service. Let a set $\mathcal{P}_j(\ell_j^P)$ of performers' service networks for service j be given. The set of suitable performers' organizations $\mathcal{F}_j$ is specified as follows. Each organization $F_j = (M_j^F, L_j^F)$ in $\mathcal{F}_j$ is a hierarchy with a nodes representing a role or position in the organization and an arc $(i,j) \in L^F$ representing a principal-agent relation with principal i and agent j. The principal-agent relations in a provider's organization are top-down oriented. It follows that any principal is a coordinator of a set of cooperating or interacting agents. The principal at the top $t(F)$ is the manager of the provider. An agent in the set $x(F_j)$ of performers who have no agents in the organization and who interact directly with the receivers, assumes a front-position in the organization and is called a *front-worker*. There is a unique path (t,f) from the manager to each front-worker $f \in x(F_j)$ with length ℓ_j^F.

Since the manager wants to support a performers' service-network in $\mathcal{P}_j(\ell_j^P)$, he will restrict the set of suitable performers' organizations to those organization-networks that are isomorphic to $\mathcal{P}_j(\ell_j^P)$. This set is denoted by $\mathcal{F}_j(\ell_j^F)$. Each organization in $\mathcal{F}_j(\ell_j^F)$ is then identified by its length $\ell_j^F = \ell(F_j)$.

The number of agents that are directly connected with a principal is called the *coordination-span* or *scope*. Since this scope is assumed to be uniform for the organization and characteristic for the stratified service, it can be denoted by κ. A second parameter characterizing the provider's organization is the X-inefficiency or the cost of the organization. This is expressed by assuming that each layer in the organization absorbs a uniform share (1-μ) of the value added, with $0 < \mu < 1$, the organizational slack. For reasons of simplicity, this value is equal for all organizations providing the same service and is given. So μ_j is the indicator of the X-inefficiency or quality of the organization.

The transaction value of the service provided is given by the performers value function (2.1), conditional on the realization of the plans of all other managed services in the economy. Although in equilibrium all these plans have to be realized, the manager has to make a decision possibly outside a state of equilibrium. The valuation of a stratified service has been based in Section 2.3 on a belief about the realization of the other stratified services. Such a belief can be transformed into an expectation if a probability can be attached to the realization of the different manageable service networks, which probability distribution has to be determined by the governance of the economy. That problem, however, goes beyond the scope of this paper, so the expected values are introduced as an assumption and become redundant in a state of equilibrium. The provider therefore knows service value function $v_j(\ell_j, \mathcal{E}(\ell_j)) = v_j(\ell_j^F | \ell_{-j}^F, \ell^R)$, where $\mathcal{E}(\ell_j)$ is the vector of expectations of the other managed services than service j, which he needs to know to decide. The expectations are realized in a state of equilibrium.

The service value function $v_j(\ell_j, \mathcal{E}(\ell_j))$ gives the external value of the provider's service, which is enhanced by the internal organization of cooperation between performers. The internal distribution of this value among the performers is determined by an allocation rule $d_j(\ell_j)$: $F_j(\ell_j) \times v_j(\ell_j, \mathcal{E}(\ell_j)) \rightarrow \mathbb{R}^{M(F(j))}$, which rule[4] assigns to any performer i in the set of performers $M^{F(j)}$ a value $d_j^i(F_j(\ell_j)\times\ v_j(\ell_j,\ \mathcal{E}(\ell_j)) = d_j^i(\ell_j^F)$. One may observe that the assumptions on the model make the distribution rule dependent only on the parameter that determines the length of the organization F_j. The distribution function is derived from a game $G(M^F, v_j)$ on the positions in the organization by extending the value function over all possible coalitions and by restricting the value of coalitions according to the specification of the principal-agent relations in the organization. The rules of the game G are invariant for the size of the organization F_j and may generate the permission value proposed by van den Brink and Gilles (1996) and van den Brink (1997), which satisfies the properties of budget neutrality, structural monotonicity, and symmetry. Budget neutrality means that the budget is balanced, that is, the sum of budgets (wages) allocated to all positions in the firm equals the value generated, $\sum_i d_j^i(\ell_j^F) = v_j(F_j(\ell_j^F)|\mathcal{E}(\ell_j))$. Structural monotonicity means that a principal does not receive a lower wage than his agent does. Symmetry means that the function assigns the same income to all positions within the same level of the organization. Let $d_j^{i,s}(\ell_j^F)$ be the *positional wage* of a performer i at level s. Homogeneity implies that all wages are equal on the same level, so the internal wage distribution is given by $d_j^s(\ell_j^F)$, s=1,2, ..., ℓ_j*, where ℓ_j^* is the optimal length for service j. The wage at the top-position $d_j^0(F_j\ (\ell_j^F))$ is the manager's income. In van den Brink and Ruys (1996) it is shown that for any constant elasticity of substitution performers' value function $v_j(\ell_j, \mathcal{E}(\ell_j))$, given the distribution game $G(M^F, v_j)$, there exists a size ℓ_j^* of the provider that maximizes the income of the top-position. This size is increasing with the degree of substitution in the value function v_j, representing the degree and the shape of internal interaction among the performers in the provider.

The optimal size ℓ_j^* is also increasing, however, with the value v_j of the service j relative to the values of the other services, as is shown in Ruys and van den Brink (1999). This effect of the external value of the service on the size of the provider, taking into consideration the internal effect, is expressed by the function $\ell_j^* = \lambda_{j,k}(p_j)$, where $p_j = v_j(\ell_j^*|\ell_{-j}^F, \ell^R)$ equals the value based on the plans of other managers in the economy. That leads to the following definition.

[4] This function can be defined for any game $(M^F,\ v_{F,v})$ with the characteristic function $v_{F,v}(S) = v(F|S)$, for $S \subset M^F$ and depends both on the network F and the value function v, where $F|S$ means that only those parts of the network F are considered in which the performers belong to the coalition S. The allocation rule may assign, for example, a Shapley-value with $d(z,\ w) = \varphi^S(M, v_{z,w})$. The distribution of a value can only take place within a governance framework.

Definition 3.1: The manager of provider k of service j *maximizes profits* by choosing the size $\ell^*_{j,k}$ of an organization in $\mathcal{F}_j(\ell^F_j, \kappa_j, \mu_j, d_j)$ such that the residual income of the top-position d^0_j is maximized. The function $\ell^*_j = \lambda_{j,k}(p_j)$, where $p_j = v_j(\ell^*_j|\ell^F_{-j}, \ell^R)$ is called the *price effect* on the optimal size of the provider. So the *supply of service j by provider k* to its receivers is given by the three set-functions of front-workers $x_{j,k}(F_j(\ell^*_{j,k})) = x_{j,k}(\lambda_{j,k}(p_j)) = x_{j,k}(p_j)$, with changing domains of definition.

When the service of front-performers is homogeneous, the set function $x_{j,k}(p_j)$ becomes a real function with a value that is equal to the value of the scope parameter κ_j taken to the power $\ell^* = \lambda_{j,k}(p_j)$. In that case, services become similar to commodities and the relation between individual receivers and individual performers can be broken. Also, the *demand of labor-input* is a real function $x_{0,j,k}(p_j)$, which is equal to the sum of labor over all the levels, $q = 0, 1, \ldots, \ell^*$; the amount of labor needed for level q is equal the value of the scope parameter κ_j to the power q.

3.2 The Procurer of a Service

An agent in the service economy becomes a *manager of a receivers' service network* if, in the process of transforming (values from) interactions desired by the receivers' service network into (values from) transactions, she establishes an organization that supports a receivers' service network, called a *procurer* of the service. The procurer aims at organizing the social choice about which stratified service to be provided (or ordered) and how it should be financed, in short, the procurement[5] of a stratified service by receivers. Let a set $\mathcal{R}_j(\ell^R_j)$ of receivers' service networks for service j be given. The set of suitable receivers' organizations $\mathcal{H}_j$ is specified as follows. Each organization $H_j = (M^H_j, L^H_j)$ in $\mathcal{H}_j$ is a hierarchy with a nodes representing a role or position in the organization and an arc $(i,j) \in L^H$ representing a principal-agent relation with principal i and agent j. The principal-agent relations in a procurer's organization are bottom-up oriented. It follows that any agent is a representative or coordinator of a set of cooperating or interacting principals. The final principals or *voters* are at the bottom of the procurer in the set $y(H_j)$, and at the top $t(H_j)$ is the *representative* agent of the procurer, who is also the manager of the procurer. A representative may of course be a Board of Representatives. There is a unique path (t,f) from the representative at the top to each voter $f \in y(H_j)$ with length ℓ^H_j.

[5] The term 'public procurement' or 'public tendering' usually means the purchase of goods and services by public or private enterprises by means of a tender or auction. Efficient public procurement involves choosing the supplier who can supply the desired goods and services at the lowest price (or, more generally, the best "value for money"). The procurement concept introduced here generalizes upon the public procurement concept.

Since the manager wants to support a receivers' service-network in $\mathcal{H}_j(\ell_j^R)$, she will restrict the set of suitable performers' organizations to those organization-networks that are isomorphic to $\mathcal{R}_j(\ell_j^R)$. This set is denoted by $\mathcal{H}_j(\ell_j^H)$. Each organization in $\mathcal{H}_j(\ell_j^H)$ is then identified by its length $\ell_j^H = \ell(H_j)$.

The number of agents that are directly connected with a coordinating agent of interacting principals is called the *coordination-span* or *scope*. Since this scope is assumed to be uniform for the organization and characteristic for the stratified service, it can be denoted by $\kappa_,$. Given the length ℓ_j of the organization and the scope κ_j, the size of the service demanded can be determined. A second parameter μ_j that characterizes the provider's organization is the X-inefficiency or the cost of the organization.

The Representative in the top position represents, facilitates and imposes the social choice made by the principals of the organization, that are, the voters. She assumes therefore two roles, the representative and the manager of the procurer. If the Representative cannot arrive at a solution or a decision, the manager may expand or contract the procurer organization until a feasible size of interaction is obtained[6].

The transaction value of the service procured is given by the receivers value function (2.2), conditional on the realization of the plans of all other managed services in the economy. Although in equilibrium all these plans have to be realized, the manager has to make a decision possibly outside a state of equilibrium. The valuation of a stratified service has been based in Section 2.3 on a belief about the realization of the other stratified services. Again, the expected values are introduced here as an assumption and become redundant in a state of equilibrium. The procurer therefore knows service value function $u_j(\ell_j, \mathcal{E}(\ell_j)) = u_j(\ell_j^H | \ell_{-j}^H, \ell^P)$, where $\mathcal{E}(\ell_j)$ is the vector of expectations of the other managed services than service j, which she needs to know to decide. The expectations are realized in a state of equilibrium.

The service value function $u_j(\ell_j, \mathcal{E}(\ell_j))$ gives the external value of the procurer's service, which the set of receivers is willing to pay for the service j. It is made cheaper by the internal organization of cooperation between receivers. The internal allocation of contributions needed to raise this value among the receivers is determined by an allocation rule or *contribution function* $c_j(\ell_j)$: $H_j(\ell_j) \times u_j(\ell_j, \mathcal{E}(\ell_j)) \rightarrow \mathbb{R}^{M(H(j))}$, which rule assigns to any receiver i in the set of receivers $M^{H(j)}$ a value $c_j^i(H_j(\ell_j) \times u_j(\ell_j, \mathcal{E}(\ell_j)) = c_j^i(\ell_j^H)$. One may observe that the assumptions on the model make the contribution rule dependent only on the parameter that determines the length of the organization H_j. The contribution function is derived from a game $G(M^H, u_j)$ on the positions in the organization by extending the value function over all possible coalitions and by restricting the value of coalitions

[6] If the rules of a Prisoners' Dilemma are understood as coming from a procurer, it shows that non-cooperative rules, which are based on separable individual decisions, usually preclude the attainment of the highest payoff for interacting agents (see Blonsky and Spagnolo, 2001).

according to the specification of the principal-agent relations in the organization. The rules of the game G are invariant for the size of the organization H_j and may generate the permission value proposed by van den Brink and Gilles (1996) and van den Brink (1997), which satisfies the properties of budget neutrality, structural monotonicity, and symmetry.

The optimal size ℓ_j^* is decreasing with (i) the degree of substitutability among receivers, and (ii) the willingness to pay expressed by the value u_j of the service j relative to the values of the other services. This effect of the external value of the service on the size of the procurer, taking into consideration the internal effect, is expressed by the function $\ell_j^* = \lambda_{j,k}(p_j)$, where $p_j = u_j(\ell_j^* | \ell_{-j}^H, \ell^P)$ equals the value based on the plans of other managers in the economy. That leads to the following definition.

Definition 3.2: The manager of procurer k of service j *minimizes contributions of voters* by choosing the size $\ell_{j,k}{}^*$ of an organization in $\mathcal{H}_j(\ell_j^H, \kappa_j, \mu_j, c_j)$ such that the residual contribution of the bottom-position c_j^0 is minimized. The function $\ell_j{}^* = \lambda_{j,k}(p_j)$, where $p_j = u_j(\ell_j{}^* | \ell_{-j}^H, \ell^P)$ is called the *price effect* on the size of the procurer. So the *demand of service j by procurer k* is given by the three set-functions of front-receivers $z_{j,k}(H_j(\ell_{j,k}{}^*)) = z_{j,k}(\lambda_{j,k}(p_j)) = z_{j,k}(p_j)$, with changing domain of definition. When labor is homogeneous, the set function $z_{j,k}(p_j)$ becomes a real function that is equal the value of the scope parameter κ_j to power $\ell^* = \lambda_{j,k}(p_j)$.

4 An Economy with Marketable Stratified Services

4.1 A Service Economy

A *service society* $\mathcal{S} = (\mathcal{N}_S, \mathcal{G}_S)$ consists of two concepts: a service opportunity structure, defined in Section 2.4:

$$\mathcal{N}_S = (M, N, (\omega_0^i)_{i \in M}, \{\mathcal{P}_j(\ell), v_j(\ell^P | \ell^R)\}_{j \in N}, \{\mathcal{R}_j(\ell), u_j(\ell^R | \ell^P)\}_{j \in N}, w),$$

and a governance structure, with the internal governance defined in Section 3:

$$\mathcal{G}_S = (M^C, N, \{M_j^F, \mathcal{F}_j(\lambda; \mu, \kappa, d)\}_{j \in N}, \{M_j^H, \mathcal{H}_j(\lambda; \mu, \kappa, c)\}_{j \in N}, \mathcal{T}(N^F \times N^H)).$$

The external governance of the service society is represented by the set of allocation mechanisms $\mathcal{T}(N^F \times N^H)$ that are suited for the type of providers and procurers active in a service sector. The service society $\mathcal{S}$ is transformed into a *service economy* $\mathcal{E}$ if all transactions are monetary transactions. In an *economy marketable stratified services (EMSS)* the only external organization is the market system. That restricts the economy drastically, of course, because a mechanism to process social demand is not present in the economy.

It excludes, for example, public services. Cooperation in supply is only realized if the provision marketable stratified services is profitable. In an *EMSS*, the length of any procurer organization equals 0 and each receiver is its own manager.

Managers in M^F and M^H assume a specialized role in the economy. Consumers in M^C are agents who integrate again their roles of performer and receiver by force of the budget constraint. The consumer also assesses the various services by the choice of a bundle of these services.

A *state* of the service society is a series of managed stratified services by providers and procurers and an allocation of services $(x^f,\ y^h)$ to individual performers and receivers of services, with $k \in M$, the set of consumers:

$$(\{F_j(\lambda), H_j(\lambda)\}_{j\in N,}, \{(x^f, y^h)\}_{f\in M, h\in M\ ,}).$$

This state is *feasible* if all agents assume a managed position and if the expectations of all managers concerning the other organizations are realized, so $F_j(\ell_j, \mathcal{E}(\ell_j)) = F_j(\ell_j^F | \ell_{-j}^F, \ell^H)$ and $H_j(\ell_j, \mathcal{E}(\ell_j)) = H_j(\ell_j^H | \ell_{-j}^H, \ell^F)$, for all services j. This implies that demand equals supply for all services. A feasible state is *optimal* if no agent can improve upon its position unless at the cost of another agent. This implies that no manager can improve upon managing its service.

4.2 General Equilibrium for an EMSS

In the economy with marketable stratified services the front performers providing the stratified services are homogeneous. The equilibrium concept is based on the formation of prices in the markets. The price vector p is defined by $p_j = v_j(\ell_j * | \ell_{-j}^F, \ell^R)$ in Section 3.2, with $\sum_j p_j = w$ if all manager's choices lead to a state of equilibrium. In case of disequilibrium, excess demand will raise this price, as expressed by the function $\lambda_j^F(p)$ and the function $\lambda_j^H(p)$, introduced in Section 3.

The assumptions introduced above simplify the various concepts drastically. Since for each service there is one representative provider that is parametrized by its length, and the representative procurer has no function any more in a market economy, the state $(\{F_j(\lambda^F),\ H_j(0)\}_{j\in N,},\ \{(x^f,\ y^i)\}_{f\in M,\ h\in M})$ of the economy is reduced to:

$$(\{t_j, \ell_j\}_{j\in N}, \{x^i\}_{i\in M,}),$$

where t_j indicates the number of providers for service j and ℓ_j the size (length) of this provider, and x^i the consumption bundle of consumer i. The manager of any provider chooses the length such that his positional budget, which is the residual income d_j^0 of the firm, is maximized over the provider set $\mathcal{F}_j(\lambda;\ \mu,\ \kappa,\ d)$. Managers are assumed to be price takers, so the optimal length $\ell_j *$ and the optimal number of providers $t_j *$ are functions of prices.

Definition 4.1: A *General Equilibrium* for an Economy with Marketable Stratified Services *(EMSS)* is a state $(\{t_j{}^*, \ell_j{}^*\}_{j\in N}, \{x^{i*}\}_{i\in M})$ and a price vector p^* of marketable services such that:

- Each performer-manager of the stratified service j chooses a size $\ell_j{}^*$ for the organization $F_j(\ell_j)$ from the set $\mathcal{F}_j(\lambda;\, \mu,\, \kappa,\, b)$, such that its profits $d_j^0(\ell_j)$ are maximized.
- Each agent i chooses a managed performer position in some $F_j(\ell_j{}^*)$ such that its positional wage $d_j^i\ (\ell_j{}^*)$ is maximized.
- Each agent i chooses a consumption bundle x^{i*} of marketable services that maximizes its utility under the income constraint.
- At price p all markets clear.

So the supply function of the homogeneous service $x_{j,jk}$ rendered by provider k of service j is:

$$S_{j,jk}\left(p\right)=\left(\kappa_j\right)^{\lambda(p)}=x_{j,jk}\left(p\right),$$

with value $p_j x_{j,jk}\left(p\right)$.

Given homogeneity of labor, the labor demand by provider k at price p is:

$$D_{0,jk}\left(p\right)=\sum_{s=1}^{\lambda(p)}\left(\kappa_j\right)^s=x_{0,jk}\left(p\right).$$

Let the income distribution within the provider organization be egalitarian[7] , so $b_j^s = b_j$, for each position on level s. Then the budget restriction of the provider reads as follows:

$$p_o x_{o,jk}(P)\leq p_j x_{j,j}(p)/t_j=p_j x_{j,jk}(p) \tag{3.1}$$

The net supply functions are composed to vector-valued functions.

$$S_j\left(p\right)=\sum_{k=1}^{t_j}S_{jk}\left(p\right)=\left(-x_{0,jk}\left(p\right),0,...,x_{j,jk}\left(p\right),...,0\right)^T.$$

Consumer i has one divisible unit of labor-service available and maximizes income by choosing a position in a provider of the service j in which his positional budget $d_j^i(\lambda_j(p)$ per unit amount of time spent is highest, irrespective of the agent's actual time spent on labor. The choice about the actual time spent on a position, which choice determines actual income, is derived from

[7] This is a strong assumption that affects the valuation function in the service structure. The distribution function should follow from the performer interaction as determined by the valuation function. Following that course, however, would make the proof very intricate.

his maximizing utility on the set of alternatives. So the agent i chooses an optimal composite service bundle $x^i = (x_0^i, x_1^i, x_2^i, \ldots, x_n^i)$, with x_0^i = 1-x^i, by maximizing $u_i(x^i)$ over his budget set:

$$\sum\nolimits_j p_j x_j^i \leq b_j^i\left(\lambda_j\left(p\right)\right) = p_0 x_{0,j}^i. \tag{3.2}$$

So consumer i^{th} net-demand function for services is, with initial labor resource ω_0^i=1, for all $i \in M$:

$$D^i\left(p\right) = \left(x_0^i - 1, x_1^i, x_2^i, ..., x_n^i\right)^T\left(p\right) = x^i\left(p\right).$$

The aggregate excess demand vector-function is:

$$Z\left(p\right) = \sum_{i \in M} D^i\left(p\right) - \sum_{j \in N} S_j\left(p\right) \leq 0.$$

This aggregate excess demand function is homogeneous of degree zero and satisfies Walras' Law, that is, $p.z(p) = 0$, for all p positive. For, summing inequality (3.1) over all providers of the same service and over all services:

$$\sum\nolimits_j p_o x_{o,j} \leq \sum\nolimits_j p_j x_j,$$

and summing inequality (3.2) over all consumers and services:

$$\sum\nolimits_j \sum\nolimits_i p_j x_j^i \leq \sum\nolimits_j \sum\nolimits_i p_0 x_{0,j}^i,$$

shows that the value of aggregate cost of provision equals the value of provision, which again equals total income. The aggregate demand function is bounded from below. Furthermore, assume that the utility functions and valuation functions are regular, implying that the demand of a service is sufficiently sensitive for a decline of its price and that the aggregate demand function is continuous. These conditions are sufficient for the existence of a Walras equilibrium.

Theorem 4.1: Let the economy satisfy the set of conditions mentioned above. Then there exists a general equilibrium for an *EMSS*.

Since all expectations are fulfilled in a *EMSS*, it is by definition effective. This property becomes interesting only in a dynamic context with a procedure comparable to the well-known tatonnement process, which procedure also leads to an effective outcome. It is also efficient, because managers can only choose from optimal groups. This property is interesting only if the managers in the model are allowed to make mistakes and have a wider set of choices than assumed here for reasons of simplicity.

Finally, one may observe that if all for all services, receivers can be separated from performers, then the service reduces to a commodity and the number of performers of that service can be identified with the quantity of that commodity. The state of the economy is then expressed in terms of quantities $\{x^i\}$, rather than relations $\{(x^f, y^f)\}$. In that case, the market economy for managed stratified services reduces to a neoclassical market economy. So the economy with managed stratified services is a genuine generalization of a neoclassical commodity economy.

4.3 The Management of Services: Conclusions

There are many open problems left in this paper. One is that a general equilibrium is not a competitive equilibrium. In order to achieve competitiveness, the organizations have to be able to split up such that the corresponding distribution functions are remunerating positions with equal requirements competitively. I think that some tools needed to further analyze the unbundling of an organization are being developed, see for example Herings e.a. (2001). This also applies to the problem of the boundary of the firm, to which problem the approach of this paper contributes. The managerial interpretation of the efficiency and the effectivity concepts improves their usefulness, in my opinion.

Value is generated only by the interaction between receivers and performers. Value is cashed by the interaction between providers and procurers. When individuals make plans for the future, they look at the roles they can play or at the positions they can assume, anticipating on the managing decision of selecting the network in which they will participate. A teenage girl may want to become a nurse without knowing a priori whom she will care, but a posteriori she will know him or her and will generate the expected value. The more effective a governance is, the more predictable is the behavior of agents and the higher is the probability of occurrence of predicted and expected outcomes. So the expectation of these values depends on a probability distribution that is eventually determined by the governance. Deriving this probability may be of great help for people making their dreams come true. The Walrasian tatonnement procedure, for example, is implicitly based on expectations. High rates of adjustment however, may lead to cycles and chaotic behavior: see Weddepohl (1999).

Another open problem is the description of the management of markets. The external organization has not been developed beyond the neoclassical assumption that prices will emerge. The main economic policy problem at the moment is: which services should fall under which external organization regime. This paper contributes to that issue, but the model for non-market regimes has to be formalized yet.

Finally, the network approach allows for introducing cultural aspects in organizations. This also promises to be a fascinating development.

References

1. Alchian, A., and H. Demsetz (1972), "Production, information costs, and economic organization", *American Economic Review, 62*, 777-795.
2. Aoki, M. (2000), "What are institutions? How should we approach them?", Stanford University Working Paper.
3. Bilbao, J.M. (2000), *Cooperative Games on Combinatorial Structures*, Kluwer Academic Publishers, Boston.
4. Bolton, P., and M. Dewatripont (1994), "The firm as a communication network", *Quarterly Journal of Economics, 109*, 809-839.
5. van den Brink, R. (1997) "An axiomatization of the disjunctive permission value for games with a permission structure", *International Journal of Game Theory, 26,* 27-43.
6. van den Brink, R., and R.P. Gilles (1996), "Axiomatizations of the conjunctive permission value for games with permission structures", *Games and Economic Behavior, 12,* 113-126.
7. van den Brink, R., and P.H.M. Ruys (1996), "The internal organization of the firm and its external environment", CentER Discussion Paper, Tilburg University
8. Calvo, G.A., and S. Wellisz (1979), "Hierarchy, ability, and income distribution", *Journal of Political Economy, 98*, 991-1010.
9. Coase, R.H. (1937), "The nature of the firm", *Economica, 4*, 386-405.
10. Coleman, J.S. (1988), "Social capital in the creation of human capital", *American Journal of Sociology, 94*, Supplement, 95-120.
11. Coleman, J.S. (1990), *Foundations of Social Theory*, Harvard University Press.
12. Dilworth, R.P. (1940), "Lattices with unique irreducible decomposition", *Annals of Mathematics, 41*, 771-777.
13. Giarini, O., and W.R. Stahel (1993), *The limits to certainty. Facing Risks in the New Service Economy*, Kluwer Academic Publishers.
14. Gilles, R.P., G. Owen and R. van den Brink (1992), "Games with permission structures: the conjunctive approach", *International Journal of Game Theory, 20*, 277-293.
15. Granovetter, M. (1985), "Economic action and social structure: the problem of embeddedness", *American Journal of Sociology, 91*, 491-510.
16. Hart, O.D., and J. Moore (1999), "On the design of hierarchies: coordination versus specialization", LSE-Sticerd Discussion Paper TE/99/375.
17. Herings, P.J.J., G. van der Laan and D. Talman (2000), "Cooperative games in graph structure", Research Memorandum 00/026, Maastricht University.
18. Holstrom, B. (1999), "The firm as a subeconomy", *The Journal of Law, Economics & Organization, 15*, 74-102.
19. Hurwicz, L. (1994), "Economic design, adjustment processes, mechanisms, and institutions", *Economic Design, 1*, 1-14.
20. Ichiishi, T. (1993), *Cooperative Nature of the Firm*, Cambridge UP.
21. Jackson, M.O. (2001), "The stability and efficiency of economic and social networks", HSS 228-77, California Institute of Technology, Pasadena.
22. Jackson, M., and A. Wolinsky (1996), "A strategic model of social and economic networks", *Journal of Economic Theory, 71*, 44-74.
23. Keren, M., and D. Levhari (1979), "The optimum span of control in a pure hierarchy", *Management Science, 25*, 1162-1172.

24. Keren, M., and D. Levhari (2002), "On the economics of managing and return to scale in communication", this volume.
25. Kranton, R.E. (1996), "Reciprocal exchange: a self-sustaining system", *American Economic Review, 86,* 830-851.
26. Ruys, P.H.M., and R. van den Brink (1999), "Positional abilities and rents on equilibrium wages", in: *The Theory of Markets*, P.J.J. Herings, A.J.J. Talman and G. van der Laan, North-Holland., 261-279.
27. Ruys, P.H.M., R. van den Brink and R. Semenov (2000), "Values and governance systems", in: *Institutions, Contracts and Organizations: Perspectives from New Institutional Economics*, Cl. Ménard (ed.), Edward Elgar, 422-446.
28. Van Zandt, T. (1997), "Real-time hierarchical resource allocation", CMSEMS DP 1231, Northwestern University.
29. Weddepohl, H.N. (1999), "Price adjustment models in a production economy", *Annals of Operations Research, 89*, 149-164.
30. Williamson, O.E. (1967), "Hierarchical control and optimum firm size", *Journal of Political Economy, 75*, 123-138.
31. Williamson, O.E. (1979), "Transaction-cost economies: the governance of contractual relations", *Journal of Law and Economics, 19,* 223-261.

Markets

Undesired Goods in Applied Demand Analysis

Anton P. Barten*

Department of Economics, Katholieke Universiteit Leuven, Belgium

Abstract. Undesired goods are goods for which the consumer has no use, which fulfill no need, like diapers for a family without children. When comparing demand by two types of households, one for which all goods have positive utility and one for which one or more goods are "undesired" one should take into account that for the undesired goods price and income fluctuations have no effect. There might be some similarity in demand between the households for the other goods. The applied demand analyst would like to exploit this similarity to pool the data for both types of households. The paper proposes to combine for this purpose the partial and conditional demand functions for the first type of consumers with the regular type of demand functions for the second type, The conditioning variables are the quantities demanded by the first type of the goods which are undesired by the second group.

Keywords: zero desirability, conditional preferences and demand, structure of demand systems, pooling data.

1 Introduction

It is a quite common feature of data sets about consumer expenditure by individual persons or households that no value, or zero, is entered for certain items. There may be various reasons for this. One is that the item in question is of no use to that particular consumption unit. Think of diapers for a family without small children. No change in relative prices or income will induce a positive expenditure. The focus of the present paper is on this situation.

Reported zero expenditure may have other causes as well. The good may have desirability, in others words, may fulfill a need, but is not purchased in the period of observation. This period is frequently a month or a year. One does not buy a pair of shoes every month or a new bed every year. It may be a matter of the indivisible nature of the item in question, but it is also possible that the time and effort for shopping is valued so much that one buys a two month supply in one go rather than visit the store every week, even if the commodity is perfectly fluid like petrol. Note that it concerns observable purchasing behaviour and not the usually unobservable consumption or use.

* The author is indebted to Frederic Vermeulen for the shared interest in the topic of this paper, the stimulating discussion and the useful comments on an earlier version. Responsibility for errors remains with the author.

Another example where no purchase is recorded for an item even though it is desirable is the one of a close substitute for a good which is purchased. Take two brands of cornflakes, say X and Y. If a box of X is bought almost certainly one will not buy a box of Y. It may be that because of difference in taste or price or income brand X is preferred over Y, which in itself remains desirable. A (small) change in the relative price of the two brands may induce the consumer to switch over to the other brand.

In this connection the classical contribution of Tobin (1958) can be mentioned. Tobin takes up the case where for a good (a consumer durable) some households register no purchase or ownership while other households report a positive amount. Within one and the same framework his model describes the decision to have or not to have the good and in the positive case, how much of it. It is an elegant blend of a qualitative response and a regular quantitative model.

In this paper, however, we will deal with the situation where some consumers do not purchase or own a particular good because it fulfills no need for them, while for other people it has a certain degree of desirability. The concept of *demographic separability* of Deaton *et al.* (1989) where some goods are only demanded by families with children is a particular case of the more general condition.

Traditional demand theory starts off from a definition of a commodity space over which preferences are defined, while satiation is excluded. When applying the theory to actual data one usually takes the dimensions of the consumption space to be the same for all consumers. This is asking for trouble. To see this, take the case where one set of consumers has a consumption set covering n goods, while the other consumers have preferences defined for the first $n-1$ goods only and have no use for the n–th good. The second set of consumers will not show any positive expenditure for the n–th good. There is no point in trying to explain this null expenditure by income/price considerations. An attempt to estimate income and price sensitivity for both sets of consumers at the same time is obviously not very meaningful.

From a theoretical point of view there is actually no problem. Each consumer taken on its own can be adequately modelled. There is no theoretical reason for different consumers to show the same responses. However, it is a matter of practical expediency to assume some similarity among consumers. One can exploit this to use data for different consumers to better estimate the income and price sensitivities. It is not uncommon to take, say, elasticities to be constant for the different data units available. Statistical efficiency requires the use of as many data as are available, as long as these are taken from the same population. This last condition is clearly not met if the set of consumers consists of different subsets of people each with a different consumption space. However, being different in one respect does not necessarily exclude similarity in other respects. The fact that one subset of consumers has no use for fish is no reason for their marginal rate of substitution for beer and cheese to be totally different from that of another subset.

So far we have reasoned in terms of a single commodity having no desirability for some of the consumers. Such a lack of desirability can occur for more than one good. In fact each consumer may have his own set of "undesired" goods. In our discussion we will concentrate on the simple dichotomy where for some consumers in our sample some of the goods are "undesired".

In the next section the issue at hand is being illustrated by a very simple demand system. The following section presents a formalization, which can serve as a basis for a generalization to arbitrary demand systems. Use will be made of concepts like partial and conditional preferences, utility and demand. The consequences of the prevalence of undesired goods for the formulation of demand systems in differential form are discusssed as well as the ensuing relations among the derivatives of such systems.The last section gives some suggestions for a further extension.

2 An Example

An extremely simple example may serve to illustrate the focus of this paper. We will consider the case of two types of consumers: a and b. The total number of different goods is 3, but consumers of type b are not interested in good 3. We will assume that the preferences of our consumers can be expressed by the well-known linear logarithmic utility function of Samuelson (1948/49). For consumers of type a one has

$$u_a(q_{1a}, q_{2a}, q_{3a}) = \sum\nolimits_{i=1}^{3} \beta_{ia} \ln(q_{ia} - \gamma_{ia}) \tag{2.1}$$

with q_{ia} being the amount of good i consumed by consumer a. The β_{ia} and γ_{ia} are fixed parameters of the utility function of a. Here, $\gamma_{ia} < q_{ia}$ and $\beta_{ia} > 0$. Without loss of generality one can set

$$\sum\nolimits_{i=1}^{3} \beta_{ia} = 1 \tag{2.2}$$

The budget constraint of consumers of type a is given by

$$\sum\nolimits_{i=1}^{3} p_i q_{ia} = m_a \tag{2.3}$$

where p_i is the given unit price of good i and m_a denotes the given total available means of consumers of type a. The quantities demanded are those that maximize utility function (2.1) while satisfying budget constraint (2.3). The system of corresponding demand equations can be written as

$$q_{ia}^* = \gamma_{ia} + \frac{\beta_{ia}}{p_i}\left(m_a - \sum\nolimits_{k=1}^{3} p_k \gamma_{ka}\right) \qquad i = 1, 2, 3 \tag{2.4}$$

Multiplying through by p_i yields the system of expenditure equations

$$e_{ia} \equiv p_i q^*_{ia} = p_i \gamma_{ia} + \beta_{ia} \left(m_a - \sum\nolimits_{k=1}^{3} p_k \gamma_{ka} \right) \qquad i = 1, 2, 3 \tag{2.5}$$

which are linear in the prices p_k and means or income m_a. The set of these expenditure equations is known as the Linear Expenditure System.

The same story can be told about the choice behavior of consumers of type b, except that indices i and k do not take on the value of 3. The utility function is now

$$u_b(q_{1b}, q_{2b}) = \sum\nolimits_{i=1}^{2} \beta_{ib} \ln(q_{ib} - \gamma_{ib}) \tag{2.6}$$

The counterparts of (2.4) and (2.5) are

$$q^*_{ib} = \gamma_{ib} + \frac{\beta_{ib}}{p_i} \left(m_b - \sum\nolimits_{k=1}^{2} p_k \gamma_{kb} \right) \qquad i = 1, 2 \tag{2.7}$$

and

$$e_{ib} \equiv p_i q^*_{ib} = p_i \gamma_{ib} + \beta_{ib} \left(m_b - \sum\nolimits_{k=1}^{2} p_k \gamma_{kb} \right) \qquad i = 1, 2 \tag{2.8}$$

respectively. Note that the counterpart of (2.2) is

$$\beta_{1b} + \beta_{2b} = 1 \tag{2.9}$$

The applied econometrician will be interested in estimating the β's and the γ's. Efficiency considerations lead him to keep the number of coefficients to be estimated as small as possible relative to the available data. He will impose certain restrictions on the coefficients. One possibility is to make these coefficients pairwise equal. Obviously, this creates a problem, because (2.2) and (2.9) are not compatible unless $\beta_{3a} \equiv 0$, which is of no practical interest. Moreover, data for consumer b do not contain information for γ_{3b},which is not defined.

One can define for consumers of type a a partial and conditional preference order over the consumption space for the goods 1 and 2 with a fixed value for q_{3a}, say, $\bar{q}_{3a}$[1]. The corresponding budget constraint is given by

$$\tilde{m}_a = p_1 q_{1a} + p_2 q_{2a} = m_a - p_3 \bar{q}_{3a} \tag{2.10}$$

and the relevant, partial and conditional, utility function can be written as

$$\tilde{u}_a(q_{1a}, q_{2a} \mid \bar{q}_{3a}) = \sum\nolimits_{i=1}^{2} \beta_{ia} \ln(q_{ia} - \gamma_{ia}) + \beta_{3a} \ln(\bar{q}_{3a} - \gamma_{3a}) \tag{2.11}$$

[1] For an early use of conditional demand functions, an implication of conditional preferences, see Pollak (1969, 1971). See also Browning and Meghir (1991).

The optimal values for q_{1a} and q_{2a}, are given by

$$\tilde{q}^*_{ia} = \gamma_{ia} + \frac{\tilde{\beta}_{ia}}{p_i}\left(\tilde{m}_a - \sum\nolimits_{k=1}^{2} p_k\gamma_{ka}\right) \qquad i = 1,2 \tag{2.12}$$

with

$$\tilde{\beta}_{ia} = \frac{\beta_{ia}}{\beta_{1a} + \beta_{2a}} \tag{2.13}$$

The conditioning value for $\bar{q}_{3a}$ is set at its optimum in the unconditional setting

$$q^*_{3a} = \gamma_{3a} + \frac{\beta_{3a}}{p_3}\left(m_a - \sum\nolimits_{k=1}^{3} p_k\gamma_{ka}\right) \tag{2.14}$$

- see (2.4). In principle this is the observed value for q_{3a},which is needed to obtain the value for $\tilde{m}_a$ - see (2.10).

For consumers of type b there is, of course, no special conditioning value. Corresponding to budget constraint (2.10) one has simply

$$m_b = p_1 q_{1b} + p_2 q_{2b} \tag{2.15}$$

Given utility function (2.6) one obtains the demand equations (2.7), given earlier, which are expressions similar to (2.13).

It is then natural for the applied econometrician to set $\tilde{\beta}_{ia} = \beta_{ib} = \tilde{\beta}_i$ and $\gamma_{ia} = \gamma_{ib} = \gamma_i$ for $i = 1,2$, and jointly estimate the conditional demand system (2.12) and the unconditional one (2.7). This yields on the basis of observations for both types of consumers estimates for γ_1 and γ_2 as well as for $\tilde{\beta}_1$ and $\tilde{\beta}_2$. Then he estimates, using data for consumers of type a only, unconditional demand equation (2.14) which gives him γ_3 and β_{3a}. As $\beta_{3a} = 1 - (\beta_{1a} + \beta_{2a})$, β_{1a} and β_{2a} can be estimated by $\tilde{\beta}_1(1 - \beta_{3a})$ and by $\tilde{\beta}_2(1 - \beta_{3a})$, respectively. This procedure achieves its purpose: logically consistent values for the coefficients of the system of demand equations.

An alternative of this two-step estimation procedure is to combine the two unconditional systems (2.5) and (2.8) and estimate them jointly employing the restriction $\beta_{ib} = \beta_{ia}/(\beta_{1a} + \beta_{2a})$, $i = 1,2$. In this way one can exploit the covariance between individual consumers at the cost of the nonlinearity in the restriction on the β's. This covariance is usually rather small and the gain in precision of the coefficient estimates negligible.

It must be granted that the example is rather specific. Utility function (2.1) represents additive preferences, which is clearly a very special, restricted case. The implied Linear Expenditure System is rather rigid in taking into account specific preference interactions. A generalization is in order.

3 A Formalization

3.1 Partial and Conditional Preferences

Let $C \subset \mathcal{R}^n$ be the $n-$ dimensional consumption set of possible vectors of quantities of commodities and services: $\mathbf{q}$. Without loss of generality one can partition this set into two non-overlapping subsets $C_A \subset \mathcal{R}^{n_A}$ and $C_B \subset \mathcal{R}^{n_B}$ with $C \equiv C_A \times C_B$ and $n = n_A + n_B$. Correspondingly, one can reorder the elements of $\mathbf{q}$ in such a way that

$$\mathbf{q}' \equiv [\mathbf{q}'_A, \mathbf{q}'_B] \qquad \mathbf{q}'_A \equiv [q_1, ..., q_{n_A}] \in C_A \qquad \mathbf{q}'_B \equiv [q_{n_A+1}, ..., q_n] \in C_B$$

with a parallel partitioning of the vector of prices $\mathbf{p} \in \mathcal{R}^n_{++}$

$$\mathbf{p}' \equiv [\mathbf{p}'_A, \mathbf{p}'_B] \qquad \mathbf{p}'_A \equiv [p_1, ..., p_{n_A}] \qquad \mathbf{p}'_B \equiv [p_{n_A+1}, ..., p_n]$$

Next, let $\mathbf{q}^0 R \mathbf{q}^1$ denote the preference order defined on C for some choice of $\mathbf{q}^0$ and $\mathbf{q}^1$. This preference order is equipped with all the usual properties - see e.g. Barten and Böhm (1982). With the partitioning into two parts one can write it as

$$\left(\mathbf{q}^0_A, \mathbf{q}^0_B\right) R \left(\mathbf{q}^1_A, \mathbf{q}^1_B\right) \tag{3.1}$$

Take the case that $\mathbf{q}^1_B$ happens to be equal to $\mathbf{q}^0_B$. Clearly, it follows from the validity of (3.1) that also

$$\left(\mathbf{q}^0_A, \mathbf{q}^0_B\right) R \left(\mathbf{q}^1_A, \mathbf{q}^0_B\right) \tag{3.2}$$

is a valid description of the preference order. The fact that the second part of the consumption bundle is the same for both full bundles in (3.2) has no consequence for its validity as a regular preference order. One can write (3.2) as

$$\mathbf{q}^0_A R_A \left(\mathbf{q}^0_B\right) \mathbf{q}^1_A \tag{3.3}$$

where $R_A\left(\mathbf{q}^0_B\right)$ denotes the *partial* and *conditional* preference order. It is *partial* because it refers to the ordering of bundles in the partial consumption set C_A. It is *conditional* because it is specific for a particular bundle of C_B, *viz.*, $\mathbf{q}^0_B$.

For fixed $\mathbf{q}^0_B$ the partial and conditional preference order (3.3) has all the usual properties of a consumer preference order. Consequently, it can be represented by a *partial* and *conditional* utility function defined on C_A:

$$\hat{u}_A\left(\mathbf{q}_A \mid \mathbf{q}^0_B\right) \tag{3.4}$$

For the purpose of this paper the set of consumers consists of two non-overlapping groups. The first one, of type a, has preferences defined for the complete consumption set C, with a regular utility function $u_a(\mathbf{q})$ and a *partial* and *conditional* utility function $\hat{u}_{Aa}\left(\mathbf{q}_A \mid \mathbf{q}^0_B\right)$.The value of $\mathbf{q}^0_B$ can be chosen freely from C_B. The second set of consumers, of type b, has a preference order defined only on C_A, with all the classical properties, including the existence of a regular utility function $u_{Ab}(\mathbf{q}_A)$.

At this stage it is useful to bring up the relation between separability of preferences and the case under discussion here. Separability of preferences between bundles from A and those from B occurs when the partial and conditional preference order (3.3) does not depend on $\mathbf{q}_B^0$. This could be the case or not for consumers of type a, but is not defined for consumers of type b, because goods from B do not appear in their preference order. Separability properties do not play an essential role for the treatment of "undesired" goods.

3.2 Partial and Conditional Demand

Demand is by definition the vector of goods and services that is most preferred among those that the consumer can afford, *i.e*, that belong to his budget set. The budget set for the consumers of type a can be written as

$$b(m_a, \mathbf{p}) \equiv \{\mathbf{q} \in C \mid \mathbf{p}'\mathbf{q} = m_a\} \tag{3.5}$$

where m_a represents the means available for a consumers. Note that the equality on the right-hand side reflects the assumption of no satiation. The optimizing bundle can then be formulated as

$$\mathbf{f}_a\left(m_a, \mathbf{p}\right) = \{\mathbf{q} \in b(m_a, \mathbf{p}) \mid u_a\left(\mathbf{q}\right) > u_a\left(\mathbf{q}^0\right) \forall \mathbf{q}^0 \neq \mathbf{q}, \mathbf{q}^0 \in b(m_a, \mathbf{p})\} \tag{3.6}$$

The counterparts of (3.5) and (3.6) for type b consumers are

$$b_A(m_b, \mathbf{p}_A) \equiv \{\mathbf{q}_A \in C_A \mid \mathbf{p}_A'\mathbf{q}_A = m_b\} \tag{3.7}$$

and

$$\mathbf{f}_{Ab}\left(m_b, \mathbf{p}_A\right) = \begin{array}{c} \{\mathbf{q}_A \in b_A(m_b, \mathbf{p}_A) \mid u_{Ab}\left(\mathbf{q}_A\right) > u_{Ab}\left(\mathbf{q}_A^0\right) \\ \forall \mathbf{q}_A^0 \neq \mathbf{q}_A, \mathbf{q}_A^0 \in b_A(m_b, \mathbf{p}_A)\} \end{array} \tag{3.8}$$

which is a regular set of demand equations.

For consumers of type a one has a partial and conditional budget set:

$$b_{Aa}^c(\tilde{m}_a, \mathbf{p}_A \mid \mathbf{q}_B^+) \equiv \{\mathbf{q}_A \in C_A \mid \mathbf{p}_A'\mathbf{q}_A = \tilde{m}_a = m_a - \mathbf{p}_B'\mathbf{q}_B^+\} \tag{3.9}$$

the condition being the value taken on by $\mathbf{q}_B^+$. The corresponding partial and conditional demands are:

$$\mathbf{f}_{Aa}^c\left(\tilde{m}_a, \mathbf{p}_A \mid \mathbf{q}_B^+\right) = \begin{array}{c} \{\mathbf{q}_A \in b_{Aa}^c(\tilde{m}_a, \mathbf{p}_A \mid \mathbf{q}_B^+) \mid \hat{u}_{Aa}\left(\mathbf{q}_A \mid \mathbf{q}_B^+\right) > \hat{u}_{Aa}\left(\mathbf{q}_A^0 \mid \mathbf{q}_B^+\right) \\ \forall \mathbf{q}_A^0 \neq \mathbf{q}_A, \mathbf{q}_A^0 \in b_{Aa}^c(\tilde{m}_a, \mathbf{p}_A \mid \mathbf{q}_B^+)\} \end{array} \tag{3.10}$$

This expression should be compared with (3.8) for the full and unconditional demands of the consumers of type b.

Partitioning the full and unconditional demands (3.6) one may write

$$\mathbf{f}_a\left(m_a, \mathbf{p}\right) \equiv \{\mathbf{f}_{Aa}\left(m_a, \mathbf{p}\right), \mathbf{f}_{Ba}\left(m_a, \mathbf{p}\right)\} \tag{3.11}$$

It can be shown that replacing $\mathbf{q}_B^+$ in (3.10) by $\mathbf{f}_{Ba}\left(m_a, \mathbf{p}\right)$ results in $\mathbf{f}_{Aa}\left(m_a, \mathbf{p}\right)$. No formal proof is given here, but the finding is intuitively obvious.

3.3 Structure of Demand Systems

The forms in which the demand functions (3.6), (3.8) and (3.10) are expressed is not very open to a further interpretation. Assuming differentiability one obtains more insight from those functions in differential form. In particular, the first-order derivatives can reveal useful information.

The regular demand system for consumers of type a can be written in differential form as

$$df_a(m_a, \mathbf{p}) = d\mathbf{q}_a = \mathbf{q}_{ma}(dm_a - \mathbf{q}'_a d\mathbf{p}) + \mathbf{K}_a d\mathbf{p} \tag{3.12}$$

Here, $\mathbf{q}_{ma}$ is the vector of derivatives with respect to total means or income and $\mathbf{K}_a$ is the Slutsky substitution matrix. According to the underlying theory the following properties hold:

$$\begin{array}{lll} \textit{Adding-up} & \mathbf{p}'\mathbf{q}_{ma} = 1 \quad , \quad \mathbf{p}'\mathbf{K}_a = \mathbf{0}' & \\ \textit{Homogeneity} & \mathbf{K}_a\mathbf{p} = \mathbf{0} & \\ \textit{Symmetry} & \mathbf{K}_a = \mathbf{K}'_a & \\ \textit{Negativity} & \mathbf{x}'\mathbf{K}_a\mathbf{x} < 0 \quad \forall \mathbf{x} \neq \alpha\mathbf{p}, \quad \alpha \in \mathcal{R} & \\ \textit{Rank} & \textit{rank}(\mathbf{K}_a) = n - 1 & \end{array} \tag{3.13}$$

In partitioned form this system can be written as

$$df_{Aa}(m_a, \mathbf{p}) = d\mathbf{q}_{Aa} = \mathbf{q}_{mAa}(dm_a - \mathbf{q}'_a d\mathbf{p}) + \mathbf{K}_{AAa}d\mathbf{p}_A + \mathbf{K}_{ABa}d\mathbf{p}_B \tag{3.14}$$

$$df_{Ba}(m_a, \mathbf{p}) = d\mathbf{q}_{Ba} = \mathbf{q}_{mBa}(dm_a - \mathbf{q}'_a d\mathbf{p}) + \mathbf{K}_{BAa}d\mathbf{p}_A + \mathbf{K}_{BBa}d\mathbf{p}_B \tag{3.15}$$

with the obvious definitions

$$\mathbf{q}_{ma} \equiv \begin{bmatrix} \mathbf{q}_{mAa} \\ \mathbf{q}_{mBa} \end{bmatrix} \qquad \mathbf{K}_a \equiv \begin{bmatrix} \mathbf{K}_{AAa} & \mathbf{K}_{ABa} \\ \mathbf{K}_{BAa} & \mathbf{K}_{BBa} \end{bmatrix}$$

For consumers of type a the partial and conditional demand functions in differential form are

$$\begin{aligned} df^c_{Aa}(\tilde{m}_a, \mathbf{p}_A \mid \mathbf{q}_{Ba}) &= d\mathbf{q}_{Aa} = \\ &\mathbf{q}^c_{mAa}(d\tilde{m}_a - \mathbf{q}'_{Aa}d\mathbf{p}_A) + \tilde{\mathbf{K}}_{AAa}d\mathbf{p}_A + \mathbf{Q}_{AX}d\mathbf{q}_{Ba}. \end{aligned} \tag{3.16}$$

Here

$$d\tilde{m}_a = d(\mathbf{p}'_A\mathbf{q}_{Aa}) = \mathbf{p}'_A d\mathbf{q}_{Aa} + \mathbf{q}'_{Aa}d\mathbf{p}_A \tag{3.17}$$

or the change in the means assigned by a for the purchase of goods from C_A. Note that , $\mathbf{q}^c_{mAa}$ and $\tilde{\mathbf{K}}_{AAa}$ display, *mutatis mutandis,* the same set of properties (3.13) as their counterparts of the complete system. One has an extra adding-up condition, viz. $\mathbf{p}'_A\mathbf{Q}_{AX} = \mathbf{0}'$. The last term of (3.16) represents the effects of the change in $\mathbf{q}_{Ba}$. If one takes its optimal value as determined by (3.15) one obtains (3.14), after some suitable manipulations - see also (3.11).

System (3.16) is one of the two building blocks, the one pertaining to consumers of type a. The other one, which describes the behavior of consumers of type b, is

$$df_{Ab}(m_b, \mathbf{p}_A) = d\mathbf{q}_{Ab} = \mathbf{q}_{mAb}(dm_b - \mathbf{q}'_{Ab}d\mathbf{p}_A) + \mathbf{K}_{AAb}d\mathbf{p}_A \tag{3.18}$$

Since (3.18) is a regular demand system the $\mathbf{q}_{mAb}$ and $\mathbf{K}_{AAb}$ satisfy the set of properties (3.13).

It is natural to pool the information on consumer behavior with respect to A goods supplied by (3.16) and (3.18). It depends on the choice of the functional form which sets of coefficients will taken to be the same across the two sets of consumers. A possibility would be to set $\mathbf{q}^c_{mAa}$ equal to $\mathbf{q}_{mAb}$ and $\tilde{\mathbf{K}}_{AAa}$ to $\mathbf{K}_{AAb}$. One should have in mind, however, that the concepts for the a consumers involve indirectly information about the preferences with respect to the B goods, which is, of course, absent in those for the b consumers.[2]

3.4 Relations among Derivatives

It is of some interest to trace the vectors and matrices of the partial and conditional system (3.16) back to those of the full and unconditional system (3.12), or of its partitioned version consisting of (3.14) and (3.15). One can derive

$$\mathbf{q}^c_{mAa} = \mathbf{q}_{mAa} - \mathbf{K}_{ABa}\mathbf{K}^{-1}_{BBa}\mathbf{q}_{mBa} \tag{3.19}$$

$$\tilde{\mathbf{K}}_{AAa} = \mathbf{K}_{AAa} - \mathbf{K}_{ABa}\mathbf{K}^{-1}_{BBa}\mathbf{K}_{BAa} \tag{3.20}$$

$$\begin{aligned}\mathbf{Q}_{AX} &= \mathbf{K}_{ABa}\mathbf{K}^{-1}_{BBa} + \left(\mathbf{q}_{mAa} - \mathbf{K}_{ABa}\mathbf{K}^{-1}_{BBa}\mathbf{q}_{mBa}\right)\mathbf{p}'_B \\ &= \mathbf{K}_{ABa}\mathbf{K}^{-1}_{BBa} + \mathbf{q}^c_{mAa}\mathbf{p}'_B\end{aligned} \tag{3.21}$$

The existence of the inverse of $\mathbf{K}_{BBa}$ is guaranteed by the set of properties (3.13). One sees that $\mathbf{Q}_{AX}$ is made up out of two parts. One, $\mathbf{K}_{ABa}\mathbf{K}^{-1}_{BBa}$, reflects the structure of preferences. The other, $\mathbf{q}^c_{mAa}\mathbf{p}'_B$, represents the fact that a change in $\mathbf{q}_{Ba}$ has to be paid for at the cost of other goods. It is a kind of income effect.

Note that relation (3.20) expresses the *Le Chatelier principle,* introduced in economics by Samuelson (1947) in the context of rationing. The matrix $\mathbf{K}_{AAa}$ is the negative definite part of the substitution matrix $\mathbf{K}_a$. To obtain $\tilde{\mathbf{K}}_{AAa}$ one subtracts the negative semidefinite matrix $\mathbf{K}_{ABa}\mathbf{K}^{-1}_{BBa}\mathbf{K}_{BAa}$. One of the consequences is that the diagonal elements of $\tilde{\mathbf{K}}_{AAa}$ are less negative than the diagonal elements of $\mathbf{K}_{AAa}$. In other words the own substitution

[2] The combination of information from a and b consumers is only meaningful if there is some similarity between the consumers. One may point out that in the specification of econometric equations equality of coefficients across different data points is current practice. This "similarity" is usually testable but not tested.

effect of the partial and conditional system is less negative than that in the original system. The price reaction is "blunted" by the presence of the condition, which restricts adjustment.

It is clear that via $\mathbf{K}_{ABa}$ and $\mathbf{K}_{BBa}$ interactions with preferences for B goods play a role, also for the partial system for the a consumers, while they are not at all involved in the case of the corresponding system for the b consumers. Setting in applied work for example $\tilde{\mathbf{K}}_{AAa}$ equal to $\mathbf{K}_{AAb}$ may imply an "heroic" assumption. However, it depends on the type of preference ordering whether this problem is serious or not.

To demonstrate this, the case of the model of Section 2 is used again. The $\mathbf{q}^c_{mAa}$, $\tilde{\mathbf{K}}_{AAa}$ and $\mathbf{Q}_{AX}$ will be expressed in terms of the β_i and γ_i of the Linear Expenditure System. The following definitions are being used:

$$\beta = \begin{bmatrix} \beta_A \\ \beta_3 \end{bmatrix} = \begin{bmatrix} \beta_1 \\ \beta_2 \\ \beta_3 \end{bmatrix} \qquad \gamma = \begin{bmatrix} \gamma_A \\ \gamma_3 \end{bmatrix} = \begin{bmatrix} \gamma_1 \\ \gamma_2 \\ \gamma_3 \end{bmatrix} \qquad \tilde{\beta}_A = \frac{1}{\beta_1 + \beta_2} \beta_A$$

while $\widehat{\mathbf{x}}$ denotes a diagonal matrix with the elements of the vector $\mathbf{x}$ on the diagonal. The subscripts a and b are deleted whenever there is no danger of confusion.

One can then establish that

$$\mathbf{q}_{mAa} = \widehat{\mathbf{p}_A}^{-1} \beta_A \qquad \mathbf{q}^c_{mAa} = \widehat{\mathbf{p}_A}^{-1} \tilde{\beta}_A$$

$$\mathbf{K}_{AAa} = -(m_a - \mathbf{p}'\gamma)\widehat{\mathbf{p}_A}^{-1} \left[\widehat{\beta}_A - \beta_A \beta'_A \right] \widehat{\mathbf{p}_A}^{-1}$$

$$\tilde{\mathbf{K}}_{AAa} = -(\tilde{m}_a - \mathbf{p}'_A \gamma_A)\widehat{\mathbf{p}_A}^{-1} \left[\widehat{\tilde{\beta}_A} - \tilde{\beta}_A \tilde{\beta}'_A \right] \widehat{\mathbf{p}_A}^{-1}$$

$$\mathbf{Q}_{AX} = \mathbf{0}$$

Note that for the β_A and γ_A being the same for both type a and type b consumers $\tilde{\mathbf{K}}_{AAa}$ is equal to $\mathbf{K}_{AAb}$, apart from the difference between $\tilde{m}_a$ and $\tilde{m}_b$. The rather complex formulation of these expressions appears to be relatively simple for this particular type of utility function.

4 Possible Extensions

In this paper the problem of the treatment of undesired goods in the context of systems of demand equations has been studied for an extremely simple set-up. Reality is much more complex. Here we have concentrated on the case of two types of consumers and a single partitioning of the consumption set. There may be multiple partitions resulting in many different partial consumption sets, each corresponding with a particular type of consumers. The case of the type of consumer with preferences defined on the full consumption set (our type a) can be presumably ruled out. What matters for the applied demand analyst is the overlap of the various sets of partial consumption sets, which can be exploited to pool data for the relevant consumer types.

It is beyond the exploratory nature of the present paper to take up the problem of undesired goods in its full dimensionality.

There are connections to be explored between the case of undesired goods as discussed here and that of rationing, where conditional systems also play a role - see also Bettendorf et Barten (1995). Another possible extension is the treatment of the arrival of new goods, where the partitioning is of a chronological nature.

In this paper no use has been made of separability assumptions, apart from those implicit in the example of the Linear Expenditure System. Separability assumptions provide a considerable amount of structure for demand models, which may be of special importance when for some consumers some of the goods are undesired.

The applied econometrician may be worried by the absence of a treatment of the random components. This is an obvious target for further work. Once the stochastics are in place there is room for significance testing. One may, like Vermeulen (1999), test whether the observed zero expenditures are due to the goods in question being undesired or to other causes, like corner solutions and the like.

References

1. Barten A.P. and V. Böhm (1982): Consumer Theory, Chapter 9 in K.J. Arrow and M.D. Intriligator (editors) *Handbook of Mathematical Economics, Volume II,* North-Holland, Amsterdam, 1982, 381-429.
2. Bettendorf, L. et A. Barten (1995): Rationnement dans les systèmes de demande: calculs des prix virtuels, *Economie & Prévision,* 121 (1995), 101-107.
3. Browning; M and C. Meghir (1991): The Effects of Male and Female Labor Supply on Commodity Demands, *Econometrica,* 59 (1991), 925-951.
4. Deaton, A.S. J. Ruiz-Castillo, and D. Thomas (1989): The Influence of Household Composition on Household Expenditure Patterns: Theory and Spanish Evidence, *The Journal of Political Economy,* 97 (1989) 179-200.
5. Pollak, R.A. (1969): Conditional demand functions and consumption theory, *The Quarterly Journal of Economics,* 83 (1969), 60-78.
6. Pollak, R.A. (1971): Conditional demand functions and the implications of separable utility, *Southern Economic Journal,* 37 423-433.
7. Samuelson, P.A. (1947): *Foundations of Economic Analysis,* Enlarged Edition, Harvard University Press, Cambridge, 1985.
8. Samuelson, P.A. (1948/49): Some Implications of "Linearity", *Review of Economic Studies,* 15 (1948/49), 88-90.
9. Tobin, J. (1958): Estimation of Relationships for Limited Dependent Variables, *Econometrica,* 26 (1958), 24-36.
10. Vermeulen, F. (1999): Do smokers behave differently? A tale of zero expenditures and separability concepts. Mimeo, Center for Economic Studies, Katholieke Universiteit Leuven, 1999.

Animal Communication

Marcel Canoy*

Centraal Planbureau, Den Haag, The Netherlands

Abstract. This is an economic analysis on regulation of telecommunications markets in the form of dialogues between a fox and an owl. At first animals use oldfashioned technology. Later innovation comes along, but new technology and markets conditions require careful regulation. Can the animals provide optimal regulatory responses or do they sell their souls to monopolistic rats?

Keywords: Competition, telecommunications, regulation, auctions

Banging Sticks

The world of animals seems so simple. The animals, led by the wise Weddy Owl and his advisor Marcy Fox, live in a forest. They cherish traditions of small scale manual work in closed communities, keeping daily life as straightforward as possible. Nevertheless, the forest is so huge that animals sometimes communicate by rhythmically banging sticks on hollow trees. This primitive but peaceful life is about to be cruelly disrupted by the visit of Maggy the Maggot from a neighboring forest.

Weddy Owl:	Hi Maggy. What brings you to our humble estates?
Maggy:	Nothing special. I enjoy untouched nature.
Weddy Owl:	We cherish our traditions. But is it so different in your place ? (*rhythmic noises of banging sticks*).
Maggy:	Well, your banging friends tell the story. You prehistorians are apparently still banging sticks. (*the fox and the owl look puzzled*)
Marcy Fox:	I recall seeing you banging them yourself not a while ago. Not unskillfully either, I might add.
Maggy:	That was *ages* ago. I almost forgot we used to have that pedestrian habit ourselves.
Weddy Owl:	And what, I might ask, has replaced the handsome handicraft routine?
Maggy:	C'mon, you *must* have heard of our Booming Buzzy system? The sewer rats invented and installed it. Smashing stuff. A simple set of tubes amplify sounds. Saves the energy of making all these trees hollow, and avoids primitive pass-through-systems when you want to reach animals on

* I thank my sister Eveline Canoy for dotting the i with the illustration.

the other side of the forest. Never getting tired of failed communication attempts when one of your sorry creatures breaks the chain? (*the fox and the owl look bemused*). Sorry, I must fly, buzzes are waiting to be answered.

Maggy the Maggot leaves in a hurry. The Owl looks disgruntled, while the Fox smiles.

Tinkly Toys

Marcy Fox: Hear, hear. Who would have thought that old Maggy uses such fashionable stuff, owly-fowly! Perhaps we can fly in some of these rats?

Weddy Owl: Listening to maggots and sewer rats these days?

Marcy Fox: Never mind these unfortunate creatures. You'd better look in the mirror. Lately you have complained quite a bit about sticks and trees.

Weddy Owl: That is true. The pigs have to perform their annual renewal of sticks and trees as soon as possible. Let us not lose our heads though. A bit of cursing doesn't fumble my forest into a tinkly toy garden for raving rats.

Marcy Fox: What if rats don't rave and toys don't tinkle?

Weddy Owl: As a traditional bird, I do not fancy perverse experiments with rats.

Marcy Fox: No experiments needed. Recall that we were about to renew the sticks and trees anyway. Anything new from these lazy pigglets lately? With your permission, I will check this ratty business myself.

Weddy Owl: Fair enough. (exit fox)

The Smell of Pigs

Not long after the exit of the fox, a pig waddles into the scene.

Weddy Owl: Ah precious piggy! I cannot wait to hear your plans on the annual sticks and trees renewal.

Pig: Plans? (*yawns*) Gosh, I was just passing by. Have I missed something? (*the owl is fidgety, drops his pipe, drops his matches while trying to pick up his pipe, puts the pipe in his bill the wrong way around*).

Weddy Owl: Ouch! Get rid of that mud in your ears and listen! Three weeks ago we agreed on a pig plan, which you pigs claimed would take you a week. I am still waiting.

Pig:	Oh that. I am afraid I have to report a delay. It took us one week to carefully select Choppy to draft the report. I just received notice of Choppy's writer's block. He claims sticks don't inspire pigs. Quite unfortunate, I admit. (*the owl looks in despair*).
Weddy Owl:	How can we agree on *anything* in this fashion?
Pig:	We do our usual stinking best, but renewal plans require careful consideration.
Weddy Owl:	But these renewals are always the same. Just take last year's plan, modify a few things and that's it.
Pig:	I guess you grossly underestimate the complexity of renewals. There are new animals, moving animals, not all of them need renewal, some insist on using old stuff. And we are not even talking about the amazing technological development of sticks. Careful consideration is the name of the game. (*the owl looks acquiescent*).
Weddy Owl:	Tell me how much time you need, but I won't accept anymore of these jellyfish excuses.
Pig:	Give us another four weeks. (*the owl grinds his bill*)
Weddy Owl:	Sign here. (*pig signs quickly*).

Adventurous Owl

The pig leaves. The owl decides to have his nap. After a while the fox re-enters the scene.

Marcy Fox:	Are you asleep?
Weddy Owl:	(*half asleep*) ....nightmare....rats....scary.....pigs....more pigs....
Marcy Fox:	The same paltry pigs that were supposed to report on renewal two weeks ago?
Weddy Owl:	(*awake now*) Four weeks! Oh my god, what did I do? (*fox looks puzzled*). I gave these pigs four unbelievable extra weeks for their plans!
Marcy Fox:	Four weeks? Two weeks!! (*it is the Owls' turn to look puzzled*). That is how much time the rats need to install Booming Buzzy.
Weddy Owl:	(*fully awake now*) Impressive, I admit. What else did you find out?
Marcy Fox:	It's cheap to use, no annual renewals, it's fast and the quality is superior. The only downside is the price of the Buzzy system itself. Bit of an investment for us I guess.
Weddy Owl:	Ah, monopoly rats! Is there nobody else who could do that job?

Marcy Fox:	Possibly yes. There are ants, bees, perhaps even pigs, who knows. We can organize an auction.
Weddy Owl:	Before you sell my forest to rats, explain some more please.
Marcy Fox:	The idea is to auction the right to install the tubes and exploit the forest locations. The one who bids highest wins a license which gives the right to offer Booming Buzzy services to the animals for a predetermined number of years. Yields a return for us as well.
Weddy Owl:	Which we subsequently lose on monopoly rat prices. I understand the name 'Booming' now.
Marcy Fox:	Not necessarily. We can outfox Maggy, by allowing for competition. Why sell our souls to rats? We might as well have ants, bees and other animals.
Weddy Owl:	I might give it some consideration. Can you produce a plan?
Marcy Fox:	No problem.
Weddy Owl:	Deal.

Head and Heart

A short while later.

Weddy Owl:	I have read your report. Impressive. If only pigs....well forget about them. Still, I am not completely comfortable with it. My head says 'Booming Buzzy', but my heart says 'Banging Sticks'. The buzzy system still smells a bit like trouble. Irreversible trouble too, I might add.
Marcy Fox:	Not more irreversible than pigs, they are simply impossible to move these animals. Think in challenges, not in problems!
Weddy Owl:	True, these pathetic pigs tried to re-re-renegotiate our last contract. Still, everybody knows that communication networks are natural monopolies. Replicating them is highly inefficient. So why are you talking about competition?
Marcy Fox:	Cost efficiency is not the same as welfare. First of all, the other animals do not have to replicate the whole network, only parts. That creates a cost inefficiency, I admit. This efficiency loss will be more than offset by creating incentives to compete, innovate and invest, things your lazy pigs are not particularly good at.
Weddy Owl:	True, but these benefits can only come about if competition works properly. I am not that convinced yet. Still, I take your point about stirring up these pigs. They claimed all those buzzy rumors were unsettling them. Ha ha! I will unsettle them more than they can image. We will have the auction!

The Auction

Pig1: Oinking Hell! Did you hear about Booming Buzzy?

Pig2: Amazing, years of stinking hard work deleted with one foxy push on the auction button. (*pigs are performing mourning oink ritual*).

Pig3: Hey! Do you see that tree over there? (*pigs look at tree*)

Pig1: The rules of the game are attached to various trees. (*reads it out to the other pigs*)

Rules

- Three licenses will be auctioned. A license gives the right to install Booming Buzzy and offer services.
- Each animal species can only bid on one license. Pigs, swine and hogs are assumed to belong to the same species.
- Bids must be delivered in a closed basket to the auctioneer, the fox. The highest three bids win.
- After the auction the three winners can announce their plans and prices.
- Animals can then choose between operators. Animals are stuck to an operator for one lunar cycle, after which they can switch, should they wish to do so.
- Bids should be delivered before the sun comes up three times. The winning animals have one lunar cycle to install their stuff. Breaking that deadline implies that the owl can reallocate the license to somebody else.

The pigs start lengthy discussions on the auction, whether they should participate and how much bids they should put in.

The Winners

Three sunrises later the animals gather together for the announcement of the winner.

Marcy Fox: All silent! On behalf of the owl, I am about to announce the winners of the auction. Before that, some facts and figures.

1. Only four legal bids were put in. An investigation on collusion will follow. Colluders will be excommunicated.
2. The bids of the pigs are considered illegal, wild boars obviously being no different species from pigs. (*loud protests from the pigs*).

3. The auction is considered to be legal and its result is binding, pending the collusion investigation.

The winners are ... ... the rats, the ants and the bees. The lice have put in a bid of one and henceforth have lost the contest. All results and consequences will be announced by leaflets before dawn. Questions?

The fox works himself bravely through a long list of questions. Later that day, he teams up with the owl again.

Assessment

Weddy Owl:	How do you interpret the result?
Marcy Fox:	Too early to assess. I sort of hoped that more animals would bid. Perhaps our notice was too short. The announcement of collusion investigation was a bit of cheap talk to scare them off. In any case, revenue was a side issue. We are the only forest to have a truly competitive Booming Buzzy before the next full moon.
Weddy Owl:	Do you think we appropriately anticipated all the risks?
Marcy Fox:	Positively. I did some auction theory when I was young. The Revenue Equivalence Theorem says that you should not bother too much on design, since all designs yield the same return. (*Owl stares pensively at the fox and fills his pipe*).
Weddy Owl:	Perhaps, but this Booming Buzzy business has more far-reaching consequences than just an auction design. My gut feeling says we have overlooked things. (*the fox and the owl depart*)

Monkey Business

The rats, ants and bees announce their terms of trade. Animals gather together to take notice and select their favorite operator.

Rat:	Here talks experience! Buy rat, now! Never trust insects! (*shouting*) No charge, no charge! (*then softly whispering*), the first period.
Bee:	Nobody buzzes better than a bee! You want your tube smelling like sewer, buy rat! Never listen to rats anyway!

Ant:	Ants are the masters of communication! Excellent industrious service all year around!
Weddy Owl:	(*to fox*) Oops, here is our first error. We did not specify the desired level of transparency in this process. This is not very pleasant for the animals. Can we stop the show?
Marcy Fox:	Not so easy I am afraid. The contracts are signed. Let the animals choose. When they are unhappy about their decision they can always switch operators.
Weddy Owl:	Very well, but I am starting to regret the adventure.

The animals all sign up with an operator in chaotic fashion. The rats obtain most customers. The owl goes for the ants, while the fox tries the bees. Several lunar cycles pass by, all operators manage to install Booming Buzzy without difficulty.

When the Shit Hits the Fan

Marcy Fox:	How does it work?
Weddy Owl:	Not bad, although I had some trouble starting the thing up.
Marcy Fox:	(*sigh of relief*) Well in the end all is right.
Weddy Owl:	Buzzy works surprisingly good. I actually start to enjoy it!

Three pigs enter the scene.

Pig 1:	We would like to file a complaint on the rats. Because of all the noises at this 'selection show' of yours, we could not hear all the details. We are trapped into unfavorable terms!
Weddy Owl:	Which unfavorable terms?
Pig 1:	Well, at first prices seem very low. There is even no charge the first period. Usage charges are also handsomely low later on. But there are other things.
Pig 2:	Such as service. My Buzzy had a buzz. It took the rats days to repair it, and they charged us hideous amounts of money for it.
Pig 3:	And they have creative entries on the bill. So I wanted to switch to these charming ants. The rats claimed that relocating Booming Buzzy is possible but expensive. 'Administrative costs'. Outrageous!
Weddy Owl:	We'll ask the fox. He knows all the answers, at least he thinks so. (*everybody looks for the fox, but he has suddenly disappeared*).

The Fox in a Box

The owl calls for the fox at his home tree.

Weddy Owl: You sneaked off quite timely yesterday.

Marcy Fox: Coincidentally, I had an urgent meeting with the rats ...

Weddy Owl: ... about whom I received an astonishing number of complaints. So the auction might have gone relatively smoothly, but we forgot to think of many other things. How are we supposed to deal with complaints? What sanctions do we have to discipline the rats? What if we do not find satisfactory solutions?

Marcy Fox: Are you sure that complaints from rancorous piggies are valid?

Weddy Owl: The pigs were the first but many others followed, including animals that do not complain easily, such as birds of prey.

Marcy Fox: The contracts are unspecified on minor details. I guess that will always be the case.

Weddy Owl: I will spell out all your *minor details*: 1. there are substantial switching costs, creating lock-in of animals, frustrating proper competition; 2. talking of which ... there are hardly any competition rules; 3. it is unclear what to do in situations for which there are no rules; 4. there are no credible sanctions if operators misbehave; 5. ... shall I go on?

Marcy Fox: (*the Fox feels guilty*) Ok, I took it too lightly, I should have listened to a couple of critical Chimps in Maggy's forest. (*whispers*) Nothing wrong with admitting a mistake, I guess.

Weddy Owl: Do you have a plan?

Marcy Fox: Not yet, I will meet the Chimps and come up with a cunning plan in one lunar cycle. No piggish scenes, I assure you, but these things really take time.

Weddy Owl: It'd better be a good plan, otherwise I will enforce banging sticks as soon as I can! (*the fox rapidly leaves for the Chimps in Maggy's forest*).

Meeting the Chimps

The fox arrives in Maggy's forest and immediately spots two giggling Chimps.

Chimp 1: What a mess you have made of that auction.

Chimp 2: Should have hired us as advisors. For a fraction of your financial disaster we would have prevented the whole lot.

Marcy Fox: (*ignoring the jeer*) Which brings me to the reason of my visit. We have to get out of this shitty mess as soon as possible. Are you prepared to advise me on a new plan?

Chimp 2: (*still giggling*) Sure, no problem, you know our fees. Which plan?

Marcy Fox: The owl cannot laugh at ratty jokes and threatens to regress into banging sticks. To convince the owl to keep Booming Buzzy, we have to come up with a cunning plan. I need your help to draft that plan.

Chimps: Deal (*sign a contract*).

A Foxy Plan

A lunar cycle has elapsed. The fox presents his plan to the owl.

Weddy Owl: I am very curious.

Marcy Fox: As I said a lunar cycle ago, one of my mistakes was not listening to the Chimps. This plan is joint work with the Chimps.

Weddy Owl: Let's hear it.

Marcy Fox: The plan is based on introducing your cousins the vultures.

Weddy Owl: Vultures? Yuck, you don't want to put your money on my awful cousins.

Marcy Fox: On the contrary, your cousins will prove to be useful.

Weddy Owl: They are only useful in cleaning up shit!

Marcy Fox: In a sense that is what they will do. In footnote 54 in the Forest Rules, it says that the Owl can decide in situations were the rules do not apply. So now all you have to do is to rule that vultures are allowed to hire capacity from the rats or the other communication suppliers.

Weddy Owl: What purpose does that serve? Surely animals will not switch to vultures.

Marcy Fox: It depends. Vultures are eager, patient and most of all persistent. And we can set the proper terms. We will forbid rats to charge these so called 'administrative costs' when animals want to switch. We determine access fees that the vultures have to pay. I think it will be quite successful.

Weddy Owl: We'll see.

The Ratbar

A short while later. The leaders of the ants, bees and rats gather together in the ratbar for their weekly 'chat amongst equals'.

Ant leader: (*nipping from fresh honey juice*). What to think of this latest foxy plan?

Rat leader: Tricky stuff. It disturbs the daily market routines.

Bee leader: It is an indecent proposal! How can we compete without being able to coordinate?

Ant leader: As usual consumers will suffer from this market disturbing entry by the vultures.

Bee leader: What worries me too is that the owl can decide what to do in the case of complaints, in situations where the rules does not apply. Honest entrepreneurs need a bit of maneuvering space, everybody knows that. (*the rat starts to vomit*).

Depressed leaders leave the ratbar.

The Highest Animal Court

Several lunar cycles have passed by since the auction. The rats did not agree with the introduction of the vultures and complained to the highest animal court. The fox and the owl are discussing the ruling of the court.

Weddy Owl: Too bad that the court has ruled in favour of the rats.

Marcy Fox: And the rats dared to question my integrity! "The fox changed the rules when the result was not to his liking". Outrageous.

Weddy Owl: An unreliable fox. Unthinkable.

Marcy Fox: It is not so clear, however, that the court ruled in favour of the rats. Think of the pigs.

Weddy Owl: Those stories about mud? Yuck!

Marcy Fox: Their slogan "muddy mobiles" was a bit one-dimensional, but a small modification will lead to a more widespread acceptance of their invention. It really works cute these mobiles.

Weddy Owl: Who would have thought that of these pigglets.

Marcy Fox: And they do look much smarter these days.

Weddy Owl: Their workout has worked out. But what do these mobiles have to do with the ruling of the Highest Animal Court?

Marcy Fox: The Highest Animal Court has ruled to re-allocate the licenses. Nothing prevents us from auctioning licenses for mobiles rather than the outdated Booming Buzzy.

Weddy Owl: Well alright, but this time we will prepare the whole thing much more thoroughly.

Muddy Mobiles

A couple of lunar cycles further. Four new licenses for mobiles are newly allocated. The pigs are among the winners. The rats withdrew.

The fox and the owl are sitting in the owl's home tree. Pigs are wallowing in the mud. Some bars of irritating muzak (vaguely resembling the Carnival des Animaux) are played. The owl grabs his mobile from underneath his feathers.

Weddy Owl: Hi cousin Eagle! (*the music has stopped*) ... I am in my tree, where are you? ... these new pig toys are an absolute s-e-n-s-a-t-i-o-n! ... What? You have a problem? Tell me ... yes, quite recognizable ... we also messed things up initially ... even our cunning little fox here did not know how to handle this (*the fox grumbles*). The auction design was very sloppily drafted and we did not have any competition rules or a proper regulatory framework ... As a consequence important animals such as birds of prey thought it too risky to compete in the bidding, so we were left with inferior species such as rats ... hold the line the fox is shouting something.

Marcy Fox: The vultures did their cleaning job, didn't they?

Weddy Owl: (to eagle) Sorry, the fox claims that the vultures saved us, but it was not particularly elegant to introduce vultures just like that. And we also violated the pigs with this sudden auction.

Pig: No offense taken.

Weddy Owl: (*fumbles his pipe*) What? ... Sorry a pig was disturbing our conversation ... these animal have to learn not to disturb mobile callers ... indeed the pigs came to our rescue with their mobile invention ... but look I have to fly. Let me finish with five lessons.

Lesson 1: Do not listen to incumbent animals. It will produce distorted signals.

Lesson 2: Take your time and communicate in a fully transparent way. This will make it likely that decent animals compete.

Lesson 3: Reserve one new license for a new animal. Otherwise competition becomes too cozy.

Lesson 4: Make licenses temporary.

Lesson 5: Install appropriate competition rules and a regulatory framework before you start anything. Even an owl cannot foresee everything.

Still Searching for Bertrand Equilibrium*

Eric van Damme

CentER, Tilburg University

Abstract. Throughout his career, Claus Weddepohl worked on models of price competition. In one of his earliest papers, he stated his motivation for this work: if economists do not succeed in constructing tractable models of price competition that generate rich insights, policy makers will assume that competition among the few will produce competitive outcomes. In this selective survey of Bertrand type models, I address the question whether price competition will produce competitive outcomes, what insights can be derived from these models, and whether a unified theory should be aimed for.

Key words: oligopoly theory, Bertrand competition, non-cooperative games.

1 Introduction

Price competition and the Bertrand model have played an important role in the work of Claus Weddepohl. In one of his first publications (Weddepohl, 1975), he writes (in my (abbreviated) translation of his Dutch):

"The theory of price formation contains three compartments. The first includes the theory of perfect competition, which is supposed to apply when there are many suppliers. The second is the theory of pure monopoly. This theory is simple and old. The third compartment is a mixed bag; here one finds theories that deal with markets with more than one, but not many competitors: oligopoly and monopolistic competition. The many theories which cover this field do not form a unity and they are complicated, some even very complicated. The consequence is that many observers believe that in cases like these, where there are "not too many" but also "not too few" competitors, price formation will more or less be as in the case of perfect competition".

In his paper, Weddepohl then continues by studying a Bertrand model in which firms produce with increasing returns to scale technologies, and he shows that the outcomes are very different from those resulting under perfect competition. In particular, his model admits many equilbria, at different price levels, and with different welfare properties. One can, hence, draw the broad conclusion that it is not wise to assume that markets in which there is competition among the few function as if they were perfectly competitive.

* The title is inspired by one of Claus Weddepohls' contributions "Op zoek naar evenwicht in een Bertrand model".

Weddepohl's article concludes by remarking that many variations of the basic model seem worthy of study and, indeed, later in his career, Weddepohl has returned to this general theme several times; several of his papers discuss variants of Bertrand's model.

Writing a quarter of a century later, I wish to return to this theme, to see what we have learned since then. The motivation for doing so is given by the final sentence from the above quotation, which I interpret as saying that, if economists do not succeed in making simple theories of competition among the few, policy makers will be inclined to assume that these markets will function as if they are perfectly competitive. The Dutch policy experience confirms this assessment of Weddepohl. Since the beginning of the 1990s, the Dutch government has embarked on a program of regulatory reform, the aim being to increase welfare, employment and growth by making markets more competitive. As I have argued at length elsewhere (Van Damme, 2001ab), policy has concentrated on reducing entry barriers, the underlying idea being that, once such barriers had been eliminated, competitive markets would arise spontaneously. The experience, however, has shown this not to be the case; where legal entry barriers were scrapped, strategic entry barriers were erected, so as to avoid competition becoming too intense. Furthermore, firms were also well aware that, in situations of tight oligopoly, it may be better to (tacitly) cooperate rather than compete, and this also contributed to the benefits of competition not being fully realised.

Returning to theory, the good news is that models are simple. The other side of the coin is that there is no unified theory; we just have a collection, a large collection, of models. Each model is reasonably easy to understand, but what may be more complicated is to exercise good judgement about which model is, or which insights are, most relevant in a certain specific case. One may expect that those who know the ins and outs of the various models, are also in the best position to judge which model fits best and which insights really apply. With this in mind, this paper provides an overview of some of the simple models of price competition, and the insights that can be derived from them.

The remainder of the paper is organized as follows. In Section 2, we discuss the basic Bertrand model together with some simple variations, including those that have been studied in the work of Weddepohl. In Section 3, we discuss issues of timing and cooperation in models of price competition, and in Section 4 we briefly turn to behavioral issues: how are price competition games played in the laboratory? While the material in Section 2 is drawn from the literature at large, the discussion in Sections 3 and 4 is mainly based on work that has been done at Tilburg University and in which I have been involved more closely. Section 5 concludes with some general remarks, inspired by the state of the art in this field and by the separation of economics and econometrics in the Netherlands.

2 Variations on Bertrand's Theme

We start with the most simple model of price competition. There are two firms that each can costlessly produce one unit of an indivisible commodity; there is one consumer that is willing to pay up to one money unit for one unit of this commodity. If the consumer buys from firm i for a price p_i, then the utility of firm i is p_i, the utility of the consumer is $1-p_i$, and the utility of firm j ($j \neq i$) is zero. The game proceeds as follows: simultaneously, each of the firms chooses its price; the consumer observes both prices and then decides to buy from one of the producers at the price quoted there, or not to buy at all.

It is remarkable how rich a set of issues can already be addressed in this setting. First of all, there are issues in pure (game) theory. Let us start by simplifying the game by eliminating the choice of the consumer, assuming that he will always choose what is in his best interest. Hence, if $\min(p_1, p_2) > 1$ the consumer does not buy; otherwise he buys at a firm charging the lowest price. In effect, we are moving from Nash equilibrium to subgame perfect equilibrium. As the reader may verify, we are eliminating some Nash equilibria in this way. (For example, consider a consumer who "wants to buy" at $\frac{1}{2}$; if both firms charge $\frac{1}{2}$, he randomizes equally among them; if exactly one firm charges $\frac{1}{2}$, he buys at that firm; and if none of the firms charges $\frac{1}{2}$, he does not buy.) Hence, one sees that the difference between Nash and subgame perfect equilibria already appears in this very simple model. Already in a simple setting like this one, one can illustrate "incredible threats" and "unreasonable equilbria".

Let us assume that the consumer indeed always behaves optimally, so that we can focus on competition between the firms. A formal definition of the game requires a specification of the strategy sets. Let us first assume that each firm can choose prices from a finite set P. As is well-known, as long as the price grid is sufficiently fine, the game is (weakly) dominance solvable, i.e. by repeatedly eliminating weakly dominated strategies, one can eliminate all prices other than the lowest, strictly positive and feasible price. Hence, one does not need an equilibrium assumption to solve the game. One may discuss several questions related to this observation: Is the procedure of repeatedly eliminating dominated strategies "justified", and under what conditions are the outcomes of such a procedure independent of the order in which one eliminates strategies? Both questions have received ample attention in the recent literature, see for example Brandenburger and Keisler (2001), and Dufwenberg and Stegeman (2001) for recent contributions.

Of course, the choice of a finite grid may be somewhat artificial. Also the case in which P is a continuum, say $P = [0, 1]$, is interesting. In this case, the unique Nash equilibrium (of the reduced game with the consumer eliminated) is $(0, 0)$, but, in a certain sense 0 is the worst strategy for each firm: it is weakly dominated by any other pure strategy in the game. Hence, here we have a paradox and an essential difference between games with a finite strategy set and games in which strategies can be chosen from a continuum. In the former case, there always exists a Nash equilibrium in undominated strategies, in the latter case apparently not. Given the fact that some Nash equilibria are unreasonable, given the need to refine the Nash equilibrium concept, given perfectness as an acceptable refinement criterion in finite games, and given that perfect equilibria of finite games are undominated, the question arises how to meaningfully define perfect equilibria for continuum games. See Simon and Stinchcombe (1995) for an answer to this latter question.

The economic message that comes from the above example is that price competition is a very competitive situation: with constant marginal cost, already two firms are sufficient to generate the competitive outcome and zero profit. To investigate how general this message is, we now study a couple of variations of the basic model. First of all, we note that, if there are more than two firms that can produce the commodity, we can observe price dispersion in equilibrium: as long as two firms each charge a price of zero, in equilibrium, all of the other firms can randomize in any way they want. (Of course, they won't necessarily make sales.) In the remainder of this Section, we first consider more general supply and demand conditions, whereafter attention is focused on the information that is available to market parties, on issues of divisibility, and on repetition of the game.

It is generally thought that, with constant returns to scale in production, the game is (almost) dominance solvable for rather general demand functions. That this, however, is not true can be seen by considering the case in which the consumer is willing to buy at any price, hence, he will always buy from the cheapest source, and the game played between the two firms is given by $\pi_i(p_1, p_2) = p_i$ if $p_i < p_j$, $\pi_i(p_1, p_2) = 0$ if $p_i > p_j$, and $\pi_i(p_1, p_2) = p_i/2$ if $p_i = p_j$. This game has multiple equilibria, in fact multiple mixed strategy equilibria. To verify this, assume that player i draws a price at random from the distribution $F(\cdot)$. Then player j's expected payoff resulting from price p is $p(1 - F(p))$. Hence, if $F(p) = 1 - K/p$ for $p \geq K$, then player j's expected payoff is K for any such price, while, furthermore, the expected payoff is lower for any $p < K$. Hence, we see that for any $K > 0$, we have such a mixed strategy equilibrium. Baye and Morgan (1999) show that a similar construction is possible for more general, downward sloping, demand functions, provided that monopoly profits are unbounded.

In his work, Weddepohl has investigated the relation between the competitive outcome and the Bertrand equilibrium for more general cost functions. Assume the consumer side is represented by a decreasing demand function, $d(p)$. Furthermore, assume that there are n identical firms that all produce with a cost function c. In Weddepohl (1975), the paper that was referred to in the Introduction, Weddepohl analyses the case of increasing returns to scale. In this case c is increasing and concave and the least costly way of serving demand is through monopoly. Weddepohl remarks that, in this case, supply functions and, hence, competitive equilibria, are not well-defined. The

Bertrand price setting game, however, is well-defined. If $n > 1$, there is competition for the market and this eliminates monopoly profits: there is an equilibrium in which one firm is active and charges price p such that $pd(p) = c(d(p))$. The other firms charge a price slightly above p, which ensures that they don't have demand and that the "active firm" cannot profitly deviate. (Note that in case the price space is a continuum, this equilibrium involves the demand side solving a coordination problem -all consumers have to go to the same firm- hence, consumers cannot simply be assigned equally to all firms with the lowest price.) Weddepohl, however, concludes that infinitely many outcomes are possible. It is not that his analysis is incorrect, he analyses a different model. Formally, he considers the following 2-stage model, which is similar to the one from Marschak and Selten (1974). In a first stage, firms decide whether or not to become active and active firms determine their prices. In a second stage, those firms that do not have the lowest price have the opportunity to switch to the lowest price. The consumer then chooses to buy at the lowest price where he equally splits his demand over all firms that do charge this lowest price. Weddepohl (1975) shows that there is a great many equilibria and that the more firms are active, the higher the price is.

In Weddepohl (1991), the paper that motivated the title of the present contribution, Weddepohl considers the case of decreasing returns to scale. Let us assume that there are n firms, that each produce with an increasing and convex cost function $c(\cdot)$ with $c(0) = 0$, and that consumers will distribute evenly across all firms that charge the lowest price. In this case, the competitive equilibrium is well-defined, but each firm charging the competitive price and producing the competitive quantity will typically not be an equilibrium of the price setting game. The reason is simple: if all firms charge the competitive price and produce the competitive quantity, each individual firm has some monopoly power over its own customers and it will have an incentive to exploit that power and to raise its price. As Weddepohl remarks, the equilibrium of the Bertrand game might depend on the fine details of the rationing rule that is used and an equilibrium in pure strategies will typically not exist. (For the special case of capacity constraints, this was already remarked in Edgeworth (1897).) Weddepohl then searches for price vectors that possess unidirectional stability (hence, either unilateral price increases or price decreases are not profitable), but also general existence of such "relaxed equilibria" cannot be established. In related work, Canoy and Weddepohl (1995), a relation "relaxation" is studied, in which firms assume that price reductions are always matched, but price increases not necessarily. We will not discuss these technically complicated aspects here, but confine ourselves to the remark that mixed strategy Nash equilibria can be expected to exist (Dasgupta and Maskin, 1986), even though they may be complicated to characterize or compute explicitly. However, it is straightforward to characterize equilibrium in case firms are not allowed to ration demand, hence, in case they are forced to serve all demand. (Of course, such an assumption may not be a reasonable one to make.) In this case, if there are m firms that charge the lowest price and if this lowest price is p_m, then each of these firms has profits equal to $\pi_m = p_m q_m - c(q_m)$ where $q_m = d(p_m)/m$, while the others have profit zero. Define p^*_m as that price that would result in zero profits if it is the minimum one in the market and it is charged by exactly m firms. These prices are well-defined under quite general conditions and, as can be easily verified, they are all equilibrium prices. Hence, we have multiple equilibria, see Dastidar (1995).

Another important modification of the basic model concerns the information condition. We have assumed that the consumer is informed about both prices and that firms know this. Imagine instead that the consumer only learns of one of these prices, with both possibilities being equally likely. After having learned p_i, the consumer may decide to buy at that price, or he may, at a cost $\varepsilon > 0$, decide to inquire information about the competitor's price p_j. After the consumer has bought the additional information, he then has to decide whether to buy the product, or not; in this case, if the consumer buys at price p_k his net utility is $1 - p_k - \varepsilon$. In a symmetric, pure strategy Nash equilibrium, the consumer expects the same prices at both locations, hence, he has no reason to switch. Since player i's best response against p_j is given by $p_i = p_j + \varepsilon$ if $p_j \leq 1 - \varepsilon$, it follows that the unique equilibrium is $(1, 1)$: each firm will charge the monopoly price. The example (originally due to Diamond (1971)) illustrates the importance of market transparency for competition: in a transparent market, the competitive price will result as soon as there are two firms; when the market is not transparent, the price will be the monopoly price, irrespective of the number of firms. Of course, in this stark form, the result may not be too relevant for actual markets. Nevertheless, this simple model may serve as a building block from which essential insights relating to market transparency, and the importance of the Internet for competition may be derived, see Van Damme and Dellaert (2001).

Of course, the situation in which n sellers compete for one client is mathematically equivalent to the situation in which n buyers compete for one object. The latter is the situation analysed in auction theory, hence, the close relation between that theory and the theory of Bertrand competition. Let us focus on the simple auction setting in which one object is available for sale with two buyers computing for it, and each of them assigning value 1 to this object. If the object is indivisible, and the seller organises a sealed-bid, first price auction, we are back in the previous case: each bidder will bid 1. What would happen if the object would be perfectly divisible with buyers assigning marginal value of 1 to each unit? In this case, each bidder i can specify a demand function $q_i(p)$ stating how many units he would like to have in case the seller charges prices p. Assume the seller organises a competition in demand functions, i.e. buyers simultaneously determine their demands and the seller then determines the price that clears the market; i.e. he sets p

such that $q_1(p) + q_2(p) = 1$. One equilibrium involves truthful revelation: $q_i(p) = \infty$ if $p \leq 1$ and $q_i(p) = 0$ if $p > 1$. However, as Wilson (1979) has shown, there are also equilibria involving demand reduction. For example, if player 1 demands $q_1(p) = 1 - p$, then player 2 faces the residual demand curve $r_2(p) = 1 - q_1(p) = p$ and, if the price is p, his net utility is $(1-p)p$. Consequently, if player 1 demands $q_1(p)$, then player 2 prefers the final price to be $\frac{1}{2}$. Now, if player 2 demands $q_2(p) = 1 - p$, then the final price will indeed be $\frac{1}{2}$, hence $q_2(p) = 1 - p$ is a best response against $q_1(p) = 1 - p$. By symmetry, both buyers demanding $q_i(p) = 1 - p$, therefore, is an equilibrium. This equilibrium, however, does not result in the competitive outcome: the seller's profit is only $\frac{1}{2}$.

Wilson (1979) concludes from the above example, and from more general ones in his paper, that share auctions may be less competitive than unit auctions. This insight is quite relevant for actual auctions, and it is wise to be on the outlook for possibilities of demand reduction. To give an even simpler example, assume there are 2 indivisible units for sale, and assume that there are 2 bidders that each assign a value 1 to each of these units. With demand exceeding supply, the competitive price is 1. The ascending, English auction, however has an equilibrium with revenue 0 for the seller. In this equilibrium, each buyer reduces his demand irrevocably to 1 unit at $p = 0$ and the auction stops immediately.

Up to now we have assumed that the underlying game is played only once, and the conclusions that have been derived depend on this assumption. Things may be completely different in a repeated setting. Imagine our simple (reduced) game being played repeatedly, at times $t = 1, 2, \ldots$, with players discounting future payoffs with a common discount factor δ. If both players are, at each time point t, fully informed about both prices that have been asked up to that point in time, there exist the well-known collusive trigger strategy equilibria: at time t choose $p_i = 1$ if previous prices have always been 1 choose $p_i = 0$ otherwise. Such equilibria, however, rely essentially on the assumption that the entire history is commonly known: with a bit of noise and private monitoring (each player only knows his own personal history of prices and sales), such equilibria may break down, see Bhaskar and Van Damme (2002) and other papers in the recent JET-Symposium on games with imperfect private monitoring. These models are quite important for the question of whether, and in which contexts, collusion may be problematic. The reader is referred to the JET-symposium for further details.

What's the moral in the story above? I have tried to show that a very simple economic situation is actually very rich. It gives rise to a great many puzzles, which provide mathematical fun and joy. The basic situation also leads to many different models and the solution may depend strongly on a specific detail of the model. Furthermore, in some situations, multiple equilibria arise naturally, which raises the general question of how to select an equilibrium. More generally, and perhaps more importantly, there is the question about how to select the model. Most models, however, contain relevant insights, and they all caution to the belief that competition among the few will lead to the competitive outcome. In any case, it seems quite clear that the search for a unified theory will be in vain, unless the theory itself is phrased in very general terms, which may not be very practical.

3 Differentiation, Timing and Cooperation

One message that comes from the models discussed in the previous section is that price competition indeed may be a very competitive situation, resulting in low profits for the firms involved. In such a situation, a firm might argue "if the game is so competitive, why would I play it?" To put it differently, would a firm have possibilities to change the nature of the game? In this section we discuss three possibilities for making the game less competitive: product differentiation, price leadership and cooperation. Our aim is to illustrate that such moves to change the nature of the game can also be analyzed by non-cooperative tools. Rather than focussing on the price game itself, in this case, one studies an enlarged (usually 2-stage) game in which the pure price game might arise as a subgame, but in which, because that game is so competitive, players decide to avoid it.

The product differentiation variant of this idea is due to Hotelling (1929). In his model, the set of consumers is modelled by the continuum [0,1], with the consumer located at x having preferences to buy a product of characteristics x. More generally, if that consumer buys a product with characteristics y for a price p, his utility is $u(y,x)-p$. In the simplest version of the model, there are two firms who play a 2-stage game. In the first stage, they simultaneously choose product characteristics, x_1 and x_2; and in the second stage, given these characteristics, they simultaneously choose prices. The simple game from the previous section arises as a special case when $x_1 = x_2$. In equilibrium, the firms have to trade-off the market demand effect (ceteris paribus, it is most attractive to be at the center of demand, $x = \frac{1}{2}$) against the competition effect (competitive pressure is lowest if the products are maximally differentiated, i.e. if $|x_1 - x_2|$ is large) and the exact nature of the solution depends on the utility function that is specified. For the linear case analyzed by Hotelling, there is no pure strategy equilibrium; for a quadratic version, the equilibrium involves maximal differentiation; for yet other specifications, firms may decide to locate at the middle. We refer to Goeree et al. (1997) for further details and other nice results. We also note that Bester et al. (1996) have shown that multiple equilibria and coordination failures may arise, if it is not imposed exogenously that firm 1 has to locate to the left of firm 2, i.e. $x_1 \leq x_2$.

An alternative way in which too intense price competition might be avoided is if players succeed in avoiding simultaneous price setting. With firms setting prices sequentially, prices will typically be higher. A good example of this in practise is price leadership, where the market leader sets a price that is then followed by the other players in the industry. Such price leadership exists, for example, in the Dutch gasoline market, where it is typically seen as one of the causes for the relatively high prices. More generally, the issue is one of timing of the moves, and the question is which firm will move when: who will be the leader? In joint work with Sjaak Hurkens (1996, 1999) I address this question. In these papers we ask the question which of the players will act as a leader and which one will follow, given that players have the possibility to move simultaneously or sequentially. Note that in a Cournot setting (quantity competition) it is well-known that it pays to commit: the leadership position is most preferred and the follower position least preferred, with simultaneous decisions taking an intermediate position. In price competition (with differentiated products), the situation is somewhat different: simultaneous price setting is least preferred; both players prefer there being a leader, but the follower profits most, hence, it is most advantageous to have that position.

In these papers, we introduce an asymmetry in cost functions to make the question "who will lead?" into an interesting one, and we show that the least cost firm will always emerge as the leader. Let us illustrate the model and the main idea in a simple asymmetric version of the 2-firm Bertrand game. Assume that the product is perfectly divisible, that the consumer has a marginal utility of 1 up to 1 unit and 0 thereafter, and that firm 2 has a capacity constraint of k. If firm 2 acts as a leader, firm 1 will undercut and firm 2's profit is zero. Firm 2's profit is also zero if price setting is simultaneous and in this case firm 1's profit is zero as well. If firm 1 commits to be leader, he will be undercut by firm 2, but given that firm 2 has a capacity constraint, firm 1 is not hurt that much by it. Firm 1 will simply commit to the monopoly price and profits will be $1 - k$ for firm 1 and k for firm 2. Hence, only in the case where firm 1 takes up the leadership position will profits be positive for each firm. It would be tempting to conclude from this that firms will coordinate on firm 1 acting as leader, however, such temptation should be resisted, players care only about their own payoffs and joint profits are not directly relevant. The papers with Hurkens derive the ordering of the moves from purely individualistic considerations. For the case of price competition, we consider the following 2-stage game. Players can choose a price in either period 1 or period 2, with player i being informed about the price p_j of his opponent if i chooses to move at $t = 2$ and j moves at $t = 1$. In the very special case considered here, this game can be solved by iterative elimination of dominated strategies and, not surprisingly, the outcome is that firm 1 commits to $p_1 = 1$ at $t = 1$. As said, Van Damme and Hurkens (1996) obtain the result that the efficient firm will lead in much more general settings. Still, some of the intuition is conveyed in this very simple example.

Above, we already briefly referred to the work of Edgeworth on Bertrand competition with capacity constraints. Edgeworth was also the one who introduced the Core as the concept that models unbridled competition. Shubik (1959) rediscovered this concept in the context of cooperative games and the close relation between the Core of the cooperative exchange game and the competitive outcome was soon discovered. Weddepohl discusses the Core equivalence theorem in his first publication, Weddepohl (1971). Cooperative games are games in which the emphasis is on coalitions and in which the analyst tries to predict the outcome (i.e. which coalition might arise and how it would divide the gains from trade) without specifying the detailed rules governing the strategic interaction between the players. The motivation to study cooperative games is threefold. First of all, cooperation really is an issue; secondly one might hope that it would be possible to predict the outcome on the basis of the little information that is contained in the cooperative game, and, thirdly, as stressed by Von Neumann and Morgenstern (See Von Neumann and Morgenstern, 1953, especially Chapter V), rational players may always be expected to attempt to violate the rules of any game when is is in their interest to do so. In our most simple price competition game (which in cooperative terms amounts to the 3-person characteristic function game given by $v(S) = 1$ if $3 \in S$ and $|S| \geq 2$; and $v(S) = 0$ otherwise, where player 3 is the consumer), the core consists of a single allocation (0,0,1), corresponding to the consumer buying from either producer for a price of 0, hence, the Core coincides with the competitive outcome, illustrating the Core equivalence theorem.

When, in the mid 1970s, I took my first courses in game theory with Stef Tijs, I considered the solution prescribed by the Core in the above game to be very natural. As a consequence, I was bothered very much by the fact that the Shapley value of this game was not an element of the Core and that it predicted a positive expected utility for each of the sellers. (As is well-known, the Shapley value of this game is (1,1,4)/6). This example and the multitude of cooperative solution concepts in general, without it being clear which one of these would apply in a specific context, induced me to turn away from Shapley value, and from cooperative theory in general, for quite a long time. It was only after I had realised that the multiplicity of solution concepts was a natural implication of trying to solve the problem at a too great level of abstraction, and that non-cooperative theory would allow one to get a better understanding of the various cooperative solution concepts, that I returned to this class of games. Of course, this position had been argued by John Nash long ago (See Nash (1953)).

Indeed, in the price competition game, coalitions arise quite naturally and, as a consequence, the Core actually provides a misleading picture. Obviously, once the sellers realize that their profits will be competed away if they do not form a cartel, they will try to form one. If they succeed in forming the cartel, they transform the situation in a bilateral monopoly in which case the negotiated price will be $\frac{1}{2}$. By symmetry, each of the sellers will get $\frac{1}{4}$ in this case. But, anticipating this, the consumer will want to form a coalition with any of the sellers, if only to prevent these sellers from entering into a cartel agreement. As Von Neumann and Morgenstern (1953) already realized, the game is really one in which players will rush to form a coalition. (See also Schelling, 1960). But then the outcome will be determined by the coalition formation process, hence, non-cooperative modelling should focus on that process.

Let us assume that the players bump into each other at random and that, if negotiations between two players are not successful (which will not happen in equilibrium), the match is dissolved and the process starts afresh. The remaining question is what price, p, the consumer will pay to the seller if a buyer-seller coalition is formed. (By symmetry, this price does not depend on which seller the buyer is matched with.) The outcome is determined by the players' outside options, i.e. by what players can expect if the negotiations break down. The next table provides the utilities players can expect depending on the first coalition that is formed

	Utility		
First Coalition	S_1	S_2	B
S_1B	p	0	$1-p$
S_2B	0	p	$1-p$
S_1S_2	$\frac{1}{4}$	$\frac{1}{4}$	$\frac{1}{2}$

For the coalition S_1B, the outside option of S_1 is $\frac{1}{3}(p+\frac{1}{4})$, while the buyer's outside option is $\frac{2}{3}(1-p)+\frac{1}{6}$. (This follows since all three 2-person coalitions are equally likely to form in the next round.) The coalition loses $\frac{1}{3}(p+\frac{1}{4})$ if it does not come to an agreement, hence, it will split this surplus evenly. It follows that the price p must satisfy

$$p = \frac{1}{3}(p+\frac{1}{4}) + \frac{1}{6}(p+\frac{1}{4}),$$

Hence, $p = \frac{1}{4}$. Since all coalitions are equally likely, the expected payoff of a seller equals $\frac{1}{6}$, while the buyer's expected payoff equals $\frac{2}{3}$. The conclusion is that expected payoffs are equal to the Shapley value of the game. Furthermore, the outcome, naturally, lies outside of the Core. Of course, the exact price will depend on the details of the matching process and different processes may give rise to different prices, hence, different cooperative solution concepts. In addition, some details in the above derivation have to be filled in. All such details can be found in the literature, we refer the interested reader to Montero (2000).

What's the lesson to be drawn from the material in this Section? I would say it is that the material shows that the simple model is even richer than Section 2 indicated it to be. While in Section 2, we focused on variations in the economic data, keeping the model of price competition unchanged, we concluded in this section that, when the game itself is very competitive, players will try to find ways around it, to change the game and make the competition less intense. Possibilities for doing so include innovation, for example, through product differentiation; strategic manipulation, for example, by influencing the timing of the game; or outright violation of the rules, by forming coalitions outside the formal rules. Clearly, one should be aware of all such possibilities, also in the policy domain. An example is provided by the Dutch UMTS-auction that took place in 2000. In that auction five licenses were for sale in a situation with five incumbents. One entrant, Versatel, participated in that auction and, just as in the above example, it had power over the incumbents: Versatel could drive up the price by staying in the auction, while it would do the incumbents a favour by stepping out; coalitions, hence, were quite natural, see Van Damme (2002).

4 Behavior

The discussion in the previous sections has been based on the assumption of rational behavior: firms and consumers maximize utility and they have no difficulty in doing so. From the experimental literature, we know that this assumption may work in some cases, but not in all: individuals may not maximize (for whatever the reason may be); or if they maximize, they need not maximize the utility function that we imputed; all kinds of other arguments might play a role. This raises the question of how well the rationality assumption works in the special, simple Bertrand game that we are considering in this paper.

Surprisingly, even though we know something about the answer to this question, we know remarkably little (See Holt (1995) for an overview.). In any case, there is some good news: the concerns for fairness that play such an important role in ultimatum bargaining experiments seem to play much less of a role in competitive price setting situations. On the other hand, there is evidence that experimental subjects realise that it is a sensible strategy to try to get out of the competitive situation. The evidence comes, among others from experiments conducted at Tilburg University, which are reported in Dufwenberg and Gneezy (1998). In what follows, we briefly describe their experiment.

The game they analyse, is based on the simple example with which we started this paper, with only the two sellers active. They bid integers from the set $\{2, 3, ..., 100\}$ and the one who chooses the lowest number, say n, gets ng where g is a parameter, a certain number of guilders, with ties being split. Clearly, in the unique Nash equilibrium of this game, each player chooses 2 and each seller receives g. In the experimental implementation of this game, there were 6 groups of 12 students and each group played 10 rounds of the game. In each round, 6 pairs of players were randomly formed, with no student meeting another student more than once. Three treatments were distinguished: in the high information treatment, H, before round $t+1$ each player was informed about all bids made in round t; in the medium information treatment, M, each player got to see all the winning bids, and in the low information treatment, L, each player was only informed about his own personal pay-off at the end of the round. Hence, the treatments correspond to various degrees of transparency about the market. For fully rational players, the amount of information should, however, be irrelevant as the equilibrium of the game does not depend on it.

How to play in such a game? What did the outcomes look like? How did the outcome depend on the information feedback? It turns out that in all sessions, behavior in round 1 is very much different from the Bertrand outcome. This actually should not be too surprising: players have to learn first to understand the game. Hence, it is more instructive to look at the data from round 10, but also then the outcomes may be different. The authors conclude: "it seems that in the treatments M and L the winning bids converged (almost) to the Nash equilibrium, whereas in treatment H there was no tendency at all of convergence to the Nash equilibrium". Formally, using the Mann-Whitney U test, the authors reject (at the 5% level) that the prices in the H-treatment are drawn from the same distribution as was observed in the M- or L-treatment. Clearly, it is encouraging that, in the L- and M-treatments, prices converge quickly to the Nash equilibrium, but what is going on in the H-treatment? The authors argue that, when loosing bids are observed, such bids may be used to signal desirable outcomes. Indeed, some of the behavior in their data is consistent with such signalling behavior. Nevertheless, the data appear somewhat puzzling: even if signalling might be a good thing during the game, it seems that there is little to gain from it in the last round. So why don't prizes suddenly drop when players know it is the last round?

Despite the above question, there is the important observation by Dufwenberg and Gneezy that a more transparent market is not necessarily more competitive. On the contrary, more transparency may make signalling easier and tacit collusion more likely. A similar message comes from other recent experimental research, such as Huck et al (1999) and Offerman et al (2001). Of course, a similar message was already conveyed by the older experimental work, such as that reported in Fouraker and Siegel (1953). Nevertheless, this message stands in stark contrast to the claim that the Internet will make markets both more transparent and more competitive (Van Damme and Dellaert (2001)) and it seems that more systematic research in this area would be worthwhile.

The main message from this section is simple. In some experiments with the Bertrand model, behavior converges quickly to the Nash equilibrium. In others, it does not. Some variables, which according to standard theory should not matter, do matter in practice. More work is to be done.

5 Conclusion

Let me try to summarize the argument that I have been trying to make. It is no doubt important to try to understand price competition among the few; will the outcome be close to the competitive outcome, or not? Some 115 years ago, Bertrand proposed a very simple model of price competition. Even though the model is simple, it is also very rich in the sense that it allows elaborations in many directions. It is also intellectually stimulating, and great fun to analyse and try to understand the variations on Bertrand's basic model. The richness of the structure has advantages and disadvantages. The great advantage is that, given an actual market, a puzzling observation or a paradox, it will be reasonably easy to find a Bertrand like model that captures some essential element of that situation, and that is at the same time reasonably well understood, so that the model may be expected to illuminate the market or puzzle. The drawback, of course, is that it will be difficult to have a unified theory, unless one is satisfied with a very general, abstract theory, or with a theory that is not neatly formalised, that is, a theory that is "just" a collection of connected insights.

Personally, I am happy with the Bertrand theory being loose, abstract and not completely formalised; in my view, the advantages mentioned above outweigh the disadvantages several times. In this sense indeed relevance may be preferred to rigour, at least if rigour is interpreted as insistence on a clear demarcation of in which contexts a specific insight will apply. Such a sharp boundary may not be possible, and, instead, one may be satisfied with relying on judgment in this respect. Of course, at the same time, one should insist on full rigour within a certain model. Reviewing the above material, I believe there are several insights there, such as the competitiveness of naked price competition, the desire of players to avoid too intense competition, hence, the incentives to change the nature of the game, the multiplicity of supply function equilibria (an insight especially relevant for the electricity market), the importance of market transparency, and the relation,up to now imperfectly understood, between transparency and competition (a question extremely relevant for Internet markets), and the insight that whether or not a Nash equilibrium is attained may depend on variables that are considered irrelevant in the standard theory.

The above point of view implies that a good economist probably should both have good powers of observation (in order to be able to detect puzzles), good judgement (so as to extract essential elements and built relevant models), and good mathematical skills (in order to structure, formalise and solve these models). In this respect, the strict separation that exists in the Netherlands between economics and econometrics is to be regretted, for it is unlikely to produce all-round economists that master all those skills. Obviously, what is unlikely is not impossible, and Weddepohl might be one of these exceptions that prove the rule. Another piece of good news is that also in the Netherlands, the sharp separation between the two fields is vanishing, with some of Weddepohl's students being active on both sides and, hence, actively contributing to tearing down the wall. The reader will probably be able to judge this himself, based on their contributions in this volume. As for Weddepohl's judgement, let me simply refer the reader back to the quote from the introduction, and let me note that indeed, because of Dutch (policy) economists being unfamiliar with oligopoly models, there has been a tendency to view oligopolistic markets as being competitive, with all the resulting disappointment.

References

1. Baye, Michael and John Morgan (1999), A folk theorem for one-shot Bertrand Games, *Economics Letters* **65**, 59-65
2. Bester, Helmut, André de Palma, Wolfgang Leininger, Elu von Thadden and Jonathan Thomas (1996), A non-cooperative analysis of Hotelling's Location Game, *Games and Economic Behavior* **12**, 165-186.
3. Bhaskar, V. and Eric van Damme (2002), Moral Hazard and Private Monitoring, *Journal of Economic Theory,* Forthcoming.
4. Brandenburger, Adam and Jerome Keisler (2000), Epistemic conditions for iterated admissibility, *Discussion Paper*, Harvard Business School.
5. Canoy, Marcel and Claus Weddepohl (1995) Alternative conjectures in a Bertrand-Edgeworth Model, *European Journal of Policitcal Economy* **11**, 577-598.
6. Damme, Eric van (2001a) Marktwerking Vereist Maatwerk *Maandschrift Economie* **65**(3) 185-208.
7. Damme, Eric van (2001b) MDW tegen het Licht, *Regelmaat* 2001/3, 87-97.
8. Damme, Eric van (2002) The European UMTS-Auctions, *European Economic Review*, Forthcoming.
9. Damme, Eric van and Sjaak Hurkens (1996), Endogenous Price Leadership, *Discussion Paper* nr. 96115, CentER, Tilburg University.
10. Damme, Eric van and Sjaak Hurkens (1999), Endogenous Stackelberg Leadership, *Games and Economic Behavior 28, 105-129.*
11. Damme, Eric van and Benedict Dellaert (2001) e-Economie: ICT en Marktwerking, in Rick van der Ploeg and Chris Veenemans (eds.) *De invloed van ICT op maatschappij en overheid*, Infodrome, 203-285.
12. Dasgupta, Partha and Eric Maskin (1986) The existence of equilibria in discontinuous games: II, Applications, *Review of Economic Studies* **53**, 27-41.
13. Dastidar, K. (1995) Comparing Cournot and Bertrand in a Homogeneous product market, *Journal of Economic Theory* ***75,*** 205-212.
14. Diamond (1971) A model of price adjustment, *Journal of Economic Theory* **3**, 156-169.
15. Dufwenberg, Martin and Uri Gneezy (1998), Price Competition and Market Concentration: An Experimental Study, *Discussion Paper* nr. 9827, CentER, Tilburg University.
16. Dufwenberg, Martin and Mark Stegeman (2001), When order matters for iterated strict dominance, *Econometrica*, Forthcoming

17. Edgeworth, Francis Ysidio (1897) The Pure Theory of Monopoly, reprinted in William Baumol and Stephan Goldfeld *Precusors in Mathematical Economics: An Anthology*, London School of Economics, 1968.
18. Fouraker, D. and Sidney Siegel (1953) *Bargaining behaviour,* McGraw Hill
19. Goerree, Jacob, Simon Anderson and Roald Ramer (1997) Location, Location, Location, *Journal of Economic Theory* **77**, 102-127.
20. Holt, Charlie (1995) Industrial Organization: A survey of Laboratory Research, Chapter 5 in John Kagel and Alvin Roth (eds.), *Handbook of Experimental Economics*, Princeton University Press, Princeton, NJ 1995.
21. Hotelling, Harald (1929) Stability in Competition, *Economic Journal* **39**, 41-57.
22. Huck, Stephan, Hans-Theo Normann and Jörg Oechsler (1999) Learning in Cournot Oligopoly: An Experiment, *Economic Journal* **109**, C80-C95
23. Marschak, Thomas and Reinhard Selten (1974), Equilibrium with Price Making Firms, *Lecture Notes in Economics and Mathematical Systems* **91**, Springer Verlag.
24. Montero, Maria (2000) *Endogeneous Coalition Formation and Bargaining*, PhD thesis, CentER, Tilburg University.
25. Nash, John (1953) Two-person Co-operative Games, *Econometrica* **21**, 128-140.
26. Neumann, John von and Oskar Morgenstern (1953) *Theory of Games and Economic Behaviour*, Princeton University Press, Princeton, NJ (First edition, 1943).
27. Offerman, Theo, Jan Potters and Joep Sonnemans, Imitation and belief learning in an oligopoly experiment, Discussion paper, CREED University of Amsterdam and CentER Tilburg University. Forthcoming in *Review of Economic Studies.*
28. Schelling, Thomas (1960) *The Strategy of Conflict*, Harvard University Press, Cambridge, MA.
29. Shubik, Martin (1959) Edgeworth market games, in A.W. Tucker and R.D. Luce (eds.) *Contributions to the theory of games IV; Annals of Mathematics Studies 40, P*rinceton University Press, Princeton, NJ.
30. Simon, Leo and Max Stinchombe (1995) Equilibrium Refinement for Infinite Normal-Form Games, *Econometrica* **63**, 1421-1443.
31. Weddepohl, Claus (1971) Concurrentie-Evenwicht en de Core van een Directe Markt, *Maandschrift Economie* **35**, 427-448.
32. Weddepohl, Claus (1975). Concurrentie in Geval van Afnemende Gemiddelde Kosten, *Maandschrift Economie* **40**, 87-102.
33. Weddepohl, Claus (1991) Op Zoek naar Evenwicht in een Bertrand Model, in Bouma, J. e.d. (eds.), *Samenspel, Opstellen aangeboden aan Prof.dr. G.F.W. Pikkemaat*, Wolters Noordhof, Groningen, 85-102.
34. Wilson, Robert (1979) Auctions of Shares, *Quarterly Journal of Economics* **93**, 675-689

Environmental Policy and Bertrand Competition: a General Equilibrium Approach*

Chris Elbers[1] and Cees Withagen[2]

[1] Department of Economics, Vrije Universiteit Amsterdam
[2] Department of Economics, Vrije Universiteit Amsterdam and Tilburg University

Abstract. Contrary to the usual approach we analyse the effects of environmental policy on competitiveness of domestic industries by means of a general equilibrium model. Attention is paid to the case where competition on the world market is perfect as well as to the case where the country under consideration is 'large'. However, we will concentrate on the case of Bertrand competition. It will be shown that under certain circumstances it is optimal from a social welfare point of view to conduct environmental policy which is lax with regard to the exporting sector of the economy compared to the sheltered sector of the economy. This result runs counter to the outcome of a partial equilibrium analysis.

Keywords: Environmental policy, competitiveness

1 Introduction

It is appealing from a partial equilibrium perspective to argue that strict environmental policy in a jurisdiction will hamper competitiveness of domestic firms: the marginal cost rise, implying an upward shift of the supply curve and therefore less exports. Several questions can be raised. First of all, is there any real world evidence for this? Second, if it is true in certain circumstances, what are the exact conditions under which it is indeed true? Third, is increased cost of production for exporters the only criterion to evaluate environmental policy? Isn't there a case to consider a broader welfare perspective, including environmental damage. Regarding the empirical relevance of loss of competitiveness, the evidence is at best mixed as well as meager. Several studies indicate that there is no harm to competitiveness, others show that the loss is only marginal, whereas still others find a reverse effect. See Mulatu et al. (2001) for a meta-analysis of international trade studies. Regarding the second question, in relation to the third one, it turns out to make a significant difference whether a partial equilibrium point of view or a general equilibrium perspective is taken and whether environmental policy is evaluated only in terms of profits to firms or that environmental externalities are taken into account. It has been shown by among others Rauscher (1994) and Elbers

* The authors wish to thank Pieter Ruys for his stimulating remarks.

and Withagen (2001) that in a two country general competitive equilibrium model, it is optimal from a social welfare point of view to impose a Pigouvian tax on the negative pollution externalities equal to marginal damage, for all sectors in the economy. There is no welfare gain in taxing the exporting sectors less than the sheltered sectors. Also the case of a country that is large on the world market has been studied. And the results indicate that is might be welfare enhancing to tax the exporting sectors more heavily than the sheltered sectors. The same result is obtained (by Elbers and Withagen (2001)) for a general equilibrium model involving two counties who are oligopolists on a so-called third market, if competition is by means of quantities.

In the present paper we investigate the case of Bertrand competition. This has been done before in a partial equilibrium setting by Barrett (1994), with emission standards set by the government as a means of precommitment. In the present paper we consider a general equilibrium model with Bertrand competition where the government uses environmental taxes. The main conclusion we reach is that contrary to the outcome of the partial equilibrium model, it might be optimal for the government to tax the exposed sector less than the sheltered sector.

The outline of the paper is as follows. In Section 2 we sketch the model. In Section 3 a description is given of the outcomes reached in the literature. In Section 4 we consider partial equilibrium, whereas Section 5 gives a general equilibrium analysis. Finally, Section 6 concludes.

2 The Model

The main ingredients of the model we employ are from the model constructed by Rauscher (1994) in his article on general equilibrium and environmental policy.

There are five commodities: three *consumer commodities*, *capital* and a *raw material.*

The first consumer commodity is produced and consumed domestically only. Production takes place in the so-called sheltered sector according to an aggregate technology described by a production function (F_1), having capital (k_1) and the raw material (y_1) as inputs. The production function is assumed to be concave, differentiable and monotonically increasing. Consumption is denoted by c_1.

The second class of consumer commodities is produced domestically in the so-called exposed sector by n ($n \geq 1$) producers, indexed by $i = 1, 2, ..., n$. Each producer i has a production function $\left(F_2^i\right)$, and employs capital (k_2^i) and a raw material (y_2^i). Part of the output is consumed domestically $\left(c_2^i\right)$, part of it is exported $\left(x_2^i\right)$ at prices p_2^i. The production functions are concave, differentiable and monotonically increasing.

The third consumer commodity can not be produced domestically. It needs to be imported. Consumption is denoted by c_3. Its given world market price is p_3. In the sequel the third commodity is usually taken as the numéraire.

Capital is immobile internationally but mobile between domestic sectors. Some empirical as well as theoretical support for the assumption of international immobility of capital can be found in Gordon and Bovenberg (1996). The economy's endowment is given by $\bar{k}$. The rate of return on capital is denoted by r.

The raw material is in principle freely available in unlimited amounts. However, processing of the raw material causes pollution. For that reason the government levies taxes τ_1 and τ_2^i $(i = 1, 2, ..., n)$ per unit of raw material used in the sheltered and the exposed sector respectively. The taxes can be differentiated between as well as within sectors. The tax revenues are recycled to the consumers in a lump sum fashion.

The income of the representative *consumer* consists of the value of the capital endowment $r\bar{k}$, the tax revenues $\tau_1 y_1 + \sum_{i=1}^{n} \tau_2^i y_2^i$, and the profits of the firms, which amount to $p_1 F_1(k_1, y_1) - rk_1 - \tau_1 y_1$ and $\sum_{i=1}^{n} \{p_2^i F_2^i(k_2^i, y_2^i) - rk_2^i - \tau_2 y_2^i\}$ for the sheltered and the exposed sector respectively, assuming for the moment that the domestic price of the second commodity equals its world market price. Under the assumption of full employment of capital (in a situation where firms maximise profits) total income boils down to (in shorthand) $p_1 F_1 + \sum_{i=1}^{n} p_2^i F_2^i$.

The consumer maximises utility, taking prices and income given. Preferences consist of two parts. First, they depend on the consumption of the consumer goods. This is represented by a utility function, denoted by $U(c_1, c_2, c_3)$, where $c_2 = (c_2^1, c_2^2, ..., c_2^n)$. The utility function is assumed to have all the usually desired properties such as concavity, differentiability and monotonicity. The use of the raw material brings along pollution (in a proportional way). That is the primary reason for the taxation. This part of the preferences is given by the (convex and increasing) damage function

$$D\left(y_1 + \sum_{i=1}^{n} y_2^i\right).$$

Finally, we impose equilibrium on the current account of the balance of payments.

3 Perfect Competition, Monopoly and Cournot Oligopoly

The first best optimum is defined as the allocation that maximises social welfare, i.e., utility of the representative agent minus the disutility of pollution damage, subject to the restrictions imposed by the technology and the condition of equilibrium on the current account of the balance of payments. In mathematical terms the first best optimum it is the solution of the maximisation of

$$U(c_1, c_2, c_3) - D(y_1 + \sum_{i=1}^{n} y_2^i).$$

subject to

$$c_1 = F_1(k_1, y_1)$$

$$c_2^i + x_2^i = F_2^i\left(k_2^i, y_2^i\right), \quad i = 1, 2, ..., n$$

$$p_3 c_3 = c_3 = \sum_{i=1}^{n} p_2^i x_2^i$$

It is easily seen that if there prevails perfect competition on the world markets of the exported commodities, the first best optimum can be realised in a decentralised setting by imposing emission taxes on the firms equal to marginal damage. Hence there is no reason to discriminate with regard to environmental policy between sheltered and exposed sectors, at least from the perspective of social welfare maximisation.

The case is different when the country is "large", so that world prices depend on the country's aggregate supply (assuming that there is no perceived interaction with other foreign players). In the literature it is usually assumed that each individual firm is small on the world markets, but that the sector producing an exported commodity, is large on an aggregate level. It can be shown that in these circumstances it is optimal to impose a tax on imports, but still to tax emissions according to marginal damage. In particular it is *not* optimal to favour producers of export goods over producers of domestic goods. But, if policy measures that cause trade distortions are not allowed, due to e.g. trade agreements, then this result no longer holds. Whether in a second-best world the exposed sector should be taxed more or less than the sheltered sector now depends on the terms of trade effect of a deviation of the emission tax away from marginal damage. Elbers and Withagen (2001) provide numerical examples which show that the optimal tax in the exposed sector is not necessarily below marginal damage and also is not necessarily below the emission tax imposed on the sheltered sector.

The case of Cournot competition on world markets has been considered by Barrett (1994) in a partial equilibrium setting, not taking into account the limited availability of capital and possible spillovers to other markets. It is shown that is it optimal to impose an emission tax below marginal damage. The model differs from the one employed here in several respects, but it is easily seen that the conclusion remains unaltered in the present framework, when a partial equilibrium is considered. Elbers and Withagen (2001) show that the conclusion no longer hold in a general equilibrium framework. This is generally the case when the domestic exporting firms play a Cournot Nash game on the world market. The government can then use its environmental policy as a precommitment device. In the next section of this essay we consider Bertrand equilibria.

4 Bertrand Oligopoly: Partial Equilibrium

In this section we consider the case where there is a Bertrand oligopoly on the world markets for the second type of consumer goods. It is common in the literature to study the oligopoly case in a partial equilibrium framework. First we summarise the work by Barrett (1994).

We cast Barrett's basic model in the format used above. In the literature attention is usually restricted to the case where the exported commodity is not consumed domestically. We adhere to that assumption. In Barrett's model domestic and foreign firms compete on the world market. Foreign and domestic firms produce varieties i $(i = 1, \ldots, n)$ of the second type of consumer good. Each variety in turn has a domestic and a foreign 'version'. In the original Barrett model each firm's costs consist of production costs, depending only on production, and abatement costs, depending on production as well as on the emission standard set by the respective governments. Since we want to stay as close as possible to the general equilibrium model set out above, and since this still serves to illustrate the main point, our model does not incorporate abatement nor emission standards. Instead we use emission taxes.

In the partial equilibrium approach the government takes as given all that occurs in the domestic sheltered sector. Moreover the government is not interested in total social welfare per se: pollution by the aggregate domestic sectors is not taken into account. Hence the government seeks to maximise export revenues minus social costs, the latter consisting of capital costs and the external damage costs caused by emissions of the exporting sectors.

The individual firm maximises profits, given the world demand schedule. The exporting firms produce heterogeneous commodities under constant returns to scale and compete with foreign firms. Without loss of generality we can perform the analysis for one variety only. The index denoting the variety is omitted for the time being. Let world demand for the domestically produced commodity be given by:

$$x = x(p, p_f)$$

Here p_f is the price charged by a foreign firm for the foreign 'version'. The domestic firm sets price p to maximise

$$px\left(p, p_f\right) - rk - \tau y$$

subject to

$$x\left(p, p_f\right) \leq F\left(k, y\right).$$

Assuming an interior solution we find as a necessary condition for optimality:

$$\{p - C(r,\tau)\}\frac{\partial x(p,p_f)}{\partial p} + x(p,p_f) = 0,$$

where $C(r,\tau)$ is the unit cost function which, due to constant returns to scale, depends only on the rate of return on capital and the emission tax. Social welfare is given by

$$W = (p - C(r,\tau))x(p,p_f) - D(y) + \tau y.$$

It consists of net profits of the firms minus damage costs, expressed in monetary value terms now, and the revenues from taxation. Now suppose that the government charges an emission tax equal to marginal damage: $\tau = D_y(y)$ where y is the profit maximising input of the polluting production factor. Next we consider the effect of a change in the emission charge, starting from the so-called optimal one.

$$dW = x(p,p_f)(dp - \frac{\partial C}{\partial \tau}d\tau) + (p - C(r,\tau))\left[\frac{\partial x(p,p_f)}{\partial p}dp + \frac{\partial x(p,p_f)}{\partial p_f}dp_f\right]$$

$$-[D_y - \tau]dy + yd\tau.$$

Using the necessary condition for profit maximisation and the fact that $y = x\frac{\partial C}{\partial \tau}$, this expression can be reduced to

$$dW = -x\frac{\partial x/\partial p_f}{\partial x/\partial p}\frac{dp_f}{d\tau}d\tau$$

The effect of a change in the emission tax depends on its impact on the equilibrium price. The second factor of the expression for dW above is negative. The third factor will be positive for a large class of demand functions (e.g., similar linear demand functions for the two varieties under consideration). Under these conditions the reaction functions are increasing functions of the competitor's prices. Since an increase in the tax rate will cause a downward shift in domestic supply (and hence an increase in the domestic price) $dp_f/d\tau$ is positive. So, an increase in taxes, starting from taxes equal to marginal damage, will increase social welfare. Henceforth the optimal taxation is larger than marginal damage. This result was also reached by Barrett in a slightly different framework. The basic idea goes back to the Brander and Spencer (1985) model where governments can take actions that constitute a commitment of domestic firms towards their competitors. In a two stage game interpretation the government makes a strategic choice before the domestic firm takes an action with regard to price setting. Price setting by the domestic firm itself is then influenced by this strategic decision. There is an extensive related literature on strategic substitutes and strategic complements (see e.g., Fudenberg and Tirole (1984)). In the case of a Cournot oligopoly with a homogeneous commodity the conclusion is the other way around: it is optimal to set emission taxes below the marginal damage for the exporting country. At least when there is a single domestic producer. So, the usual outcome that Cournot competition and Bertrand competition lead to different policy conclusions is reached here as well.

We provide an example to illustrate the result. Suppose that the world demand schedule is linear:

$$x = 1 - ap + bp_f$$

and that the same demand function applies for the commodity produced abroad. Assume also that the production function is Cobb-Douglas with a production elasticity of capital equal to β. Profit maximisation then leads to:

$$1 - 2ap + bp_f = a(\frac{\tau}{1-\beta})^{1-\beta}(\frac{r}{\beta})^{\beta}.$$

In this case the domestic reaction function is upward sloping in the price charged by the foreign competitor. Moreover, an increase of the emission tax will cause a downward shift of the reaction curve. Hence, given the foreign reaction curve, it is optimal to set an emission tax higher than marginal damage.

Finally, a more realistic case is the one where domestic consumption of the exported commodity is allowed for. It has received only minor attention in the literature. According to Ulph (1997), if there is domestic consumption and the domestic market is oligopolistic, there is an incentive for the government to impose less stringent taxation in the case of Cournot competition because imperfect competition will in general lead to too low output levels from the point of view of social welfare. Hence production for the domestic market should be stimulated, reinforcing the optimality of low emission taxation. In the case of Bertrand competition the introduction of domestic markets will have a countervailing effect.

5 Bertrand Oligopoly: General Equilibrium

In this section a general equilibrium model is constructed with domestic producers who act as oligopolists on world markets. For the time being we abstract from domestic consumption of the traded commodities, as in the previous section.

The motivation for analysing oligopoly in a general equilibrium setting was originally given by Rauscher (1994), who argued that the strategy of increasing domestic production (in his case by relaxing the standards) has an effect also on the allocation of capital in the economy through the rate of return. This effect is neglected in a partial equilibrium setting, but still it might be important. Of course the same general equilibrium considerations pertain in the case of Bertrand competition.

The social optimum is the solution of the following optimisation problem: maximise

$$U(c_1, c_3) - D(y_1 + \sum_{i=1}^{n} y_2^i)$$

subject to

$$c_1 = F_1\,(k_1, y_1)$$

$$p_3c_3 = \sum_{i=1}^{n} p_2^i x_2^i(p_2^i, p_f^i) \qquad i = 1, 2, ..., n,$$

$$\bar{k} = k_1 + \sum k_2^i$$

$$x_2^i(p_2^i, p_2^f) = F_2^i(k_2^i, y_2^i), \qquad i = 1, 2, ..., n,$$

where the home government takes the prices of the foreign firms as given. Necessary conditions for a social optimum are

$$U_{c_1} = \varphi_1$$

$$U_{c_3} = \varphi_3$$

$$\varphi_1 F_{1k} = \mu; \quad \varphi_1 F_{1y} = D_y$$

$$\varphi_2^i F_{2k}^i = \mu; \;\; \varphi_2^i F_{2y}^i = D_y, \;\; i = 1, 2, ..., n,$$

$$\phi_3 x_2^i(p_2^i, p_2^f) + (\phi_3 p_2^i - \phi_2^i)\frac{\partial x_2^i(p_2^i, p_2^f)}{\partial p_2^i} = 0, \qquad i = 1, 2, ..., n.$$

The multipliers φ_1, φ_2^i $(i = 1, 2, ..., n)$, φ_3 and μ correspond to the first consumer commodity, the exported commodities, the imported commodity and capital, respectively. The subscripts c_1, c_3, y and k denote partial derivatives with respect to raw material and capital inputs.

The optimum is realised in a decentralised setting if $p_1 = \varphi_1/\varphi_3$, $p_3 = 1$, $r = \mu/\varphi_3$ and $\tau = D_y/\varphi_3$. This can be seen by inserting these parameters in the optimisation problems of all individual agents in the economy. An alternative route to this result is to calculate the welfare impact of a deviation from setting emission taxes equal to marginal damage, taking the following into account:

- utility maximisation on the part of the consumer implies $U_{c_1}/p_1 = U_{c_3}$;
- profit maximisation in the competitive sheltered sector gives $p_1 F_{1k} = r$ and $p_1 F_{1y} = \tau_1$;

- profit maximisation in the exporting sector yields: $r = \psi_2^i F_{2k}^i$, $\tau_2^i = \psi_2^i F_{2y}^i$, where ψ_2^i is a Lagrange parameter denoting marginal costs of production. Moreover:

$$[p_2^i - \psi_2^i] \frac{\partial x_2^i(p_2^i, p_f^i)}{\partial p_2^i} + x_2^i(p_2^i, p_f^i) = 0;$$

- the given total availability of capital.

Then in case of emission taxes equal to marginal damage:

$$dW = U_{c_3} \sum_{i=1}^{n} (p_2^i - \psi_2^i) \frac{\partial x_2^i(p_2^i, p_2^f)}{\partial p_2^i} dp_f^i.$$

Therefore, in case of a Nash equilibrium on the world markets the impact of a deviation of a tax rate from marginal damage is zero: it is optimal from a social welfare point of view to have all emission taxes equal to marginal damage. Hence there is no reason to discriminate against specific sectors or to favour sectors.

The more interesting case is where the domestic firm can be made to act as a Stackelberg leader in equilibrium. This is achieved through the government's policy as in Brander and Spencer (1985). In order to perform the analysis we assume that the other country is a Stackelberg follower but is otherwise identical to the country under consideration, except possibly in size, measured by means of total available capital. We have to derive the other country's reaction functions and to incorporate them into the objective function of the leaders. We will perform a numerical analysis. In order to keep things simple we restrict our attention to the case of a single domestic exporting firm. We specify the functions involved as follows.

$$U(c_1, c_3) = \ln c_1 + \ln c_3$$

$$F_1(k_1, y_1) = k_1^{\alpha} y_1^{1-\alpha}$$

$$F_2(k_2, y_2) = k_2^{\beta} y_2^{1-\beta}$$

$$D(y_1 + y_2) = [y_1 + y_2]^2$$

$$x(p_2, p_f) = 1 - ap_2 + bp_f, \qquad a, b > 0.$$

The imported commodity serves as the numéraire: $p_3 = 1$. National income equals $p_1 F_1 + p_2 F_2$. In view of the utility functions and in the absence of domestic consumption of the exported commodity, it follows from utility maximisation on the part of the consumers that:

$$p_1 c_1 = 1/2 [p_1 F_1 + p_2 F_2], \qquad c_3 = 1/2 [p_1 F_1 + p_2 F_2]$$

Equilibrium on the current account implies that $c_3 = p_2 F_2 = p_2 x$. Hence $p_1 c_1 = p_2 F_2 = p_2 x$. Profit maximisation in the sheltered sector requires that the equilibrium prices lie on the factor price frontier given by:

$$p_1 = \left(\frac{r}{\alpha}\right)^{\alpha} \left(\frac{\tau_1}{1-\alpha}\right)^{1-\alpha} = g_1(r, \tau_1)$$

The exporting firm is maximising profits. This implies:

$$1 - 2ap_2 + bp_f = a\left(\frac{\tau_2}{1-\beta}\right)^{1-\beta} \left(\frac{r}{\beta}\right)^{\beta} = ag_2(r, \tau_2)$$

Factor demands are given by

$$k_1(r, \tau_1, c_1) = (\frac{r}{\alpha})^{\alpha-1} (\frac{\tau_1}{1-\alpha})^{1-\alpha} c_1$$

$$y_1(r, \tau_1, c_1) = (\frac{r}{\alpha})^{\alpha} (\frac{\tau_1}{1-\alpha})^{-\alpha} c_1$$

$$k_2(r, \tau_2, x) = (\frac{r}{\beta})^{\beta-1} (\frac{\tau_2}{1-\beta})^{1-\beta} x$$

$$y_2(r, \tau_2, x) = (\frac{r}{\beta})^{\beta} (\frac{\tau_2}{1-\beta})^{-\beta} x.$$

Taking into account that $c_1 = p_2 x(p_2, p_f)/p_1$ and $x = x(p_2, p_f)$ the mathematical problem faced by the government can be stated as: maximise

$$\ln p_2 x(p_2 + p_f)/p_1 + \ln p_2 x(p_2 + p_f) - 1/2\,[y_1 + y_2]^2$$

subject to

$$k_1 = k_1(r, \tau_1, p_2 x(p_2, p_f)/p_1)$$

$$y_1 = y_1(r, \tau_1, p_2 x(p_2, p_f)/p_1)$$

$$k_2 = k_2(r, \tau_2, x(p_2, p_f))$$

$$y_2 = y_2(r, \tau_2, x(p_2, p_f))$$

$$g(r, \tau_1) = p_1$$

$$1 - 2ap_2 + bp_f = ag_2(r, \tau_2)$$

$$k_1 + k_2 = \bar{k}.$$

This problem cannot be solved in such a way that the sensitivity with respect to the parameters can be determined analytically, as one would wish to do in order to find out how to set optimal taxes in different circumstances. Therefore a number of numerical calculations have been executed.

In the calculations the capital endowments of the home and foreign countries are set equal to 10. The parameters in the world demand function are $a = b = 0.2$. The production elasticities of capital are $\alpha = 0.4$ and $\beta = 0.2$.

First we computed the social optimum where the government takes the price set by the other country as given. We take $p_f = 1$. The policy maker sets the export price p_2 and the emission tax rates $\tau_1$and τ_2. The optimal values are

$$p_1 = 0.93, r = 0.089, p_2 = 3.27, \tau_1 = \tau_2 = 1.46, x_2 = 0.55.$$

As already derived theoretically, emission taxes equal the monetary value of marginal damage. The value of the social welfare function is 0.90.

Next we calculated the so-called Nash trade equilibrium. It is a set of prices (p_2, p_f) such that given p_f the welfare maximising strategy of the home government yields, through the emission taxes, an export price p_2. And, given p_2, the foreign country's optimal policy generates p_f. In the case of two identical countries the result is:

$$p_1 = 2.25, r = 0.22, p_2 = p_f = 3.92, \tau_1 = 3.45, \tau_2 = 2.18, x_2 = 1.$$

The value of the social welfare function is 1.53. The equilibrium emission tax in the sheltered sector equals the monetary value of marginal damage. Therefore, the example shows a case where in general equilibrium, the export sector should be taxed less than marginal damage, contrary to what is optimal in a partial equilibrium setting.

Finally, we have calculated what we call the Brander Spencer equilibrium, with the home country as leader. This equilibrium is generated as follows. A domestic export price induces a foreign export price through welfare optimisation on the part of the foreign country. Taking this mechanism into account the domestic government can calculate the optimal domestic export price and of course also the optimal emission tax profile yielding this optimal price. The optimal solution now reads:

$$p_1 = 2.26, r = 0.22, p_2 = 3.97, p_f = 3.93, \tau_1 = 3.51, \tau_2 = 1.84, x_2 = 0.99.$$

Social welfare is 1.53. One observes that the deviation is only marginal. Again, the tax rate for the sheltered sector equals the monetary value of marginal damage, and the exposed sector pays less than marginal damage.

Figure 1 depicts the reaction curves for the previous two cases. The steep curve is showing the foreign firm's response. In the curve the foreign country's optimal reaction has already been incorporated. The lower of the two flatter curves is the home country's reaction curve; the intersection with the foreign

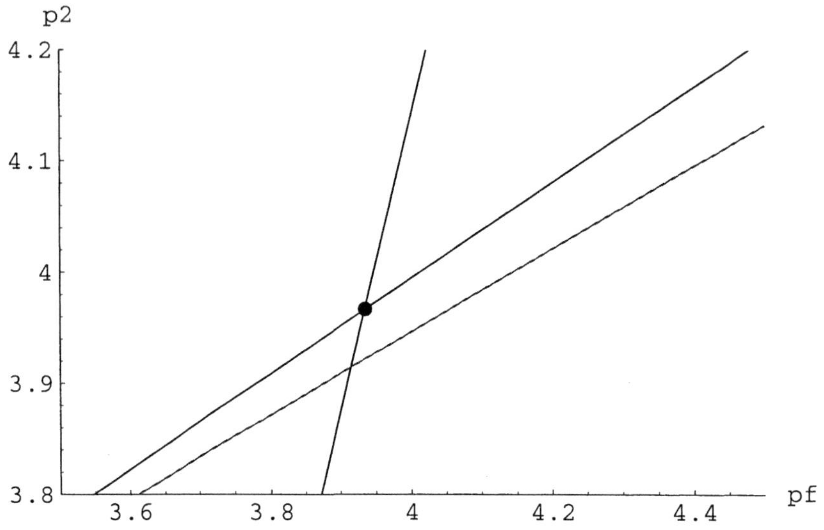

Fig. 1. Brander Spencer (dotted) and Nash trade equilibria.

curve is the Nash trade equilibrium. The upper reaction curve results if the government anticipates the foreign response and leads to the Brander Spencer equilibrium.

6 Conclusion

In the design of environmental policy and with regard to the desirability of a stringent environmental policy it is crucial to be aware of the situation in which the economy finds itself. In particular this holds for the 'competitiveness' issue which is brought into the discussion by many lobby groups. For optimal environmental policy it makes a difference whether the economy finds itself in a situation of perfect competition on the world market or not. The reason is that the environmental policy of a 'large' country will have an impact on the terms of trade. When there is oligopoly in the world market and a terms of trade effect can be expected, it again makes quite a difference whether competition takes place by means of quantities (Cournot oligopoly) or prices (Bertrand oligopoly). Finally, optimal policies may differ according to whether a partial equilibrium analysis is appropriate or a general equilibrium analysis is necessary. In the present paper we have concentrated on the case of Bertrand competition. Two main outcomes arise. It has been confirmed in our setting, with emission taxes rather than standards, that environmental policy should be stricter for the exporting sectors than for the sheltered sectors when a partial equilibrium perspective is taken. But, in a general equilibrium setting this results no longer holds true. We have provided examples where the exposed sector should be taxed *less* than the sheltered sector.

References

1. Barrett, S. (1994): "Strategic environmental policy and international trade", *Journal of Public Economics* 54, pp.325-338.
2. Brander, J. and Spencer, B. (1985): "Export subsidies and international market share rivalry", *Journal of International Economics* 18, pp. 83-100.
3. Culioli, C. (1996): "Optimization with *Mathematica*", Chapter 3 in H. Varian (ed.), *Computational Economics and Finance,* Springer-Verlag, New York.
4. Elbers, C. and Withagen, C. (2001): "Environmental regulation and international trade: a general equilibrium approach", mimeo Free University Amsterdam, forthcoming in A. de Zeeuw and J. List title to be inserted.
5. Fudenberg, D. and Tirole, J. (1984): "The fat cat effect, the puppy dog ploy, and the lean and hungry look", *American Economic Review, Papers and Proceedings,* 74, pp. 361-368.
6. Gordon, R. and Bovenberg, L. (1996): "Why is capital so immobile internationally? Possible explanations and implications for capital income taxation", *American Economic Review* 86, pp. 1057-1075.
7. Mulatu, A., Florax, R. and Withagen, C. (2001): "Environmental regulation and competitiveness: a meta-analysis of international trade studies", *mimeo* Free University Amsterdam.
8. Rauscher, M (1994): "On ecological dumping", *Oxford Economic Papers* 46, pp. 822-840.
9. Ulph, A. (1997): "International trade and the environment: a survey of recent economic analysis", in H. Folmer and T. Tietenberg (eds.), *The International Yearbook of Environmental and Resource Economics 1997/1998,* Edward Elgar, Cheltenham, pp. 205-242.
10. Wolfram, S. (1996), The Mathematica book, Cambridge University Press, Cambridge.

Retiring Rich*

Jacob K. Goeree

CREED, Faculteit der Economische Wetenschappen en Econometrie, Universiteit van Amsterdam

Abstract. To raise money for his retirement, Claus intends to auction the original handwritten version of one of his classic papers among his caring advisees. We describe Claus' discovery that "standard" winner-pay auctions yield surprisingly low revenues in this context because of the positive externality bidders forgo if they top another's high bid. By requiring losers to pay their bids, Claus is able to raise much more money and secure the kind of retirement he deserves and desires.

Keywords: Altruistic bidders, winner-pay auctions, all-pay auctions, revenue dominance.

Claus has been looking forward to his retirement for some time. Like other retirees to be, he intends to visit exotic places, read the books he accumulated over time, and even take some art classes. Fun plans are costly, however, and Claus wonders how to pay for them once his pension is his sole source of income. He doesn't like the idea of selling the house (or the proverbial boat) and is determined to find alternative ways of raising money for his pension.

One day, while cleaning up the attic, it hit him. He had just come across the original handwritten version of his *Econometrica* paper "An Equilibrium Model with Fixed Labor Time," which was published back in 1979. Didn't his most recent graduate advisees, Mark, Jack, and John, tell him on numerous occasions they would love to acquire this masterpiece? Indeed, was it not Mark, or John, or maybe Jack, who had asked him about its whereabouts not too long ago? Claus decided it was time for his former advisees to put their money where their mouth was. He would organize a contest between Mark, Jack, and John and have them compete for the handwritten manuscript. Thrilled by this prospect, Claus hurried downstairs to tell his wife Ank who liked the idea but wondered "what type of contest will you organize?"

Claus realized the answer to his wife's question was not obvious. First of all, what would be the objective of the contest? Raising as much money as possible, sure. But as an economist, Claus disliked the idea of employing an

* Dutch title: "Pensioneren met het Zwitserleven Gevoel." This paper is based on "How (Not) to Raise Money," by Jacob K. Goeree and John L. Turner. Financial support from the Bankard Fund at the University of Virginia is gratefully acknowledged.

inefficient sales mechanism just to raise more revenue. He considered it imperative that the manuscript go to whomever valued it the most. "Wait," he thought, "isn't this what auctions do best? Aren't auctions praised for their efficiency while generating lots of money for the seller?" But what auction should he use? There were so many to choose from: English, Dutch, Japanese, ...

Claus decided to give the matter some serious thought since his retirement was at stake. Armed with pencil and paper he withdrew into his study, or the "serious room" as his family jokingly called it. Seated in his favorite chair he entered a state of deep thought, filling the air with the fumes from his infamous pipe.

1 The Usual Suspects

Claus started by reading through his *Auction 101* book, acquired when he was a freshman in college. It made a distinction between "single-shot" and "multi-round" or "dynamic" auctions. An example of the latter is the English auction, widely used in the sale of art. In the English auction bidders are allowed to raise their bids after they have observed others' bids. In contrast, in a single-shot auction, all bidders write a bid on a piece of paper, put it in an envelope, seal the envelope and hand it over to the seller. The seller then determines the highest bidder who gets the auction's prize, a payment is exchanged and the auction is done. Claus liked the simplicity of single-shot auctions, which were also called "sealed-bid" auctions for obvious reasons. He decided to employ a sealed-bid format because he preferred a speedy process and he deemed it less prone to strategic manipulation by bidders.

He continued reading *Auction 101*, which talked about two different sealed-bid formats: the first-price and second-price sealed-bid auction. In the former, the highest bidder wins and pays his own bid, while in the latter the highest bidder only has to pay the second-highest bid. "Hmm," pondered Claus, "if there are first and second-price auctions, why is there no mention of a third-price auction?" He made a note to investigate the third-price auction and figure out why the *Auction 101* book was silent on the topic. (Claus' curiosity proved invaluable, as we will see shortly.[1]) Claus felt relieved he had been able to make a selection of formats to study and was happy his list was a short one: first-price, second-price and third-price sealed-bid auctions. Due to the popularity of the first two, Claus nicknamed them his "usual suspects."

Claus was ready for some modeling. He lit his pipe, sharpened his pencil, and started describing the information bidders possess in the auction. "Mark, Jack, and John probably have different values for the manuscript," Claus

[1] Probably *Auction 101* ignored the subject because the third-price auction is too difficult to analyze for most undergraduates, as we will discover shortly. This, of course, did not stop Claus who liked the mathematical challenge more than anything else.

argued, "and their values are likely to be independent." Claus had no idea what the range of values could be, but he didn't see this as a problem. He argued he could just normalize units such that values were between 0 and 1.[2] Moreover, he saw no reason why some values would be more likely than others. He wrote his findings thus far in mathematically concise notation: $v_i \sim U[0,1]$ for $i = 1, 2, 3$, which simply means that Mark, Jack, and John draw their values from a uniform distribution on the unit interval.

Next he turned to the payoffs that bidders derive from bidding in the auction. Here he stumbled on a puzzle. The *Auction 101* text simply stated that the winner gets his value and makes a payment (e.g. the highest or second-highest bid) while the losers get nothing and pay nothing. Claus did not think this description was adequate for the contest he was planning. He knew his advisees felt a great deal of gratitude and were quite fond of him. "All three of them will derive pleasure seeing how much money is raised for my pension," Claus felt, "so even the auction's losers will receive a warm-glow feeling knowing that I receive its proceeds. But how should I model this warm-glow feeling?" Claus decided to keep things simple and assume the additional utility was simply a constant, α, times the auction's revenue. The α constant, he thought, could be interpreted as a measure of altruism, or as a "coefficient of sympathy" as Edgeworth had noticed long before him. When $\alpha = 1/4$, for instance, his advisees would value 1 Euro given to him the same as 25 Euro cents for themselves. Claus made the natural assumption that α was restricted to lie between 0 and 1.

Claus' next step was to predict Mark's, Jack's, and John's optimal bids given their values for the manuscript. Of course, as the seller, Claus would not know their values before the auction took place, but predicting how they would bid for arbitrary values would help him guess how much money would be raised on average. Claus wondered if the problem had been studied for auctions where bidders benefit from the revenue they raise themselves. He searched the literature and found only a few papers, which proved very useful, however.

Claus read the 1994 paper by professor Engelbrecht-Wiggans who claimed that, with three bidders, the optimal bids in the first-price auction are

$$B_1(v) = \frac{2v}{3-\alpha}.$$

Claus checked his *Auction 101* book, which listed the answer for the standard case without altruism, i.e. when $\alpha = 0$: $B_1(v) = 2v/3$. "Aha," Claus cried out, "Engelbrecht-Wiggans' answer is correct in the limit case, and it makes sense that bids are higher when $\alpha > 0$ because of the warm-glow utility that bidders derive." Convinced Engelbrecht-Wiggans' paper wasn't a waste of time, Claus decided to do the math and verify the professor's claim more thoroughly. He

[2] With 0 corresponding to the lowest possible valuation for the manuscript, say 10 Euros, and 1 corresponding to the highest possible valuation, say 10,000 Euros.

was relieved to find that Engelbrecht-Wiggans had been entirely correct.[3] In the same paper, Claus found the optimal bids for the second-price auction, which were given as

$$B_2(v) = \frac{v+\alpha}{1+\alpha}.$$

Claus convinced himself that the answer reduced to bidding one's value when $\alpha = 0$, and that bids were shifted upward when $\alpha > 0$. He trusted that Engelbrecht-Wiggans had again found the correct result.

Now that Claus knew how bidders' values translated into bids, he could calculate the amount of money raised on average. He argued that with three draws from a uniform distribution, the highest and second-highest draws were 3/4 and 1/2 on average. Hence, the revenue from a first-price sealed-bid auction would simply be $R_1 = 2\cdot\frac{3}{4}/(3-\alpha) = 3/(6-2\alpha)$ and that of a second-price would be $R_2 = (\frac{1}{2}+\alpha)/(1+\alpha) = (1+2\alpha)/(2+2\alpha)$. "Great," Claus told himself, "when $\alpha = 0$ both formats yield the same revenue of 1/2, which is William Vickrey's famous revenue equivalence result." Excited he tried a few other values of α, and to his surprise, the revenue of the second-price auction exceeded that of the first-price auction for all $0 < \alpha < 1$, while the formats were equally profitable for $\alpha = 0$ and $\alpha = 1$. Claus concluded his work had not been in vain: already he had discovered that he should prefer the second-price to the first-price auction when selling the manuscript.[4]

Suddenly, Claus remembered the little note to himself, suggesting to also look at the third-price auction. He searched the literature again but found no papers written on this case. "Finally, something new," he thought, "I'll crack this one myself. But first, let me pour myself a nice glass of *Château Margaux*." Having enjoyed a few sips of his wine, Claus went back to work. He soon discovered why no one had dealt with this case before: it was far more complex than the first and second-price auctions. Challenged, Claus worked all night on the problem until he finally solved it at dawn. He glanced at the empty wine bottle and then at his answer and wondered how he ever figured it out:

$$B_3(v) = \frac{2(v-\alpha)}{1-\alpha} + \frac{\alpha}{2(1-\alpha)}\left(1+\sqrt{1+8/\alpha}\right)(1-v)^{\frac{1}{2}(\sqrt{1+8/\alpha}-1)}$$

with corresponding revenue

$$R_3 = \frac{1-\alpha+3\alpha^2\left(3-\sqrt{1+8/\alpha}\right)}{2(1-\alpha)(1-3\alpha)}.$$

[3] Consider a bidder with value v who bids as if he has value w and who faces rivals that bid according to $B_1(\cdot)$. The expected payoff is: $\pi^e(B_1(w)|v) = (v-(1-\alpha)B_1(w))w^2 + \alpha\int_w^1 B_1(y)dy^2$. It is easy to verify that the first-order condition for profit maximization yields: $\partial_w\pi^e(B_1(w)|v) = 2(v-w)w$, so it is optimal for a bidder with value v to bid $B_1(v)$.

[4] See also Engers and McManus (2001).

Claus checked that also the third-price auction generated 1/2 in revenue when $\alpha = 0$, as dictated by William Vickrey's famous theorem. He was intrigued to see that revenue equivalence also held at the other extreme when $\alpha = 1$: again revenue was 3/4. Claus' main interest, however, was how the third-price auction performed relative to the other two formats for intermediate values of α. This time he tried more than a few values. "Let me graph the revenue results I've found so far," Claus thought, "it'll make it easier to explain them to Ank." Satisfied with his graph (see Figure 1), Claus retreated from his study to go to work. He took one final look at the graph and chuckled "What do you know, a third-price auction to fund my retirement."

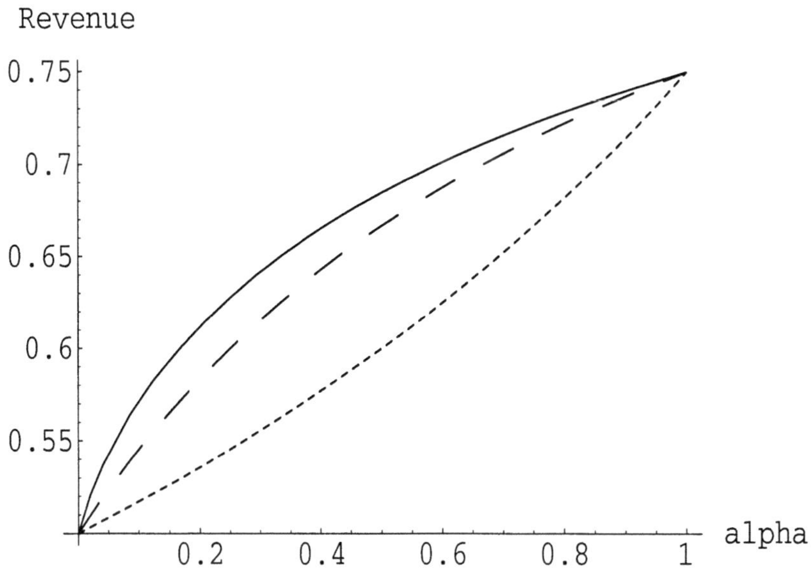

Fig. 1. *The revenue from a first-price (short dashes), second-price (long dashes), and third-price (solid line) auction for* $0 \leq \alpha \leq 1$.

When he returned from work later that day, Claus felt less satisfied with the results obtained. It was interesting to see that the third-price auction was superior in terms of raising money, but revenues still seemed rather low. Indeed, in the extreme case when $\alpha = 1$, all three formats yielded revenues of 3/4 even though bidders were indifferent between keeping an Euro for themselves or giving it to Claus! Surely, one should be able to raise more money in this extreme case. "When $\alpha = 1$, I can expect more money by asking for voluntary contributions without having to sell the manuscript," Claus later complained to Ank. She agreed and asked "but what's your intuition why auctions perform so poorly?" Claus realized his wife had asked the right question again and decided he would spend another night in the serious room.

Somewhat frustrated, Claus sat down at his desk. "What suppresses bids in these auctions," he asked himself, "why don't they bid infinity when $\alpha = 1$?" When he heard himself asking the question out loud, he realized how naive he'd been. "Of course," he cried, "opportunity costs! When someone overbids the highest bid of the others, he forgoes the warm-glow utility derived from that bid." He turned to the first-price auction to make his ideas more precise, knowing that it would work the same for the other two formats. He conjectured that when $\alpha = 1$ it would be optimal to bid one's value in the first-price auction (a curious result, since it is normally the second-price auction that results in "truth telling"). "Let's see," Claus argued, "suppose two bidders bid their values, then I have to show that the third one also bids his value." This was easy enough: if the third bidder would bid less than his value and lose the auction because of it, he would only receive the highest of the others' bids instead of his value and thus be worse off. Likewise, if the third bidder would increase his bid above his value and win the auction because of it, he would get his value but forgo the highest of the others' bids and again be worse off.

Having shown that bidders would bid only as high as their values, Claus concluded the auctions he had tried were inferior for raising money. "Revenues are extremely low because of the positive externality that is lost when one bidder overbids another's high bid," he reasoned, "what I need is a different format." He decided to inform Ank of his latest findings and then go to bed early. She was impressed by his progress and suggested "maybe it shouldn't be only the winner that has to pay; maybe all of them should pay even when they lose. I'm sure they won't mind since they all get their warm-glow feeling from contributing to your pension." Startled by her insight, Claus decided this would be the first thing to try in the morning. But first some sleep ...

2 Everyone Pays, Claus Gains

Well rested, Claus entered his study early Saturday morning. He was determined to find a better method to raise money and believed the way to do it was making the losers pay. He had read about this possibility in *Auction 101*, which referred to such auctions as "all-pay auctions." Claus decided he would analyze a first-price all-pay auction where everyone pays their own bid, a second-price all-pay auction where losers pay their own bid and the winner pays the second-highest bid, and a third-price all-pay auction where all three bidders pay the third-highest bid. He knew these formats had not been studied before with altruistic bidders, so he started from scratch.

"One way to determine the optimal bids," Claus thought, "is to look at the marginal cost and marginal benefit of increasing one's bid, which have to be equal in equilibrium." Consider, for instance, the first-price all-pay auction. The positive effect of increasing one's bid from $B(v)$ to $B(v+\epsilon) \approx B(v) + \epsilon B'(v)$ is twofold. First, it might lead one to win the auction that otherwise would have been lost. This occurs when the highest of the others' values was between v and $v+\epsilon$, which happens with chance $2\epsilon v$. The expected gain therefore is equal to $2\epsilon v^2$. Second, an increase in one's bid raises the warm-glow utility by $\epsilon \alpha B'(v)$. Combined, the two yield an expected marginal benefit equal to $\epsilon(2v^2 + \alpha B'(v))$. Since all bidders have to pay their bids, the marginal cost is simply $\epsilon B'(v)$. Equating marginal cost to marginal benefit results in a differential equation for the bidding function that can easily be integrated: $B_1^{AP}(v) = 2v^3/(3-3\alpha)$. More importantly, the resulting revenue is

$$R_1^{AP} = \frac{1}{2-2\alpha}.$$

"Finally," Claus thought, "the revenue of the first-price all-pay auction does diverge when the altruism coefficient, α, tends to 1. In addition, it raises more money than the first-price auction for all positive α." Claus also discovered that for some values of α, the first-price all-pay auction was less profitable than the third-price auctions. He wasn't discouraged by this, however, and conjectured that the second-price all-pay auction would always outperform the second-price auction, and, similarly, the third-price all-pay auction would always yield higher revenues than the third-price auction. Claus suspected a third-price all-pay auction would perform best of all, which is why he turned to that format next.

"Let me see," Claus murmured, "what are the marginal costs and benefits this time?" Now, an increase in one's bid from $B(v)$ to $B(v) + \epsilon B'(v)$ only yields more warm-glow utility if the bid is the lowest, which happens with probability $(1-v)^2$ and then the gain is $3\epsilon\alpha B'(v)$. The expected marginal benefit is thus $\epsilon(2v^2 + 3\alpha B'(v)(1-v)^2)$. Moreover, the marginal cost is only $\epsilon B'(v)$ when others bid higher, so the expected marginal cost is $\epsilon B'(v)(1-v)^2$. The optimal bids again result by equating marginal costs and benefits, and the corresponding revenue is[5]

$$R_3^{AP} = \frac{1}{2-6\alpha}.$$

"Wow," Claus yelled ecstatically, "if my advisees have an α close to 1/3, I can generate a lot of money this way." This result was intuitive, Claus felt, "in the limit case when $\alpha = 1/3$, bidders know that if they pay an amount X, a total amount of $3X$ is contributed and so all three get back $3\alpha X = X$ in warm-glow utility. Hence they should be willing to bid an infinite amount."

[5] The optimal bids are $B_3^{AP}(v) = \frac{2}{1-3\alpha}\int_0^v \left(\frac{x}{1-x}\right)^2 \mathrm{d}x = \frac{2}{1-3\alpha}(2\log(1-v) + v(2-v)/(1-v))$.

Claus was pretty sure the third-price all-pay auction was his best choice, but he felt inclined to analyze the second-price all-pay action nonetheless. He was getting faster at it. "Now, the marginal cost is $\epsilon B'(v)$ only when $B(v)$ is not the highest bid, which happens with probability $(1-v^2)$," Claus figured, "and the marginal benefit is simply $\epsilon\big(2v^2 + \alpha B'(v)(1-v^2) + 2\alpha B'(v)v(1-v)\big)$, where the last term occurs when $B(v)$ is the second-highest bid." Claus quickly integrated the resulting differential equation to derive the optimal bids and the corresponding revenue:[6]

$$R_2^{AP} = \frac{1}{2(1-3\alpha)^5}\Big\{(1-3\alpha)(1+23\alpha-45\alpha^2-3\alpha^3) \\ +48\alpha(1-\alpha)^2(1-2\alpha)\log\left(\tfrac{1-\alpha}{2-4\alpha}\right)\Big\}.$$

"Shiskebab," Claus said, "who could have anticipated this expression." He carefully took the limit when α tends to 1/3 and found that revenue was finite (6/5) in this case. "Maybe revenue now diverges when $\alpha \to 1/2$," Claus wondered, "since it is a second-price all-pay auction so two bidders have to pay the same price." When he took the limit, however, he found that the revenue was 7! "I want to understand that result better," Claus promised himself, "but first let me make a graph of the revenues and show it to Ank."

Ank was amazed by Claus' findings. "So if they're altruistic enough to value one Euro for you as 33 Eurocents for themselves," she recapitulated, "which, incidentally, I don't think is unrealistic, then you can generate a pension that will last forever?" She sat down to let the news sink in. "I've heard talk," she said, "that the first publication in a top journal raises lifetime wealth by roughly 200,000 Euros, but you're telling me its value is limitless." Claus felt overwhelmed with joy as he heard his wife talking about their bright financial future, and almost forgot there were a few loose ends to tie up. He promised Ank he would not be gone in the serious room for long, and that when he returned they would open a bottle of *Château Petrus*, 1989.

3 Afterthoughts and Future Plans

Claus' goals were twofold: he wanted to comprehend the curious revenue result for the second-price all-pay auction and he wanted to make sure the third-price all-pay auction was indeed the best format to use. To get a better understanding of the revenue results, Claus started by carefully studying Figures 1 and 2 from which he derived the following ranking of revenues:

[6] The optimal bids are given by $B_2^{AP}(v) = \frac{(1-\alpha)^2}{(1-3\alpha)^2(1-2\alpha)}\log\left(\frac{1-\alpha+(1-3\alpha)v}{1-\alpha}\right) + \frac{1}{1-2\alpha}\log\left(\frac{1}{1-v}\right) - \frac{2v}{1-3\alpha}$.

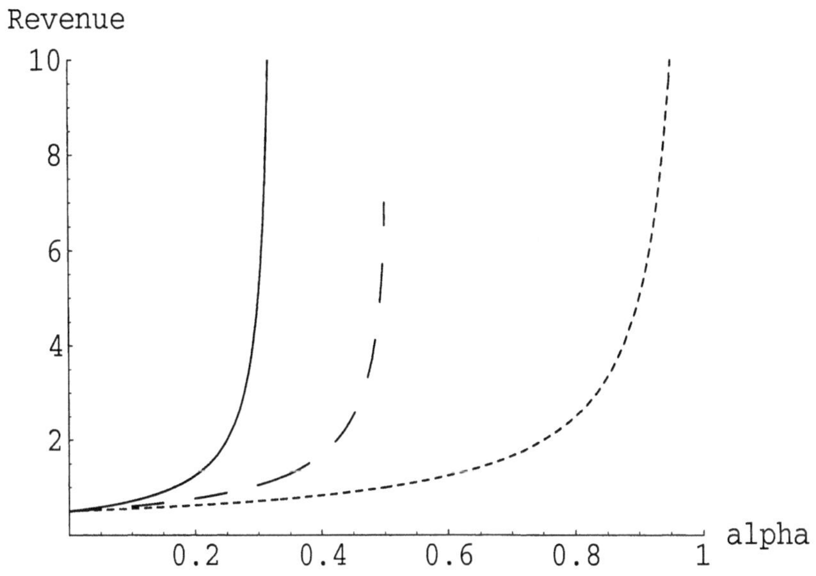

Fig. 2. *The revenue from a first-price (short dashes), second-price (long dashes), and third-price (solid line) all-pay auction for* $0 \leq \alpha \leq 1$.

$$\begin{array}{ccccc} R_3^{AP} & > & R_2^{AP} & > & R_1^{AP} \\ \vee & & \vee & & \vee \\ R_3 & > & R_2 & > & R_1 \end{array}$$

So among the auctions studied, the third-price all-pay action generated the most revenues. Claus wondered if this was true more generally, i.e. was the third-price all-pay auction the optimal sales mechanism?[7] Claus decided to follow a somewhat more general approach to investigate this issue. "In any auction the total surplus generated is divided between the seller and the bidders," Claus reasoned, "and in the case at hand, the total surplus is simply the highest of three value draws, 3/4, plus the bidders' warm-glow utility, $3\alpha R$, where R is the auction's revenue." In order to derive the seller's and bidders' share of the total surplus, Claus first considered the expected equilibrium

[7] See also Maasland and Onderstal (2001).

profits of a bidder with value v. Employing a standard Envelope Theorem argument, Claus calculated this profit to be

$$\pi^*(v) = \pi^*(0) + \frac{1}{3}v^3,$$

where $\pi^*(0)$ denoted the expected equilibrium profit of a bidder with the lowest possible value 0. "In most auctions the lowest-value bidder would have zero expected profits," Claus realized, "but not necessarily in the case where bidders derive warm-glow utility from the auction's proceeds." Next, Claus averaged the bidder's expected equilibrium profit over all possible values and multiplied the result by three

$$\pi_{bidders} = 3\pi^*(0) + \frac{1}{4}.$$

"What is left for me," Claus claimed, "is the total surplus net of what the bidders get." Claus scribbled down a nifty formula that summarized this claim: $R = 3/4 + 3\alpha R - \pi_{bidders}$. Solving for R he derived

$$R = \frac{1/2 - 3\pi^*(0)}{1 - 3\alpha}.$$

"Aha," Claus gasped, "when $\alpha < 1/3$, I'm better off the lower the expected profits of the bidders, while for $\alpha > 1/3$ our interests are aligned." Lower values of α seemed more realistic to Claus, and he told himself to compute $\pi^*(0)$ for the third-price all-pay auction. "But first," he murmured, "I want to understand why the second-price all-pay auction yields a revenue of 7 when $\alpha = 1/2$."

"Let me consider extreme value draws of 0 and 1," Claus started. He argued that the zero-value bidder loses for sure and only derives warm-glow utility. The lowest of the others' values is then 1/3 on average, so both others pay $B_2^{AP}(1/3)$ and $\pi^*(0) = 2\alpha B_2^{AP}(1/3) = B_2^{AP}(1/3)$ since $\alpha = 1/2$. Similarly, a bidder with a value of 1 wins for sure and receives 1. Moreover, the lowest and highest values of the other two bidders are then 1/3 and 2/3 respectively, so he also receives $\alpha B_2^{AP}(1/3) + 2\alpha B_2^{AP}(2/3)$ and pays $B_2^{AP}(2/3)$. Hence, $\pi^*(1) = 1 + \frac{1}{2}B_2^{AP}(1/3)$, or $\pi^*(1) = 1 + \frac{1}{2}\pi^*(0)$. Claus remembered that a simple Envelope Theorem argument established an alternative link between the expected profits of the lowest and highest-value bidders: $\pi^*(1) = \pi^*(0) + 1/3$. Taken together, these implied $\pi^*(0) = 4/3$ and a resulting revenue of $R = (1/2 - 4)/(1 - 3/2) = 7$. "Excellent," Claus concluded, "the magic 7 explained."

Claus' final task was to compute the expected profit of a zero-value bidder in a third-price all-pay auction. "Easy," Claus argued, "this bidder loses for sure and his warm-glow utility is simply $3\alpha B_3^{AP}(0)$ because he himself determines the lowest bid. Since $B_3^{AP}(0) = 0$ the zero-value bidder has zero expected payoffs!" This demonstrated that the third-price all-pay auction was indeed optimal for $\alpha < 1/3$ since $\pi^*(0) = 0$ was the lowest possible equilibrium expected profits that a rational bidder would accept. Moreover, when $\alpha > 1/3$, a third-price all-pay would obviously yield an infinite revenue and no other format could do better than that.

Satisfied, Claus returned to the living room and sat down on the wooden couch next to his wife who had opened a bottle of *Petrus* to let it breathe for a while. "Imagine," Claus kidded after a few glasses, "soon we'll be scuba-diving in the Red Sea while Mark, Jack, and John are bankrupt but happy nevertheless once they receive our postcard from paradise."

References

1. Engelbrecht-Wiggans, R. (1994), Auctions with Price-Proportional Benefits to Bidders, *Games and Economic Behavior*, **6**, 339-346.
2. Engers, M. and B. McManus (2001), Charity Auctions, mimeo, University of Virginia.
3. Goeree, J.K. and J.L. Turner (2001), How (Not) to Raise Money, mimeo, University of Virginia.
4. Maasland, E. and S. Onderstal (2001), Optimal Auctions with Financial Constraints, mimeo, CentER, University of Tilburg.
5. Weddepohl, H.N. (1979), An equilibrium model with fixed labor time, *Econometrica*, **47**, 921-938.

Scope for Defined Contribution*

Paul van den Heuvel

OPTAS NV, Rotterdam

Abstract. This paper is a contribution to the discussion which pension scheme is better, defined benefit, in which the employee has no investment risk, or defined contribution, where the premiums are fixed. In contrast with defined benefit, in a defined contribution scheme the employee has to deal with uncertainty of the pension level. In the Netherlands the discussion is dominated by the proponents of defined benefit. The critique to defined contribution is that there is a significant probability of a lower than aspired pension result. In particular, intergenerational solidarity ensures a lower risk and higher expected outcome for defined benefit. The aim of this paper is to mitigate these claims.

Keywords: pension, investment, defined contribution, defined benefit

1 Introduction

The Netherlands has a long tradition of capital-funded company pensions, either at the enterprise or at sector level. *Defined benefit* is very predominant. In a *defined benefit* scheme payments after retirement are determined by the wage of the member during the career. The most common form of defined benefit is final pay, in which only the wage at the end of the career is relevant. *Defined benefit* requires a high degree of mutual solidarity among the participants of the pension plan.

A different form of capital funding is *defined contribution.* In a defined contribution scheme premiums are determined by the firm as a portion of the wage of the insured or member[1]. The net contributions after deduction of administration cost and risk premiums are added to an individual portfolio of investments. This portfolio is the source of pensions after retirement. As a consequence, there is an investment risk involved for the individual member. A major concern with respect to defined contribution schemes is that at the age of retirement the investment portfolio is insufficient, a state called "shortfall of coverage".

* This paper is based on joint research with Jennifer Baker, Alain Verberkmoes and Johan de Vries.

[1] In the context of an insurance company we use the word "insured", while "member" is associated to a pension fund. In this paper it is indifferent if we have to deal with an insurance company or a pension fund and both terms are used. Hence we mostly use the neutral term "employee".

For the insured (members) *defined contribution* implies more freedom of choice with respect to the premium, risk coverage and investment of the contribution. The level of solidarity is much lower than with *defined benefit*. An insured does not need to pay for risk coverage for risks that are not relevant for him.

There are several options to form the individual investment portfolio. The choice of investments may be either the responsibility of the employee, or of the insurance company/pension fund. The latter situation may be accompanied by a total return guarantee. In that case the insurer covers the risk of lower tail results, and a minimum value at the retirement age is assured. In case the insurance firm constructs the investment portfolio, the case considered here, it is relevant to find an optimal investment path over the relevant career.

Recently, in the Netherlands several institutions have commented on the desirability of *defined contribution*, among which the Scientific Council for Government Policy (WRR), see Boender, van Hoogdalem, Jansweijer and van Lochem [2000], and research institute NYFER, see NYFER [2000]. These studies concentrate on the importance of solidarity, in particular, amongst generations. Solidarity implies transfers of benefit and cost. This solidarity leads to the possibility of a smoothing of investment risk in time. Furthermore, it is sometimes claimed (see eg. NYFER [2000]) that in a collective portfolio it is possible to increase the average investment time horizon, giving room for an increase in investment risk premium.

Given wage development *defined benefit* gives more certainty with respect to pension level that can be attained. It is suggested by NYFER and WRR that via intergenerational solidarity a higher outcome can be obtained, since the risk premium of the collective portfolio is higher due to a higher equity exposure. The *defined benefit* portfolio presupposes the financing of a substantial "buffer", reserve relative to liabilities. This buffer is financed by the parties involved, company and employees, either by a guarantee by the employer, additional contribution, or accumulation of profit sharing.

The question is if employees will remain in support of this form of solidarity. A problem with pension fund oriented studies in the Netherlands is that they start with a dynamic set of employees with a long career in the same firm or sector. Employees, however, form a heterogeneous group, and the tendency in society is that heterogeneity is getting more important. Besseling and Van de Ven [2001] also make this comment. Homogeneity of the working population is gradually disappearing. Firstly, labour mobility has increased. Secondly, the individual employee has an increased preference for custom-made insurance. Differences in preferences can pertain to the family situation. Another aspect of heterogeneity is the personal wealth situation.

In order to give more room to this heterogeneity we start on a more fundamental level, with the problem of consumption and savings planning for the individual. In contrast to Boender et al. [2000], a utility function is formulated. Thomson [1998] and Booth and Yakoubov [2000] follow a similar approach, in particular aimed at the determination of the optimal investment path over the life cycle.

2 Model

We start with a general framework. In the formulation of the model assumptions are made with respect to the utility function, time preference and wage development. For a more extensive description of the model, see Baker et al. [2001]. Eventually, we will restrict the model to the perspective of the insurance firm.

2.1 Utility Function and Investment Portfolio

Let the function U represent the utility of the size of the investment portfolio S at the retirement age. The aspiration level of the value of the portfolio is denoted as A. This aspiration level depends on the wage before the retirement age. In this way the aspiration level can be considered to be analogous to a defined benefit scheme, where the aspiration level is, for example 70% of the last wage.

The utility function is given by

$$U(S) = \min\{\frac{S}{A}, 1\}^{\mu}/\mu.$$

This particular type of utility function is chosen in order to impose that a pension result higher than the aspiration level generates no additional utility. In fact, a critical level of consumption is linked to a percentage of wage income. Above the critical level of consumption, marginal utility is zero. By this specification of the utility function an asymmetry is built in around the aspiration level of the pension. Not reaching the aspiration level relatively gives a penalty, while there is no reward corresponding to an excess value. It can be observed that without the threshold level A, this function would have constant relative risk aversion.

F_x is the probability of reaching age x, given present age x_0. Maximum age is x_{max}. Furthermore,

$$F_{x_{max}-1} > 0, \quad F_{xmax} = 0 \quad \text{and} \quad F_{x_0} = 1.$$

Net contributions are added to the individual investment portfolio, starting from the first contribution. This portfolio consists of several asset categories. The asset mix determines the expected return and risk of the portfolio.

Important for the portfolio value are:

- contribution, depending on wage development
- return of the portfolio in turn determined by the asset mix.

- survivor bonus F_x / F_{x+1}

If S_x is the value of the portfolio the development can be given by

$$S_{x+1} = (S_x + c_x)(1 + R_{x+1}) \frac{F_x}{F_{x+1}},$$

where the stochastic variable R_x is the total return of the portfolio over the age period $[x, x+1]$. The net contribution is denoted by $\mathbf{c}_x$.

Return of the Portfolio

The assumptions on the total returns of the portfolio have to be elaborated further. Five asset categories are distinguished, equities, real estate, euro bonds, non-euro bonds, and euro price index-linked bonds. In the table below the assumptions on returns of asset categories are summarised. Moreover, for convenience it is assumed that serial correlation is absent.

The assumptions are expressed in terms of *net real returns*. Due to the problems of trends, mean reversion and non-stationarity the results are always questionable. Therefore, we restrict ourselves to an eclectic set of assumptions, and did not do an attempt to estimate these data from time series.

Net real returns, standard deviations and correlations (base currency: euro)					
	equities	real estate	euro bonds	non euro bonds	euro index bonds
average	7.1%	5.6%	3.1%	3.1%	2.9%
st.deviaton	15%	12%	7.5%	9.5%	8.0%
correlation	equities	real estate	euro bonds	non euro bonds	euro index bonds
equities	1.00				
real estate	0.80	1.00			
euro bonds	−0.15	−0.20	1.00		
non euro bonds	0.75	0.50	−0.25	1.00	
euro index bonds	0.10	0.30	0.10	0.20	1.00

Moreover, the portfolio is supposed to be on the efficient frontier: Every expected return is attained with minimal risk. The efficient frontier of return/risk combinations is a monotonically increasing function of expected real return to risk. An increase of expected real return could only be obtained by a shift to higher risk categories.

The aspired level of the portfolio (at or after the pension date) depends on the consumption path after the pension date. At a young age the (e.g. Macaulay [1938]) duration of these outgoing cash flows is very large. Reaching the aspired level would not be a problem for the employee, if he could completely immunize the cash flows after the pension date by, for instance, long index linked bonds.

However,

- the younger he is, the more future net contributions have to be added to the portfolio and (even if this path is assumed) the future returns of these contributions are random;
- complete matching of his future liabilities is impossible, since the life span is uncertain, and the right investments are unavailable, implying unavoidable risk.

As a consequence, the aspired level of the portfolio on the pension age cannot be reached with a minimum risk portfolio. In fact, in order to attain the aspired level a significant amount of high risk has to be incorporated. However, when the employee is close to the pension date and has a sufficiently large portfolio to reach the aspired level, a low risk strategy is attractive.

Under an autoregressive conditional heteroskedastic model the optimal path is also time dependent (see eg. Siegelaer [1992]).

Survivor Bonus

Within the framework of an insurance company, the portfolio of a deceased employee is transferred to the insurer instead of to the inheritors in the form of bequests. As a compensation, the employee receives a compensation for not passing away, the survivor bonus, which is added to the portfolio. The survivor bonus is quantified by equalising the expectation of, on the one hand, the benefit for the insurance company of the decease of the employee, and, on the other hand, with the benefit for the employee of not dying. The probability of dying is determined by a mortality table.

The survivor bonus is in essence a form of solidarity between employees of the same age. This solidarity is relevant for *defined contribution* in contrast with the self-funding of old age by an individual.

2.2 Optimisation Problem

Optimisation can be carried out with the help of dynamic programming, cf. Vigna and Haberman [2001]. The asset mix is optimised simultaneously. Since we change the perspective from the employee to the insurance company, we consider consumption, wages and hence net benefit before the pension date as given. Our focus is the wealth that can be generated by the portfolio of an individual employee and the resulting pension that can be obtained.

Firstly, it was assumed that the risk attributed to the portfolio over the career follows a pattern of monotonically decreasing risk (in terms of real return). Since real return and risk have a one to one relationship for portfolios on the efficient frontier this means decreasing return over the life cycle. Real return is specified as a monotonically decreasing function of time, starting with maximum risk at the time of entry within the scheme, and ending with a minimum at the pension age. This can be specified, for instance, by a piecewise linear function.

Secondly, we allowed a feedback with regard to the intermediate size of the portfolio. If the portfolio is sufficiently large in order to attain aspired level at the pension age with minimum risk, then the strategy changes to a minimum risk strategy. With our particular utility function this is an attractive strategy. Upside potential is exchanged for more certainty.

3 Results

Asset mix optimisation leads to a sequence of asset mixes in terms of the five asset categories over the career of the employee. We refrain from presenting these sequences, since being on the efficient frontier only the expected return and risk matter for the final result.

Optimisation leads to a high-risk mix at younger ages. Risk tolerance is declining after a certain age. Then, lower risk bonds take the place of higher risk equities and real estate. The age independent asset mix strategy, which is one of the alternatives viewed by Boender et al. [2000] is inferior, as a result of the threshold in the utility function. Furthermore, we allow a feedback with respect to the volume of the portfolio. The main characteristic of the latter strategy is that if the portfolio is large relative to the aspiration level, then a low risk asset mix is adopted. Since there is no gain in excess value portfolios, it is preferred to "ensure" the aspiration level instead of risking extreme returns. This latter strategy is our main focus.

Intertemporal behaviour and the development of the asset mix largely depend on the form of the utility function. This finding reflects that there is a great diversity in employees' characteristics.

The form of the career is important for pension choices. A very important observation is that the relative advantages of defined benefit are pushed aside by career steps leaving the solidarity collective. We have done a number of simulations for basically two wage development paths, "steep" and "flat". It is a property of final pay schemes that there is a transfer of employees with flat wage development to employees with steep wage development, especially at the end of careers. Interestingly though, in our model the gradient of the wage development path hardly matters for both the asset mix and the *probability of shortfall of coverage.*

Feedback with relation to the portfolio size, on the other hand, is relevant. In the table below the results for the feedback strategy are summarised. Usually Dutch final pay plans aim at 70% of final pay. In our results the 70% final pay benchmark is attained in 69% of all 10,000 scenario simulations. The success number increases to 97%, if the aspiration level is decreased to 50% of final pay.

Probability of shortfall of coverage for the feedback strategy	
Percentage of final pay	Probability pension < percentage of final pay
70	31
65	17
60	10.5
55	6
50	3

A disadvantage usually attributed to *defined contribution* is that the probability distribution of pension outcomes is wider, so that there is a substantial chance that the aspiration level is not reached. However, our finding that only 3% of pension outcomes are worse than 50% of final pay is more favourable than the results reported in e.g. Boender et al. [2000]. These improvements are obtained by dynamically optimising the asset mix, with a feedback with respect to the size of the portfolio.

A further possible remedy against lower tail results is an interest guarantee, provided by the insurance company of pension fund, in exchange for a fee, which may be a modest percentage of the savings portfolio. Such a guarantee represents a put option on the underlying assets (see Blake [1998]). For *defined contribution* schemes the risk of an insufficient portfolio size at the age of retirement can be covered by the insurance company at the price of the relevant put options.

4 Discussion

For companies that have a short life cycle, both company and employees might be better of with *defined contribution.* For instance, consider technology sectors and smaller companies in innovative fields. The typical employee in such a firm will not be 40 years active within the same firm. The life span of products is typically short. Moreover, employees are often young, at the beginning of a career, and are not worried about their pensions. For firms in sectors like technology capital mainly consists of human resources. However, this form of capital tends to have a high mobility. Products often have a strong relationship with individual employees.

If a young employee leaves the company this employee is out of sight for the employer. The latter is not concerned with the employee's pension anymore. A *defined contribution* plan is simply transferred to the insurance company or pension fund of the new employer. In a *defined benefit*-plan on the other hand, the employee and his widow will be probably in the pension plan for the rest of their lives. Because of the solidarity mechanism the resigned has contributed more than proportional to his "rights". Hence, in a *defined benefit* the employee who quits has a relative disadvantage, even if his rights are indexed.

Boender et al. [2000] compare the setting of intergenerational solidarity with a single individual investing for his own pension. An important advantage of *defined contribution* within the framework of an insurance company or pension fund in comparison with the self-investor is the addition of the survivor bonus to the portfolio. This bonus is in fact a form of solidarity as well between the employees within the same scheme. It can be noticed that it replaces bequests (inheritance) of the portfolio to descendants that would be the case for a single investor. The survivor bonus has a growing importance as people get older. For a 52-year old 50 basis points are added to the portfolio additional to investment return. For a 65-year old this number is over 2 percentage-point. After the pension date the portfolio can remain in existence as a source for payments. The contribution of the survivor bonus increases as the insured person gets older. In addition, it is recognised that it is not necessary to cut back all equity exposure from the portfolio after the pension date. After all, according to the recent mortality tables people still have quite a long life span after the pension date, so that a substantial investment horizon is allowed.

One had the impression that the WRR study sees it as an advantage that the solidarity collective has a longer investment horizon in comparison with the individual, allowing for a higher-risk portfolio. However, this argument is denied by one of the authors, Jansweijer (Jansweijer and Winde [2001, p.81]) who stresses that the mechanism of solidarity between generations is much more effective than choosing the right asset mix. Hence, the advantage claimed is more that solidarity avoids that individual pensioners run out of money before they pass away. We have shown, however, that the probability of this "bankruptcy" phenomenon in a defined contribution framework does not have to be significant. At the retirement age a lifelong annuity can be purchased for the employee. There are a number of ways to add price indexation to the mechanism, either on an individual basis or by means of a collective pool.

One of the aspects that are largely ignored by Boender cs. is the opportunity cost of buffers (see also Besseling and Van de Ven [2001]). For providing *defined benefit* pension schemes reserves, buffers are required. Boender concludes that these buffers have to be relatively high to ensure that the goals of the scheme are reached. That implies that premium holidays and retribution to the companies are rare. Boender cs. illustrate that if assets exceed 150% of liabilities, only a third of the excess can flow back to the firms. In his opinion it is short run policy to opt for a refund from the buffer, since the return of the pension portfolio has a higher expectation than that of the investment in the firm. However, it should be noted that for a large number of firms this short-term consideration is sensible anyway, if the continuation of the company is at stake. These firms would prefer not to build a pension buffer at all and hence be better of with *defined contribution.*

It can be remarked that the view of Boender cs is seen from a national perspective. In the Netherlands historically solidarity was considered to be an important fundament. In America, for example, the individual role has more weight.

In other countries the pension system is predominantly "pay as you go". Pay-as- you-go is an extreme form of intergenerational solidarity; see eg. Fabel [1994] for an overview.

Pay-as-you-go is considered the first pension pillar (in the Netherlands "AOW"). Pay-as-you-go is also open for the part of the population without a career in a firm, institution, or government. The ageing of the population, however, which is an important phenomenon all over Europe, handicaps intergenerational solidarity, since the possibility to transfer the pension burden to later generations is more limited.

5 Concluding Remarks

In a world of homogeneous and stationary working forces *defined benefit* may lead to a result in terms of the probability distribution of pension outcomes that is more favourable than the results with defined contribution schemes. In this respect intergenerational solidarity plays a role.

Nevertheless, the above only holds under stringent assumptions, and there are a number of situations were the comparison is in favour of *defined contribution.* We have also tried to make it more likely that pensions under *defined contribution* schemes may be more prudent than often depicted in the Netherlands.

References

1. Baker, Jennifer, Paul van den Heuvel, Alain Verberkmoes, and Johan de Vries [2001], "Dispersion of defined contribution results", OPTAS NV Rotterdam.

2. Besseling, Paul, and Martijn van de Ven [2001], "Wat is een verzekering van het inflatierisico waard?", in Jansweijer and Winde ed.
3. Blake, David [1998], "Pensions schemes as options on pension fund asset: implications for pension fund management", Insurance: Mathematics and Economics 23, 236-286.
4. Boender, Guus, Sacha van Hoogdalem, Roel Jansweijer, and Evelyn van Lochem [2000], "Intergenerationele solidariteit en individualiteit in de tweede pensioenpijler: een scenario analyse" Wetenschappelijke Raad voor het Regeringsbeleid.
5. Booth, P., and Yakoubov, Y., [2000]. "Investment policy for defined contribution pension scheme members close to retirement: an analysis of the "lifestyle" concept". North American Actuarial Journal 4 (2), 1-19.
6. Elbers, L., and Claus Weddepohl [1986], "Steady-state equilibria with saving for retirement in a continuous time overlapping generations model", Zeitschrift für Nationalökonomie-Journal of Economics 46, 253-288.
7. Fabel, Oliver [1994], "The economics of pensions and variable retirement schemes", John Wiley and sons, Chichester.
8. Jansweijer, R. and A. Winde ed. [2001], "Intergenerationele solidariteit en individualiteit in de tweede pensioenpijler: conferentieverslag", Den Haag, Wetenschappelijke Raad voor het Regeringsbeleid.
9. Kritzman, M. and D. Rich [1998], "Beware of dogma: the truth about time diversification", Journal of Portfolio Management,24, p.66-77.
10. Macaulay, F.R. [1938], Some theoretical problems suggested by the movements of interest rates, bond yields, and stock prices in the United States since 1856, Columbia University Press, New York.
11. NYFER [2000], "Marktwerking in de pensioensfeer, een recept met bijwerkingen".
12. Siegelaer, Gaston [1992], "Risicoperceptie en optimale beleggingsportefeuilles" Financiering en belegging, stand van zaken anno 1992, deel 15, 135-147.
13. Thomson, R.J. [1998], "The use of utility functions for investment channel choice in defined contribution retirement funds". In: Trans. 16th Conf. Int. Assoc. Cons. Act.
14. Vigna, Elena, and Steven Haberman [2001], "Optimal investment strategy for defined contribution pension schemes", Insurance: Mathematics and Economics 28, 233- 262.

Dynamics

On the Possibility of Economic Dynamics

Frank Hahn

Abstract. After discussing a number of examples of dynamic analysis, I discuss whether it is a reasonable enterprise. Dynamics inevitably uses mathematics. Mathematical analysis of mappings in time is extremely sensitive to the exact form posited. I express some doubt that we are anywhere near a real understanding, although the rough qualitative outline of behaviour is probably pretty safe.

Keywords: equilibrium, disequilibrium, dynamics.

Like a good many others, including the honorand of this volume, I have spent some time in an attempt to fashion an economic dynamics. Looking back I now believe that none of us engaged in this enterprise have been conspicuously successful. Is it possible to draw some conclusions from this lack of success other than our feeble intellectual powers?

The general answer is actually fairly clear: except for the case of intertemporal optimisation under full information, there is no fundamental theory which can help in the construction of dynamical models. This has been disputed by arguing that our universal postulate of the rational agent provides all the theoretical underpinning which is required. But this is certainly not the case for the learning agent nor when we consider an aggregate of agents. All of this will be spelled out below. But at the moment we may usefully make the distinction between *equilibrium and disequilibrium dynamics.*

The distinction I believe is self explanatory as is the problem with the last. That is that equilibrium conditions place some restrictions on the models we can build while in the last category the choice is wide open. In these cases we would not expect to get much help from rationality. Much of the dynamical studies in economics are of equilibria and so of no help in asking whether there are forces making for equilibrium. This last remark begs the question of whether we have short period equilibrium or long period in mind, or whether the two are useful distinctions. Again this as well as other related matters will be considered below.

1 An Example of Equilibrium Dynamics

It is convenient with Solow's famous growth model (Solow, 1956). Take the macro-economic equilibrium condition of savings equal to investment, and think of it as savings per man and investment per man. Make the usual assumption that savings per man is a constant fraction (s) of output per man, so that that has to equal investment per man. Investment per man in turn can be thought of as the change in capital per man plus the growth in capital per man: $\{gk + (d/dt)k\} = i = sy$. Since we are doing an equilibrium dynamics this equation must hold all the time, and it is easy to prove that the well-behaved production function with constant returns to scale converges to the steady state, $k^* =$ a constant (so that the growth rate is g).

Here the assumption of continuous equilibrium in all markets at all t is very helpful in deriving the dynamics with its conclusion. Of course the special forms of the relationships assumed, is also important for the conclusion. But we must not deduce from this example that all models of equilibrium dynamics have agreeable outcomes. For instance the equilibrium dynamics of an overlapping generations model with outside money may converge to an autarchic equilibrium which is the pessimal state of the economy.

But the best example I can think of are models which at every moment of time have markets clearing but also are optimising for the economy as a whole. Let us call them Ramsey models. This means more than Pareto efficiency although of course that efficiency is necessary for the optimum and so one also has market clearing at all times. These models can most conveniently be expressed as the maximisation of an integral of the utility of a representative agent with average consumption, possibly discounted subject to the constraint that at all times average consumption of the representative agent plus gross investment per head does not exceed output per head.

It is this last equation which together with the Euler-Lagrange conditions which constitute the "laws of motion" of the economy. But there are many paths satisfying these and one needs to ensure that the chosen path goes through the initial given value of the control (here k) and that as time goes to infinity the value of capital is bounded(by zero). This is so well known and the literature is so voluminous that I eschew details.

The point of these examples is not really the outcome of the dynamics but rather that the latter emerges as a by-product of the model of agents who are at all times undertaking the optimum of the actions available to them (Hence markets also clear). The necessary conditions of the optimum plus the intertemporal constraints yield two differential equations in k and in c. On the other hand these conditions may be satisfied for an arbitrary finite time and yet observation of the path of the economy would not confirm (or disconfirm) its optimality. That is so because we have seen that one of the necessary conditions involves infinite time. Once again one needs to note that while in the case I have just discussed there is convergence to steady state this is not the case for many other cases. Recall the literature on "Real Business Cycle".

There is no doubt that equilibrium dynamics does not leave one quite as much at sea as one would be with models which allow disequilibrium. However I must enter a caveat: as far as I can see there is nothing in the axioms of rationality which require modeling in continuous time. If we carry out the previous exercise in discrete time e.g. with the maximand $\sum U(c_t)/N$ and the constraint suitably adjusted the dynamics may not be as well behaved as earlier and certainly will be different. So even here we have some unwelcome freedom in modeling.

One may ask whether these difficulties are peculiar to disciplines dealing with human behaviour in which at least some element of foresight is involved since that behavior is purposeful. When one engages in equilibrium dynamics one is hoping that one has a full and accurate description of the agent and of a group of these interacting so that there is then no difference between the theory of the agent and of say a ball rolling down a hill and hitting another one. The belief that agents have rational expectations is very helpful in an enterprise designed to find a unique dynamics!

But of course we all know some modern physics, and we know that the behaviour of certain physical objects is stochastic e.g. that of the electron or photon. Physicists seem to rely on the laws of large numbers which in our case suggests that we ought to study macro economics. For a number of reasons this is not perhaps good advice for economics since being a social animal there is too little independence between the actions of individuals as to allow us to reach for a law of large numbers. But so far there is too little macroeconomics based explicitly on a large number theorem to make a definite judgment. I do not know of any work, except on distribution, where the econometrics is adapted to test some large number theorem applied to heterogeneous agents. But theorists seem to have opted for the representative agent who obeys the standard axioms and given him rational expectations.

This procedure has had great success with economists especially American economists. The examples which I have briefly discussed show that this procedure not only makes life simple but automatically can produce the appropriate dynamics. (This point has a family resemblance to the early Samuelson's suggestion that it is second order conditions of maximisation which help to decide the stability of equilibrium). There is no reason to think that we make more serious mistakes when we adopt equilibrium dynamics. But because it is by and large so simple, a stability result will be crisp and welcome when instead it should be treated with reservation.

2 Disequilibrium Dynamics

We are now going to consider dynamic "error correction" mechanisms, where the error is deviation from equilibrium. It seems to me that this is the only way in which there is a hope of establishing that the path an economy seeks when out of equilibrium is one whose limit is an equilibrium. It has a long history in economic theory: Adam Smith and Walras are of course the most prominent early instances. Adam Smith's views were not in the form of formal propositions, while Walras proposed some which are still, albeit decreasingly, studied.

Indeed it is tatonnement which has attracted most modern attention. The error here is a non-zero vector of excess demands to which prices respond, by the rate of change of every price being of the same sign as that of the corresponding excess demand. However at every price vector thus generated the households are all in equilibrium and the real state of the economy is unchanged. There is no production or exchange allowed except in equilibrium. The excess demand vectors are "notional" in that they are in the agent's minds (or are announced).

This then is not a complete disequilibrium approach to dynamics. In the 1960's a good deal of technical work was done on this Walrasian method. This was much helped by the economists' discovery of Liapounoff's second method of providing qualitative solutions to systems of differential equations (Arrow and Hurwicz, 1958). This enabled the theorist to go beyond the first order expansion of excess demands to create a linear model and provide an analysis in the small. The latter had been used to good effect by Hicks and his result was easily extended to stability in the large. The impressive feature of these results was that for the case of pure exchange they could be derived in a straight forward manner from the micro-theory of choice. In this instance where agents' income effects cancelled one was left with the negative definite substitution matrix which is well understood. For global analysis one takes as the Lyapounoff function $\sum(p_i - p_i^*)^2$. Since by Walras law $\sum pz = O$ for all p, and by the law of price change $z_i dp_i/dt \geq 0$ one finds that the time differential of our Lyapounoff function is either 0 if $z(p^*) = 0$ or $-p^* z(p) < 0$ (assume $z_{p*} = 0$) where the inequality follows from the axiom of revealed preference between z_p and z_{p*} (Hicks' proof considered the properties of the Jacobian of z_{p*}).

More cases were analysed in the sixties. Hicks himself considered an adjustment process which he christened as one which could achieve *perfect stability* which considered markets finding equilibrium one after another. But many other cases were studied of which that of gross substitutes was a favourite and it was no longer necessary to confine the analysis to pure exchange (see Hahn 1962). Arrow and Hurwicz (1958) were at the forefront of these explorations. But the enterprise was doomed not to be capable of reaching general conclusions in the Walrasian setting. A theorem not directly connected with dynamics did the damage. It was shown that any set of continuous excess demand functions which obeyed Walras law could be generated by the maximisation of well behaved utility functions of a minimum number of agents and an endowment distribution between them (See Sonnenschein 1972, Debreu 1974 and Mantel 1974). It now followed that many cases of instability could be constructed since many elements of the excess demand Jacobian could be chosen to yield instability without fear of going outside the model. In fact in the absence of an explicit theory of out of equilibrium behaviour there was no reason to expect stability. But this is speaking purely theoretically: as Keynes noted a long time ago relative prices do not behave in a manner that suggests instability. If that is a correct observation there must be some hope for theory to catch up.

Apart from technical problems there are also serious economic ones. The model is one of perfect competition, everyone takes prices for given. Who then is supposed to change them? The only escape in the Walrasian model is the auctioneer. This suggests that search for realism will lead away from perfect competition towards at least some monopoly power of some firms. If that is not done we are in the situation of looking for a theory of dynamics applied to the auctioneer and since he is a fiction and we have agreed that under perfect competition no agent active in the economy changes price, so we reach a dead end. Of course all this occurred to Arrow some years ago (Arrow, 1959).

The trouble now is that there exist only rudimentary models of GE when competition is not perfect. Also such models as have most recently been constructed are largely game theoretical. There are quite a few attempts to model possible paths to an equilibrium of the game thought to be most appropriate. There are also well known attempts to understand the dynamics as evolutionary. I shall not be discussing this literature since it would take me too far off the path I wish too pursue. In any case I know of no game-theoretic discussion of price dynamics. There seems to be a need to consider alternating moves which I don't like much.

Disequilibrium dynamics can be carried out at the level of the agent where one looks for a process in which the agent is correcting mistakes of foresight. These may consist in giving too low a probability to an event now observed or not considering it at all at the time of decision. There are many plausible ones to choose from. There is no agreement nor deep theoretical reasons to model decisions under uncertainty one way rather than another except that we are used to one particular one. The belief has arisen that Bayesian updating is the only rational procedure but it is not (see Binmore (1987)). Moreover it does not remove all arbitrariness (e.g. in the choice of priors). We simply do not have an agreed way of proceeding in these cases, which presumably explains why we stick to equilibrium dynamics.

So I feel free to make some suggestions which however are no more based on any rigorous theory than any other in this field. It seems reasonable to suppose that the past experiences of an economy influence its reaction to mistakes. Thus an economy which has had the experience of persistent inflation is likely to consider it as part of the normal state of the economy. By 'normal' I mean "not surprising"; if on the other hand prices have been steady for some long periods and rarely falling then that will be counted as normal; and of course the economy may have experienced both. In fact one would expect both to be the typical case. Notice that it will not be too extravagant an assumption to suppose that most agents have the same 'normal' expectations. Moreover the normally expected prices may include mistakes –after all it is 'normal' to have wrong expectations so that the observed prices are what they are because some agents had false expectations.

But more is needed. There is in the theory being outlined a price index (P) and its normal value is P_t^* at date t. Let $d_{it} = p_{it} - P_t^* k_i^* = p_{it} - p_{it}^*$. Here d_{it} is the error made by producers of good i when they take k_i^* to be the normal ratio of their price to the normal price index. There is no harm in thinking of P_t^* as the rational expectation of the price index for t. But it is k_i^* which is changed when d_{it} is the realised error. We have considerable latitude in choosing the form of the adjustment which is consistent with the price index being rationally expected. That means the agents are able to make rational forecasts of the aggregate behaviour of others but not of these others individually. I do not deny that these are strong assumptions but they seem far less so than RE. One could illustrate by choosing some functional forms for the updating process such that it does not conflict with the rational expectations of the price index. Evidently there are here opportunities for pleasant and unpleasant outcomes, But one cannot rely on these being "real" and so I shall leave it to others who might find this version of expectation formation of interest. My own conclusion is that rational price index expectations is not as silly a hypothesis as rational price expectations and is enough to rule out wildly unstable equilibria. Such situations as Keynes noted have not been observed.

Going back to an auctioneer and an economy without production we can improve on some tatonnement results by allowing exchange out of equilibrium. The benefit is that we no longer have to assume special forms of excess demand functions to get convergence; the cost is that initial conditions are restricted and the process of exchange is efficient (Hahn-Negishi (1962), Uzawa (1962), Hahn (1962)).

Here is a quick sketch of the theory: Let V(p,pe) be the indirect utility function of an agent. Then dV/dt = (-x + e)dp/dt since (de/dt)p =0. So the derivative of the indirect utility function for every agent is negative (non-positive) provided that private excess demand for a good is not of opposite sign to the aggregate excess demand at any date including zero, (x.dp/dt is non-negative all t), i.e. exchange is efficient and has been before the story starts. Since what has been said is true of every agent the sum over agents of their indirect utility functions is a Lyapounov function.

There are other ways of proving that a process of exchange converges, for instance the assumption that exchange will take place between any two agents if it is Pareto improving. Here the sum of utility functions is non decreasing during exchange and since maximum utility of every agent is bounded by what it would be if the whole endowment were given to him the sum of utilities is positive and increasing during exchange and so is a Lyapounov function (see Uzawa (1962) who called this the Edgeworth process).

When production is included we are back in familiar difficulties, although Fisher (1983) has suggested some ingenious ways of bypassing these. I am very sympathetic to the view that every day experience should count for more than it has done but I also agree that it is an unreliable guide. The model of the economy is far too abstract to allow straight forward translation of experience. For instance one clearly wants the sequential version of G.E. and the question of modelling inventories arises. There is also the difficulty of grafting money onto the story, and that is bound to affect the story of exchange and production. There is nothing in economic theory to prevent for instance chaotic dynamics but there may be in the empirical evidence; I consider that to be a very lame conclusion. I conclude that the search for a rigorous dynamics of the price mechanism will be in vain as long as we have only highly abstract and not entirely logically coherent models of the economy. That is why a common sense assertion that the price mechanism 'works' will not do since a general demonstration of this claim is beyond us.

3 Disequilibrium Dynamics in the Theory of Growth

We have already discussed the leading theory of equilibrium dynamics in the theory of growth (Solow 1956). Like many models of growth it is highly aggregated and not well supported by the evidence which apparently suggests cyclical growth even for macro series. In any case since there plainly is a question of how this ideal comes about, I shall spend a little time on the subject of the heading to this section.

I fear that I have already written on this matter extensively and will confine myself here to some general conclusions.

Let us consider the Solow model once again. In that model if planned savings is always realised, there may have to be some capital deepening. In a strictly Keynesian model there may have to be unwanted accumulation of inventories or of money which is not in the model. If nonetheless we consider the latter then we predict a fall in interest rates and in prices since it is only real money stocks which can change. This in turn will affect the real rental of capital (lowering it) and so raise the capital/labour ratio i.e. lead to deepening. But not instantaneously, so that we could apply the usual Samuelson formulation of dynamics: $dp/dt = f(\text{investment - savings})$; $dk/dt = g(\text{marginal product of k-rental})$. In equilibrium dynamics one assumes that all of these adjustments are simultaneous. That is unlikely. On the other hand the above formulation needs justification in delays.

But while this example makes the problem of disequilibrium dynamics clear, it is not one which one should use for other than expository purposes. Note that its dynamical conclusions may be the same as those of Solow but they need not be. Convergence to the steady state will depend on relative reaction speeds of the two processes. But it is all too highly aggregated and of course the assumptions, as usual are arbitrary. It is very clear where we run into difficulties: it is lack of a proper theory of behaviour when agents are not already at their final optimum and anything plausible goes in modelling. The difficulties are not mathematical although complexity difficulties are bound to arise.

None the less one feels that this is too pessimistic. I am reluctant to leave matters thus unresolved; it may well be that the approach is too reductionist and that one ought to distinguish between central tendency and noise. That suggests some kind of averaging. That in turn leads one to think of 'herd behaviour' as perhaps a good route to the central tendency. For instance if the notion of the 'normal' is strongly held as it should surely be if there has been long persistence. Then the variables which are regarded as having 'normal' values will not long deviate from these - unless they do and it is time to change expectations of normal values. It could also be that there is a belief that 'normally' some variable fluctuates in value. One is on the way to constructing a socio-psychological theory and I for one am out of my depth. But it is along such lines that more plausible theories may be buildable, they are not axiomatically but empirically based. On the other hand recent work in evolutionary game theory has made the dynamics out of deviations of one's own strategy from the average and succeeds in showing convergence.

I showed some years ago in a von Neuman model that if the maximal rate of growth is regarded as 'normal' as is the associated interest rate and if price adjustments depend on the difference between actual minimum unit cost and 'normal' unit cost, there will be convergence to Neuman prices. In that paper I did not allow non zero excess demand to affect prices since the linear production model made cost determined prices plausible. The model distinguished between consumption goods and capital goods and modelled demand for the latter as logistic in changes in demand. Adding a 'plausible' expectation function did the trick of ensuring convergence. I cannot claim to have a high regard for this model and I mention it here to give support to Lucas's view that disequilibrium dynamics requires a good deal of "ad hockery".

Now an ad hoc assumption is one not derived from theory but from the facts or from intuition derived from these. Results thus derived are often all we can hope for. The question is whether they are worth having or indeed may be dangerous. I do not know but I strongly hold the view that finessing the issue and concentrating on equilibrium dynamics is at least as dangerous. Moreover it is also ad hoc since just to take one aspect, it tells us very little or nothing about lags and since it is quite complicated so far it has been almost always a linear theory, (log-linear). As I have said these authors are blameless since it is very hard to deal with systems of non-linear equations. But they are not blameless for taking this work so seriously, as indeed containing the truth. I fancy that there will be a rude awakening.

4 A Brief Digression

In recent years there has been a great development of dynamic analysis and games. It turns out that the evolutionary analogy has been found useful. The list of authors is far too large to state here, but Kandori Mailath and Rob (1993) and (1995) is an important reference, as is Maynard Smith (1982), *Evolution and the Theory of Games*, and a splendid book, *Individual Strategy and Social Structure - An Evolutionary Theory of Institutions*, by H. Peyton Young, (1998). One should also note Weibull, *Evolutionary Game Theory* (1995) and K. Binmore and L. Samuelson, (1994), *An Economist's Perspective on the Evolution of Norms.* The evolutionary doctrine that traits which are relatively unsuccessful will be eliminated and those whose success is demonstrable will survive. Of course the "fittest" here means the fittest in reproduction. In games one must think of strategies as being fit when they lead to higher pay-offs. It will be noted that there is a by-product from this approach, it provides a theory of learning and it removes the ultra-rationality of ordinary economics.

There are many examples (replicator dynamics is one and stochastic gradient climbing is another) that converge to a Nash equilibrium or a risk-dominant equilibrium. These results and methods seem very promising, but they are also relatively useless when it comes to shorter term problems than the formation of institutions. For instance, I doubt whether the path of the economy following an interest rate change could usefully be modelled in this way.

But there does seem to be a possibility that our reactions to changes in economic indicators are routinised and that routinisation is the outcome of some social evolutionary process. As I have already noted, the evolutionary approach need not assume strict rationality: for instance, satisficing may be an evolutionary inheritance. But as I said, none of this is suitable for discussion here.

5 Summing Up

I can be very brief since this paper has done no more than use examples from the literature to persuade the reader that most of the work on stability analysis is either ad hoc or not usable for a study of the stability of equilibrium (Game theoretical analysis being exempted). What is needed is not mathematical techniques since most contributors to the literature are fully up to what is required. What is needed is more theory and more applied and empirical foundations of how agents and the economy react to mistakes. Every one knows that a second order equation gives a different path than does a first order or third order one. It seems very unlikely that theory alone will decide which is appropriate. At the same time while I deplore fake "scientific" air which pervades this subject, (I have also sinned) I have admitted that Keynes was right when he warned against models of violent fluctuations; the world does not seem to be like this. But what is it like?

September 19, 2001

References

1. Arrow, K.J. (1959): "Towards a theory of price adjustment" in *Allocation of Economic Resources*, ed. M. Abramovitz et al, Stanford University Press, 41-51.
2. Arrow, K. J. and L. Hurwicz, (1958): "On the stability of the competitive equilibrium I", *Econometrica* 26, pp. 522-52.
3. Binmore, K. (1987): "Modeling Rational Players I", *Economics and Philosophy* 3, 179-214.
4. Binmore, K. and Samuelson, L. (1994): "An Economist's Perspective on the Evolution of Norms", *Journal of Institutional and Theoretical Economics* 150: 45-63.
5. Debreu, G. (1974): "Excess Demand Functions", *Journal of Mathematical Economics* 1, pp.15-23.
6. Durlauf, S. N. (1993): "Nonergodic Economic Growth", *Review of Economic Studies* 60, pp.336-349.
7. Fisher F.M. (1983):*Disequilibrium Foundations of Equilibrium Economics*, Cambridge University Press.
8. Hahn, F. (1962): "On the stability of a pure exchange equilibrium", *International Economic Review* 3, pp. 206-214
9. Hahn, F. and Negishi, T. (1962): "A Theorem of Non-Tatonnement Stability", *Econometrica* 30, pp 463-9.
10. Hahn, F. (1963): "On the Disequilibrium Behaviour of a Multisectoral Growth Model", *Economic Journal* 73, pp.442-57.
11. Kandori, M. Mailath G. and Rob, R. (1993): "Learning, Mutation, and Long-Run Equilibria in Games", *Econometrica* 61:29-56.
12. Mantel, R. (1974): "On the Characterisation of Aggregate Excess Demand", *Journal of Economic Theory* 7, 348-353.
13. Maynard Smith, J.(1982): *Evolution and the Theory of Games*, Cambridge University Press.
14. Solow, R. M. (1956): "A Contribution to the Theory of Economic Growth", *Quarterly Journal of Economics*, 70, pp. 65-94.
15. Sonnenschein, H. (1972): "Market Excess Demand Functions", *Econometrica* 40, 549-563.
16. Uzawa H. (1962): "On the Stability of Edgeworth's Barter Process", *International Economic Review* 3, pp. 218-232.
17. Weibull, J. (1995): *Evolutionary Game Theory*, Cambridge: MIT Press.
18. Young, H. Peyton (1998): *Individual Strategy and Social Structure - An Evolutionary Theory of Institutions*, Princeton University Press.

A Bifurcation in an Overlapping Generations Model

Jan Bogers*

Department of Quantitative Economics, University of Amsterdam

Abstract. In a Diamond type overlapping generations model stability properties of steady state perfect foresight equilibria are analyzed under changes in the growth rate of the population. In contrast with the model of Diamond it is assumed that production is carried out according to decreasing returns to scale. Consequently profits are no longer zero and are uniformly distributed among young consumers. The stability analysis is carried out by analyzing an example with Cobb-Douglas utility and production functions. It is shown that under assumptions on the production coefficient of capital goods the set of perfect foresight steady state equilibria changes if the growth rate of the population changes. Also it is shown that the stability properties of the equilibria change under changes in the growth rate. This implies that there exists a saddle node bifurcation point.

Keywords: Overlapping generations, population dynamics, stability, bifurcation.

1 Introduction

The model presented in this contribution is the model that Claus and I developed for my thesis during the years 1985 to 1990. The model was presented and investigated in Bogers [2]. The goal of the research we carried out during these years was to investigate the consequences of a changing growth rate of the population on economic equilibria. This was and still is a very important problem.

The model was the framework for the *pension models* we investigated extensively in the dissertation. The model, which is based upon an overlapping generations model, has a dynamical structure. We concentrated on temporary equilibria and on perfect foresight steady state equilibria.

In general equilibria need not exist or in case they exist they need not be stable. The emphasis of the dissertation was on the stability properties of perfect foresight steady state equilibria and on the existence of bifurcations. These properties are investigated by analyzing the Cobb- Douglas version of the model.

Basically it is the same type of model as developed by P. Diamond in 1965 [4]. In contrast with the model of Diamond however, we assume that

* The author is indebted to A.J.J. Talman and H.N. Weddepohl for helpful suggestions and correction of the earlier versions of the paper.

production is carried out according to decreasing returns to scale and that taxes are levied by the government, not in order to redistribute income by means of lump sum transfers, but in order to finance government spending on the consumption good.

In Section 2 we describe the basic features of the model. In Section 3 we discuss existence of the equilibria and finally in Section 4 we investigate the existence of bifurcations in case the model is specified by Cobb-Douglas utility and production functions.

The idea for the research came from Claus. He originally planned a more empirical approach to the problem but during our research we gradually moved to a more theoretical approach.

2 The Model

In the model there is a consumption sector, a production sector, and a public sector. The time space of the model is $\{0, 1, 2, ...\}$. An element t of $\{0, 1, 2, ...\}$ is a period. Let $T = \{2, 3, ...\}$. There are three tradable goods: a consumption good, labor, and debt certificates. The debt certificates have the function of a claim in this model. A debt certificate is a promise to pay one unit of account, the next period. Consequently there are always two parties involved in transactions with debt certificates: one party who *emits* the certificates at time t and *receives* in exchange of a certificate *units of accounting* at a certain discount rate, and a party who *receives the certificates* at time t. By definition units of account can **not** be transferred from period t to period $t + 1$.

Consumers and producers are assumed to make decisions over two periods. Decisions are based upon prices of the current period and price expectations about the next period.

For consumers the demand for debt certificates is equivalent to savings (supply of units of accounting, saving), for producers the emission of debt certificates is equivalent to the purchase of capital goods (demand for units of accounting), and for the government the actions concerning debt certificates are equivalent with saving or spending, depending on whether the government is buying debt certificates or selling them. The relevant prices in period t are $p_c(t) = 1$, the price of the consumption good, $w(t)$, the real wage rate, and $p_m(t)$, the price of debt certificates, where $p_m(t) = 1/(1 + r(t))$ with $r(t)$ the nominal rate of interest at time t. All decision makers have perfect foresight in period t about prices in the next period: $p_c(t + 1) = 1$ is their expected price of the consumption good and $w(t + 1)$ is the expected wage rate in period $t + 1$. All prices are positive, i.e., $w(t), p_m(t) \in \mathbb{R}_{++}$.

In every period t with $t > 0$ there are two types of production units in the economy, old units and new units. Old units decide at time t about production given an amount of capital goods resulting from an investment decision in period $t-1$, new units decide in the current period about the amount of capital goods needed in period $t+1$. Since consumption goods can be transformed into capital goods, the production of capital goods for the next period leads to an investment demand for consumption goods in the current period. The units of accounting necessary for this demand are acquired by the *emission* of debt certificates.

In every period t with $t > 0$ there are old and young consumers in the economy. Old consumers effectuate the debt certificates they possess from the previous period and spend the units of accounting acquired in this way.

In period t young consumers decide about their consumption in period t and about consumption in period $t+1$. This decision about consumption in period $t+1$ leads to a demand for debt certificates. In period t old consumers effectute their consumption plans.

The public sector emits in every period a fixed amount of debt certificates per unit of labor, in the same period it collects taxes, the receipt from these debt certificates and taxes are spent on government consumption under a closed budget.

The structure of the model is that of an overlapping generations model. The introduction of debt certificates as a medium of transferring value from one period to the next was an idea of Claus.

Production

The production side of the economy is specified as follows. At time t there are β^t old production units and β^{t+1} new production units in the economy, for some given $\beta \in (0, \infty)$. Old units decide about production, while new units decide about investment. We assume that new units have a lifetime of two periods.

The production of consumption goods in every period is specified by a function $f : \mathbb{R}^2_+ \to \mathbb{R}_+$. The production of consumption goods takes place by means of capital and labor. In period t old units have a given capital stock resulting from the investments in period $t-1$. Production of capital goods takes one period and capital goods are produced from consumption goods. Let $c(t)$ denote the production of consumption goods and $\ell(t)$ the demand for labor by old units at time t, and let $k(t)$ denote the capital stock of old units at time t. Furthermore, let $k(t+1)$ denote the amount of capital produced by new units in period t, and let $c^i(t)$ denote the demand for consumption goods by new units at time t. We assume that $k(t+1) = c^i(t)$. Regarding the production function f, we assume that $f(0,0) = 0$, $f \in C^2(\mathbb{R}^2_{++}, \mathbb{R}_+)$, f is strictly concave, $\forall k \in \mathbb{R}_+$, $f'(\cdot, k) \in C^1(\mathbb{R}_{++}, \mathbb{R}_{++})$, $\forall \ell \in \mathbb{R}_+$, $f'(\ell, \cdot) \in C^1(\mathbb{R}_{++}, \mathbb{R}_{++})$, $f'(1,k)k \to \infty$ if $f'(1,k) \to 0$, and $f'(1,k)k \to 0$ if $f'(1,k) \to \infty$.

In order to pay for $c^i(t)$ it is necessary to emit debt certificates. The emitted amount of certificates $m^p(t)$ follows directly from the budget equality

$$p_m(t)m^p(t) = c^i(t).$$

The amount of debt certificates to be emitted is equal to

$$m^p(t) = k(t+1)/p_m(t).$$

The debts certificates emitted at time t have to be repaid in period $t+1$. The capital goods produced in period t are available at time $t+1$ and their lifetime is assumed to be one period. At time t new units decide about investment. This decision consists of maximizing expected *net* profits:

$$\begin{aligned} &\max\ c(t+1) - w(t+1)\ell(t+1) - k(t+1)/p_m(t), \\ &\text{such that}\ \ c(t+1) \leq f(\ell(t+1), k(t+1)). \end{aligned} \tag{2.1}$$

The investment decision problem leads to an optimal plan $k^*(t+1)$, given $w(t+1)$ and $p_m(t)$. The optimal demand $c^{i*}(t)$ follows from $k^*(t+1)$. This $c^{i*}(t)$ leads to $m^{p*}(t)$ as discussed above. The relevant results from the investment problem at time t are: $k^*(t+1)$, $c^{i*}(t)$, and $m^{p*}(t)$. The relevant results from the investment problem at time $t+1$ are $c^*(t+1)$ and $\ell^*(t+1)$. At time t an old unit decides about production given the amount of capital resulting from the investment decision of the previous period. As we assumed perfect foresight this decision leads to the optimal allocations $c^*(t)$ and $\ell^*(t)$ from the investment decision of period $t-1$ *given* $k^*(t)$. The debt certificates emitted at time $t-1$ and cashed at time t, are paid from the profits at time t. The realized net profit of each old production unit at time t therefore is equal to

$$\pi(t) = c^*(t) - w(t)\ell^*(t) - m^{p*}(t-1),$$

where $m^{p*}(t-1)$ is the amount of debt to be repaid by each old unit. The total amount of profits resulting from the production activities of all old units is given by $\beta^t\pi(t)$. This amount is uniformly distributed among young consumers. Realized net profits cannot be negative as decisions are made under perfect foresight.

From standard arguments it follows that the decision problem (2.1) has a unique and bounded solution, and furthermore this solution is a C^1 function of the prices $p_m(t)$, $w(t+1)$. It is obvious that the optimal solution of the old units are C^1 functions of the prices $p_m(t-1)$, $w(t)$.

The reader can see that even production is carried out according to an overlapping generations structure. This idea was the result of the discussions I had with Claus in the year 1986.

Consumption

All consumers are identical. At time t, there are β^t young consumers and β^{t-1} old consumers. Actions of young consumers are indicated with a superscript ^{1}, actions of old consumers are indicated with a superscript ^{2}. Consumers live no longer than two periods. We assume that during period t *no* consumers die. At $t = 0$ only young consumers are present in the economy. The lifetime preferences of the young consumers are represented by a utility function $U : \mathbb{R}^2_+ \rightarrow \mathbb{R}_+$. The amount of labor to be supplied by a young consumer in period t equals 1. An old consumer is not supplying any labor. At time t a young consumer receives a profit equal to $\pi(t)$. We assume that at time t each young consumer pays taxes over his wage income to the public sector, equal to $\tau_l w(t)$, for some fixed $\tau_l \in [0,1)$. Also at time t each young consumer pays taxes over his profit income, being equal to $\tau_\pi \pi(t)$, for some fixed $\tau_\pi \in [0,1)$. Regarding the utility functions we assume that $U(0,0) = 0$, $U \in C^2(\mathbb{R}^2_{++}, \mathbb{R}_+)$, U is strictly concave, and strictly monotone increasing i.e., $\forall c^2 \in \mathbb{R}_+$, $U'(\cdot, c^2) \in C^1(\mathbb{R}_{++}, \mathbb{R}_{++})$, and $\forall c^1 \in \mathbb{R}_+$, $U'(c^1, \cdot) \in C^1(\mathbb{R}_{++}, \mathbb{R}_{++})$. The decision of a young consumer at time t consists therefore of maximizing utility under restriction of his budget constraint, i.e.,

$$\begin{aligned} &\max \ U(c^1(t), c^2(t+1)), \\ &\text{such that } \ c^1(t), c^2(t+1) \in \mathbb{R}_+, \text{and} \\ &c^1(t) + p_m(t)c^2(t+1) \leq (1-\tau_l)w(t) + (1-\tau_\pi)\pi(t). \end{aligned} \tag{2.2}$$

The demand for certificates by a young consumer follows from $m^{1*}(t) = c^{2*}(t+1)$. Problem (2.2) results into $c^{1*}(t)$ and $m^{1*}(t)$. From the perfect foresight assumption it follows that at time t old consumers demand $c^{2*}(t)$ resulting from the solution of (2.2) at time $t-1$. Again as with production it can be proved by standard arguments that problem (2.2) has a unique bounded solution being a C^1 function of the prices $p_m(t-1)$, $w(t)$, $p_m(t)$. Obviously $c^{2*}(t)$ is a C^1 function of the prices $p_m(t-2)$, $w(t-1)$, $p_m(t-1)$.

The formulation of the consumption sector follows directly from the overlapping generations model of Diamond 1965 [4]. The introduction of taxes was my own idea.

The Public Sector

We assume there is one public sector in the economy. At time t former debts are repaid, taxes are collected, new debt certificates are emitted, and there is a demand for the consumption good. Taxes are imposed on wage income and on profit income. Total tax income of the public sector at time t equals

$$\beta^t[\tau_l w(t) + \tau_\pi \pi(t)].$$

In the context of this model we use debt certificates as a medium of transferring wealth by consumers: $m^g \in \mathbb{R}_+$ is the fixed amount of debt certificates emitted per young consumer by the public sector in every period. The *total* amount of debt certificates emitted by the government at time t equals $\beta^t m^g$. Consequently the repayment of former debts at time t equals $\beta^{t-1} m^g$. Public demand for consumption goods follows from the net income of the public sector at time t,

$$c^{g*}(t) = \beta^t[\tau_l w(t) + \tau_\pi \pi(t)] - [\beta^{t-1} - \beta^t p_m(t)]m^g.$$

From standard arguments it follows that $c^{g*}(t)$, is a C^1 function in the prices $p_m(t-1)$, $w(t)$, $p_m(t)$.

This is a different formulation of the public sector than in the model of P. Diamond [4]. In our model there are no lump sum transfers to consumers. Instead we introduced a demand for consumption goods. In the discussions with Claus we decided to incorporate government debt in the model as a supply of debt certificates.

Excess Demand Function

In order to analyse equilibria it is necessary to state the situation on the markets. Therefore we define the excess demand function.

Definition 2.1 At time $t \in T$ the excess demand function $z_t : \mathbb{R}^6_{++} \to \mathbb{R}^3$ with arguments $p_m(t-2)$, $w(t-1)$, $p_m(t-1)$, $w(t)$, $p_m(t)$, $w(t+1)$ is given as

$$z_t^1(p_m(t-2), w(t-1), p_m(t-1), w(t), p_m(t), w(t+1)) =$$

$$\beta^t c^{1*}(t) + \beta^{t-1} c^{2*}(t) + c^{g*}(t) + \beta^{t+1} c^{i*}(t) - \beta^t c^*(t),$$

$$z_t^2(p_m(t-1), w(t)) = \beta^t(\ell^*(t) - 1),$$

$$z_t^3(p_m(t-1), w(t), p_m(t), w(t+1)) = \beta^t m^{1*}(t) - \beta^t m^g - \beta_{t+1} m^{p*}(t).$$

We are now ready to define the equilibria.

Equilibria

Definition 2.2 A perfect foresight equilibrium is a solution $\{w^*(t+1), p_m^*(t)\}_{t \in T}$ to the following problem

$$z_t(p_m^*(t-2), w^*(t-1), p_m^*(t-1), w^*(t), p_m^*(t), w^*(t+1)) = 0, \forall t \in T,$$

$$\textit{with } p_m^*(0), w^*(1), p_m^*(1), w^*(2) \in \mathbb{R}_{++} \textit{ given.}$$

Definition 2.3 A perfect foresight steady state equilibrium is a solution w^*, p_m^* such that $\{w^*, p_m^*\}_{t \in T}$ is a perfect foresight equilibrium with $p_m^*(0) = p_m^*(1) = p_m^*$ *and* $w^*(1) = w^*(2) = w^*$.

In the dissertation [2] we introduced a much more complicated version of the model. Already at this point I got the advice from Claus to put real effort in developing a version which would be easier to analyze and which was not so complicated.

3 Existence

Perfect Foresight Equilibria

In this section we analyze the existence of the equilibium concepts defined in Definitions 2.2 and 2.3.

For convenience of denotation we will delete the time index t as much as possible. We will write (-1) for $(t-1)$, $(+1)$ for $(t+1)$ and nothing for t also we will delete the superscripts *. From standard arguments we can derive the *factor-price frontier* denoted by ψ, which gives a relation between w and $p_m(-1)$ such that the labor market is in equilibrium, i.e., $z_t^2(p_m(-1), w) = 0$.

Lemma 3.1 *For any* $t \in T$, $p_m(-2)$, $w(-1)$, $p_m(-1)$, *if* $w^* = \psi(p_m(-1))$, *and* $z_{t-1}^3(p_m(-2), w(-1), p_m(-1), w^*) = 0$ *then* $z_t^3(p_m(-1), w^*, p_m^*, w^*(+1)) = 0$ *for some* $p_m^* \in \mathbb{R}_{++}$ *and* $w^*(+1) = \psi(p_m^*)$ *is a necessary and sufficient condition for a perfect foresight equilibrium to exist.*

From Lemma 3.1 we can derive the following proposition.

Proposition 3.1 *Given any* $p_m^*(0)$, $w^*(1)$, $p_m^*(1)$, $w^*(2) \in \mathbb{R}_{++}$ *with* $w^*(t) = \psi(p_m^*(t-1))$, *for* $t = 1, 2, \ldots$ *there exists a perfect foresight equilibrium.*

Perfect Foresight Steady State Equilibria

In case we analyse perfect foresight steady equilibria we can prove the following.

Lemma 3.2 *If* $w = \psi(p_m)$ *then* $z_t^3(p_m, w) = 0$ *for any* $t \in T$, *is a necessary and sufficient condition for a steady state perfect foresight equilibrium to exist.*

Proposition 3.2 *Let* $w = \psi(p_m)$. *If for* $p_m \to 0$, $c^2 * (+1) \to \infty$, *and for* $p_m \to \infty$, $c^{2*}(+1) \to 0$, *then there exists a steady state perfect foresight equilibrium.*

In trying to prove existence I got into real trouble. Again I got the advice to strip the model and to present only a core version. According to Claus this would be complicated enough! After a struggle of almost a year I could prove existence only under very stong assumptions. The main results of my struggle are stated in the lemma's above.

4 Bifurcations

In this section we analyze changes in the set of perfect foresight steady state equilibria and changes in the stability properties of these equilibria if the growth parameter β changes. As the values of β changes a so-called bifurcation may occur. At such a bifurcation there is a qualitive change in the behaviour of the model. For example, there might be a bifurcation value of β, say β^*, such that for $\beta < \beta^*$ a perfect foresight equilibrium exists, but for $\beta > \beta^*$ such an equilibrium does not exist (or a completely different steady state equilibrium emerges). Popularly speaking this means that if there exists a bifurcation point, then an economy which is in a steady state equilibrium, need *not* converge to another steady state equilibrium if the growth rate of the population suddenly changes beyond that point. It might end up in *chaos.*

In the mathematical literature we find several classifications of bifurcations in non linear dynamical systems. See for instance Lauwerier [7], Broer and Verhulst [1], and Wiggins [9]. In the economic literature we find studies of changes in the set of equilibria as a consequence of changes in the model parameters, for instance see Dierker [5], Hildenbrand [6], and Debreu [3]. Their concept of regular economies and non regular economies corresponds to the concept of saddle-node bifurcations. Here we use the classification of Wiggins.

Before we can discuss dynamical properties of the model we have to make a statement about the existence of a dynamical process in terms of p_m. This follows from the following theorem.

Theorem 4.1 *Given $p_m^*(-1) \in \mathbb{R}_{++}$, if $w^* = \psi(p_m^*(-1))$, and $w^*(+1) = \psi(p_m^*)$, then at a perfect foresight equilibrium there exists a function $F \in C^1(\mathbb{R}_{++}, \mathbb{R}_{++})$ such that $p_m^* = F(p_m^*(-1))$.*

Proof. From Proposition 3.1 it follows that the manifold defined by

$$z_t^3(p_m^*(-1), w^*, p_m^*, w^*(+1)) = 0,$$

is not empty. Substituting $w^* = \psi(p_m^*(-1))$ and $w^*(+1) = \psi(p_m^*)$ gives

$$m^1(p_m^*(-1), \psi(p_m^*(-1)), p_m^*) = m^g + \beta m^p(p_m^*, \psi(p_m^*)).$$

This can also be written as

$$G(p_m^*, p_m^*(-1)) = 0.$$

From the differentiability of m^1, m^p, and ψ it follows that $G \in C^1(\mathbb{R}_{++}^2, \mathbb{R})$. The theorem follows now directly from the implicit function theorem.

At the time I wrote the dissertation bifurcation theory was not common in economic literature. Claus and I introduced this element in our model to get a better grip on the consequences of changes in the growth rate of the population.

The Cobb-Douglas Version

Finally I took the advice and moved from the general formulation of the model to a version that would lend itself better for the analyses of the changes in the growth rate of the population, which was the original purpose of our research!

Let the utility function of young consumers be given by

$$U(c^1, c^2(+1)) = \alpha \log c^1 + (1-\alpha)\log c^2(+1), \alpha \in (0,1). \tag{4.1}$$

The production technology is specified by

$$f(\ell, k) = (\ell)^{\nu}(k)^{\eta}, \nu, \eta \in (0,1), \nu + \eta < 1. \tag{4.2}$$

Let $\omega = 1 - \upsilon - \eta$. The supply and demand functions are given by

$$m^1(p_m^*(-1), w^*, p_m) = (1-\alpha)[(1-\tau_l)w^* + (1-\tau_\pi)\pi(w^*, p_m^*(-1))]/p_m,$$

$$m^p(p_m, w^*(+1)) \quad = [\upsilon/w^*(+1)]^{\upsilon/\omega}[\eta p_m]^{(1-\upsilon)/\omega}/p_m.$$

$z_t^3(p_m^*(-1), w^*, p_m^*, w^*(+1)) = 0$ implies

$$(1-\alpha)[(1-\tau_1)w^* + (1-\tau_\pi)\pi(w^*, p_m^*(-1))]/p_m^* - m^g =$$

$$\beta[\upsilon/w^*(+1)]^{\upsilon/\omega}[\eta p_m^*]^{(1-\upsilon)/\omega}/p_m^*.$$

Let $k: \mathbb{R}_{++} \to \mathbb{R}_{++}$ be defined as

$$k(p_m) = [\eta p_m]^{1/(1-\eta)}.$$

We write k^* for $k(p_m^*)$, and k_{-1}^* for $k(p_m^*(-1))$. Furthermore, let

$$A = (1-\alpha)[(1-\tau_l)\upsilon + (1-\tau_\pi)\omega]/\beta \quad \text{and} \quad B = m^g/(\eta\beta).$$

Together with

$$\pi(w^*, p_m^*(-1)) = \omega[\upsilon/w^*]^{\upsilon/\omega}[\eta p_m^*(-1)]^{\eta/\omega},$$

and

$$w^* = \psi(p_m^*(-1)) = [\upsilon(\eta)^{\eta/(1-\eta)}][p_m^*(-1)]^{\eta/(1-\eta)},$$

then $z_t^3(p_m^*(-1), w^*, p_m^*, w^*(+1)) = 0$ can be written as

$$A(k_{-1}^*)^\eta - k^* - B(k^*)^{1-\eta} = 0, \text{with } A \in \mathbb{R}_{++}, B \in \mathbb{R}+. \tag{4.3}$$

A steady state perfect foresight equilibrium is defined by a fixed point k^* of (4.3), hence

$$A(k^*)^\eta - k^* - B(k^*)^{1-\eta} = 0. \tag{4.4}$$

Existence of Steady State Equilibria in the Cobb-Douglas Version

It turns out that even under this simple version of the model existence cannot be shown easily Without proof we state the existence theorem for the Cobb-Douglas version. Let C be

$$C = [\eta/(1-\eta)]^{[(\eta-1)/(2\eta-1)]} - [\eta/(1-\eta)]^{[-\eta/(2\eta-1)]}.$$

Proposition 4.1 *In case the utility and production function are specified by* (4.1) *and* (4.2) *then there exists* **one** *steady state perfect foresight equilibrium if* $\eta < 1/2$ *or if both* $\eta = 1/2$ *and* $A > B$ *or if both* $\eta > 1/2$ *and* $[A^\eta B^{\eta-1}]^{1/(2\eta-1)}C = 1$, *and there exist* **two** *steady state perfect foresight equilibria if both* $\eta > 1/2$ *and* $[A^\eta B^{\eta-1}]^{1/(2\eta-1)}C > 1$.

From this proposition we can see that we have to analyze three cases $\eta < 1/2$, $\eta = 1/2$ and $\eta > 1/2$. To study the dynamical behaviour of the model under changes in the growth rate of the population we have to investigate the stability properties of the steady state perfect foresight equilibria.

Stability of Steady State Perfect Foresight Equilibria in the Cobb-Douglas Model

From (4.3) we can see that we cannot derive the function F as meant in Theorem 4.1 explicitly. Most authors solve this type of problem by stating the dynamical process *backward looking.* We use the implicit function theorem to analyze stability. This results into the following proposition.

Proposition 4.2 *If* $\eta < 1/2$ *then the steady state perfect foresight equilibrium is asymptotically* **stable**. *If* $\eta = 1/2$ *and* $A > B$ *then the steady state perfect foresight equilibrium is asymptotically* **stable**. *If* $\eta > 1/2$ *and* $[A^\eta B^{\eta-1}]^{1/(2\eta-1)}C = 1$ *the steady state perfect foresight equilibrium is asymptotically* **unstable** *and if* $\eta > 1/2$ *and* $[A^\eta B^{\eta-1}]^{1/(2\eta-1)}C > 1$ *then one equilibrium is asymptotically* **stable** *and the other equilibrium is asymptotically* **unstable.**

Figure 1 illustrates the case that $\eta > 1/2$ and $[A^\eta B^{\eta-1}]^{1/(2\eta-1)}C > 1$. In this case the set of steady state perfect foresight equilibrium prices equals $\{p_m^{1*}, p_m^{2*}\}$. The equilibrium price p_m^{1*} is asymptotically *unstable* whereas the price p_m^{2*} is asymptotically *stable.*

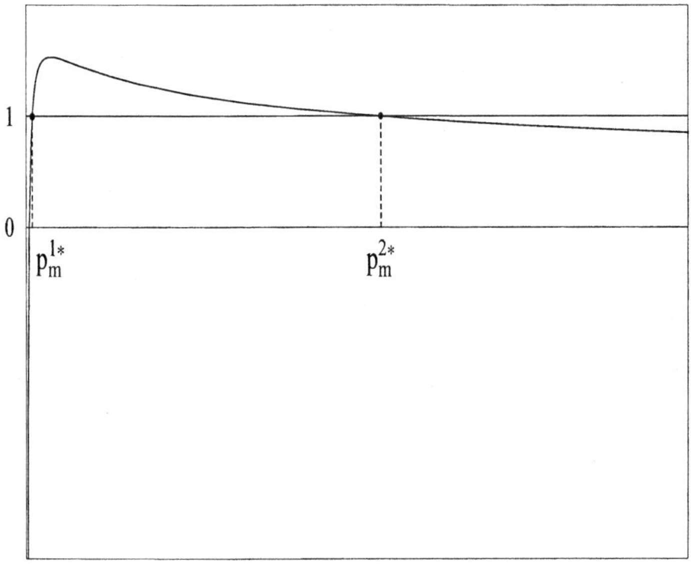

Fig. 1. For the case $\eta > 1/2$ and $[A^\eta B^{\eta-1}]^{1/(2\eta-1)}C > 1$ the set of steady state perfect foresight equilibrium prices equals $\{p_m^{1*}, p_m^{2*}\}$. The equilibrium price p_m^{1*} is asymptotically *unstable* whereas p_m^{2*} is asymptotically *stable.*

Bifurcations

For a complete description of the classes of bifurcations, see [9]. The type of bifurcation found in this example is a *saddle node bifurcation.* This means that we can find a value of β, say β^*, such that for $\beta < \beta^*$ there exist *two* equilibria one asymptotically stable and one asymptotically unstable, whereas for values of $\beta^* > \beta$ there exists *no* equilibrium. For $\beta = \beta^*$ there exists only *one* asymptotically unstable equilibrium. Let in the sequel

$$A' = (1-\alpha)[(1-\tau_l)\upsilon + (1-\tau_\pi)\omega],$$
$$B' = m^g/\eta.$$

Proposition 4.3 follows from Proposition 4.2.

Proposition 4.3 If $\eta \leq 1/2$ *then no bifurcation points exist. If* $\eta > 1/2$ *then there exists a saddle node bifurcation with* $\beta^* = [(A')^\eta (B')^{\eta-1}]^{1/(2\eta-1)} C$ *as bifurcation value.*

Proof. From the definition of A and B it follows that if $\eta \in (1/2, 1)$ then

$$d[A^\eta B^{\eta-1}]^{1/(2\eta-1)} C / d\beta < 0.$$

From Proposition 4.1 it follows that when β *decreases* the number of equilibrium prices changes from zero to one and finally to two. Furthermore from Proposition 4.2 it follows that in case there exists one equilibrium then this equilibrium is asymptotically unstable and in case there exist two equilibria then one of these equilibria is asymptotically stable and one is asymptotically unstable. This now implies that a *saddle node bifurcation point* exists. The bifurcation value of β is given by the value of β for which there is *one* equilibrium price, i.e.,

$$\beta^* = [(A')^\eta (B')^{\eta-1}]^{1/(2\eta-1)} C.$$

In the table below we summarize the results from propositions 4.1 to 4.3.

Table 1. Summary

	$\eta \in (0, \frac{1}{2})$	$\eta = \frac{1}{2}$	$\eta \in (\frac{1}{2}, 1)$
Equilibrium set	$\{p_m^*\}$	$\{p_m^*\}$if $A > B$	$\{\emptyset\}, \{p_m^*\}$or$\{p_m^{1*}, p_m^{2*}\}$
Stability	p_m^* stable	p_m^* stable	p_m^* unstable p_m^{1*} unstable p_m^{2*} stable
Bifurcations	No	No	Saddle node

From the analysis above the reader can see that even this version of the model is quite complicated. The advice that I got from Claus was good and necessary.

5 Conclusions

- The Cobb-Douglas version of the model has a saddle node bifurcation point.
- Sudden changes in the growth rate of the population can lead to fundamental changes in the set of equilibria *and* in the stability properties of the equilibria.
- In case $\eta \leq 1/2$ no bifurcation point exists and equilibria are stable.
- In case $\eta > 1/2$ a change in the growth parameter β causes a change in the set of equilibria. An *increase* of β causes a *decrease* in the number of equilibria from two to one and finally to zero. Also the stability properties undergo changes as β changes. If there exists one equilibrium then this is unstable, if there exist two equilibria then there is one stable and one unstable equilibrium.
- In terms of regular economies the case $\eta > 1/2$ means that the economy defined by a specific β^* is *not regular.*
- Welfare and policy issues are not investigated in this research. The main purpose was to get an impression of the consequences of a change in the growth rate of the population.

References

1. Broer H., and Verhulst F., Dynamische Systemen en Chaos een Revolutie vanuit de Wiskunde, Epsilon Uitgaven, Utrecht, The Netherlands, 1990.
2. Bogers, J.W.T., The Influence of Changes in a Steady State Demographical Development on Economic Equilibria: a General Equilibrium Approach, Doctoral Thesis, University of Amsterdam, The Netherlands, 1992.
3. Debreu, G., "Regular differential economies", American Economic Review 66, 1976, pp 280-287.
4. Diamond, P.A., "National debt in a neoclassical growth model", American Economic Review 55, 1965, pp 1126-1150.
5. Dierker, E., Topological Methods in Walrasian Economics, Lecture Notes in Economics and Mathematical Systems 92, Springer Verlag, Berlin, Germany, 1974.
6. Hildenbrand, W., Core and Equilibria of a Large Scale Economy, Princeton University Press, Princeton, NJ, USA, 1974.
7. Lauwerier H. A., "One dimensional iterative maps", in Chaos, Holden A. V., ed., Princeton University Press, Princeton, NJ, USA, 1986, pp 39-57.
8. Samuelson, P.A., "An exact consumption-loan model of interest with or without the social contrivance of money", Journal of Political Economy 66, 1958, pp 467-482.
9. Wiggins, S., Introduction to Applied Nonlinear Dynamical Systems and Chaos, Texts in Applied Mathematics 2, Springer Verlag, Berlin, Germany, 1990.

Heterogeneous Beliefs and Routes to Complex Dynamics in Asset Pricing Models with Price Contingent Contracts*

William A. Brock[1] and Cars H. Hommes[2]

[1] Department of Economics, University of Wisconsin, Madison
[2] Center for Nonlinear Dynamics in Economics and Finance (CeNDEF), Department of Quantitative Economics, University of Amsterdam

Abstract. This paper discusses dynamic evolutionary multi-agent systems, as introduced in Brock and Hommes (1997). In particular, the heterogeneous agent dynamic asset pricing model of Brock and Hommes (1998) is extended by introducing derivative securities by means of price contingent contracts. Numerical simulations suggest that in a boundedly rational heterogeneous evolutionary world futures markets may be destabilizing.

Keywords: heterogeneous beliefs, bounded rationality, arrow securities, evolutionary dynamics.

1 Introduction

This volume in honour of Claus Weddepohl evolves around the themes *Equilibrium, Markets* and *Dynamics.* Our contribution focuses on a dynamic equilibrium asset pricing model with evolutionary selection of heterogeneous forecasting rules or strategies. Our framework fits into equilibrium theory in the sense that there is market clearing. However, expectations are in 'disequilibrium' in the sense that agents may deviate from fully rational expectations or perfect foresight. One might describe our evolutionary adaptive belief systems (ABS) as an 'approximate rational expectations equilibrium'.

Before getting into details let us discuss the main issue we wish to deal with here. Brock and Hommes (1997, 1998), henceforth BH, attempt to develop a positive and normative equilibrium theory of endogenous dynamic belief formation in intertemporal markets, especially asset markets. Agents can choose from a finite set of competing forecasting rules or trading strategies. Agents are boundedly rational in the sense that they tend to use strategies that have performed well in the recent past. BH show that a *rational route to*

* We would like to thank an anonymous referee for detailed comments on an earlier draft of this paper. This research was supported by NSF under grant SES-9911251, the Vilas Trust, and by the Netherlands Organization for Scientific Research (NWO) under a NWO-MaG Pionier grant. None of the above are responsible for errors, opinions, and shortcomings in this paper.

randomness, that is a bifurcation route to local instability, limit cycles and chaos, may arise when evolutionary pressure for strategy selection increases. Stated differently, as the traders' sensitivity to differences in strategy performance increases, complicated and unpredictable asset price fluctuations may arise; see e.g. Brock (1997) or Hommes (2001) for recent reviews. However, BH (1998) did not consider the impact on belief heterogeneity of trading derivatives. The basic issue we wish to study here is the impact of introduction of a plentitude of "derivative" securities upon the dynamics of belief heterogeneity. *Does the introduction of derivatives in the asset market stabilize the rational routes to randomness?*

A common critique of any theory of diverse beliefs for economies with well developed asset markets is the following. The number of price contingent contracts (PCC's) (this terminology follows Kurz (1997)) such as derivative securities is rich enough in well developed markets, that enough agreement is forced by trading such objects that the quantitative effect remaining of initial belief diversity after several rounds of trading such securities is likely to be minor. Anyone who has visited an options exchange after opening rotation is completed is aware of how fast trading quiets down after the initial burst at the open (Brock and Kleidon (1992)). To put it another way, trading volume is typically higher at the open and at the close in comparison with volume during the rest of the day. This well known stylized fact is consistent with the view that several rounds of trading reduces belief heterogeneity, at least on time scales consistent with the securities being traded. The initial burst of trading at the open is consistent with the hypothesis that belief heterogeneity increased during the closed period before the opening. The rapid decrease in trading observed after the open is consistent with the hypothesis that trading decreases belief heterogeneity.

A corollary of this argument is that one should focus on researching relatively minor variations around some kind of natural focal baseline such as the Rational Expectations baseline (e.g. noisy rational expectations, adaptive learning centered at a rational expectations baseline, and so forth). Indeed, we have heard arguments that the presence of a plentitude of derivative securities in real markets should remove so much of the belief heterogeneity that the BH (1998) dynamics of belief heterogeneity would be "crushed" quite quickly into belief homogeneity. Magill and Shafer (1991) review General Equilibrium theory with Incomplete markets, hereafter called "GEI" theory. A basic idea of GEI theory is this. The more ways there are for agents at date t to move income across states at date t+1 via trading of different securities at date t, the less disagreement in beliefs across agents about date t+1 economic states. Intuitively, expected marginal rates of substitution across different agents across date $(t, t+1)$ event pairs are equated by trading of contracts at date t that pay one if event E_i occurs at date t+1 and zero otherwise. The more of these securities there are that enable the young to move income across states in old age by using available securities while young, the smaller the disagreement in beliefs about the future should be. It would seem that if there is a security for each state or if there is "spanning" (an equivalence), then disagreement in beliefs about future returns would vanish and that's the end of BH (1998) complex dynamical possibilities. Extreme versions of this type of argument suggest that the entire research program that studies the dynamics of belief heterogeneity in markets is essentially a waste of time. This paper argues that matters are rather more subtle and presents a numerical example suggesting that in a boundedly rational, heterogeneous evolutionary world price contingent contracts may actually *destabilize* the asset market.

The ultimate fate of dynamics of heterogeneous beliefs depends upon the dimension of the rule set relative to the number of derivatives or PCC's being traded among other things. Let us consider what Kurz (1997) has to say about belief heterogeneity even in contexts with a plentitude of PCC's. Despite the existence of such PCC's in real markets Kurz (1997) still gives a vigorous argument for quantitative importance of his alternative to Rational Expectations Theory (REE theory), which he calls Rational Beliefs Theory (RBE). He calls a belief, a "Rational Belief", provided that it can not be contradicted by intertemporal observed data. He works with a class of stochastic processes rich enough so that even though each Rational Belief is consistent with all possible limiting time averages generated by the economy, convergence to the true equilibrium stochastic process does not occur. Kurz works with an *exogenously fixed* set of RB's, but he imposes a strong consistency requirement with the data. Kurz (1997) stresses the importance of endogenous uncertainty created by asset markets not only for understanding the volatility patterns in returns data but also for policy purposes.

The objective of BH (1998) theory is similar to Kurz in that both theories wish to give a coherent and data disciplined approach to endogenous uncertainty and belief formation, but BH works with an *endogenous* set of beliefs that co-evolve over time. In the BH (1998) setting there is an additional evolutionary dynamics on the fraction of belief types as well as on prices. This additional dynamics adds extra complexity. BH theory pays the price that most of the beliefs will *not* be consistent with the data at all times, but the data are allowed to determine the set of "surviving" beliefs via the force of evolutionary selection based upon how well each belief type does in trading against the set of belief types in the economy as a whole. Structural rational expectations is always an equilibrium of the BH system if agents have free access to such structural information. If the information is expensive however, a dynamical tension develops which tends to create phases of "random" length where the economy runs close to rational expectations for a while then gradually drifts away from RE as some agents gradually realize they can get by with using naive predictors and not pay to get RE predictors. The size of departures away from RE and how long such excursions away from RE can be becomes a complicated function of underlying parameters of the economic system.

In BH financial modeling, the bond price (i.e. the return on bonds) was assumed to be exogenously fixed. We give a treatment of endogenous bond pricing in two period Overlapping Generations setups below. We restrict attention to a two periods OG setup, to avoid tractability problems that arise in T period settings of a heterogeneous agents economy when the horizon, T, is large. The BH-model with endogenous bond pricing is a true general equilibrium model whereas the original BH financial work is partial equilibrium. In a two period setup analytical problems are simplified for several reasons. First, the issue of wealth in a heterogeneous expectations BH setup and its book keeping problem to keep track of wealth accumulated by different traders that have switched strategies many times is simplified because the OG system is "re-initialized" each period as the young become old and pass from the scene. Brock (1990) generalizes a simple version of Lucas's Asset Pricing Model to OG setups. We can do the same here with the BH program. Second, "bubble" solutions are automatically crushed in OG models. "Up bubbles" are killed by the endowment of the young bounding the price of the asset from above. "Down bubbles" are killed off by limited liability. So, unlike BH mean variance theory, no exogenously specified "outside" force as used by Gaunersdorfer and Hommes (2000) is needed to crush bubbles. Bubbles are crushed by the economics of the models. Third, it is easy to insert Arrow securities (securities sold at date t a unit of which pays 1 if price plus dividend, $p_{t+1} + y_{t+1}$, next period takes a particular value and zero otherwise) into this OG model and explore their impact on forcing agreement of BH type heterogeneous expectors. We will propose arguments later that BH type learning should apply to the "p" part, i.e. the expected future price, of an Arrow security but perhaps algorithmic learning like Marcet/Sargent should apply to the "y" part, i.e. the expected future dividend part. See Evans and Honkapohja (2001) for an extensive recent treatment of algorithmic learning. The argument will revolve around Nature selecting and creating the y part, but Society co-creating the p part. We shall argue that it is not automatic that a full set of Arrow securities with learning in a BH framework gets rid of the dynamical complexity of BH learning. Hedging arguments against BH heterogeneity are not so easy because BH heterogeneity itself is unobservable and fluctuates unlike sunspots. So derivative security sunspot irrelevance type arguments do not automatically apply to BH heterogeneity. See Brock (1990) for discussion of Arrow securities in OG models. See Guesnerie's recent book (Guesnerie (2001))for a recent review of the sunspots literature.

Our contribution to this Festschrift should be viewed as a sketch of ideas illustrated by some first, exploratory numerical simulations that will be developed more rigorously in future work. The paper is organized as follows. Section 2 presents a general two period OG-setup of the BH-model with heterogeneous beliefs and introduces PCC's in the model. Section 3 discusses an example with three belief types and two PCC's and presents numerical simulations of the dynamical behaviour. Finally, section 4 gives some concluding remarks and briefly discusses some future work.

2 The Model

Consider an OG model with one risky asset, e.g. a stock, and one risk free bond with $h = 1, 2, ..., H$ different beliefs about the value of the risky asset. Assume agreement on the bond. Assume utility functions are the same across agents. Let u, v denote utility functions while young and old. Let p_t, y_t denote price and earnings of the risky asset at date t and let R_t denote gross return on the bond at date t. Following Brock (1990) let w_y, w_o denote endowments of young and old respectively. First let us assume homogeneous beliefs and solve for rational expectations equilibrium in order to establish a useful baseline. Assume the young solve (1) below at each date t,

$$\text{Maximize}\{u(w_y - b_t - p_t z_t) + E_t[v(w_o + (p_{t+1} + y_{t+1})z_t + R_t b_t)]\}, \tag{1}$$

where b_t and z_t are the demands for the bond and the risky asset respectively. First order necessary conditions for maximum z_t, b_t are given by,

$$\begin{aligned} p_t u' &= E_t[(p_{t+1} + y_{t+1})v'] \\ u' &= R_t E_t(v'). \end{aligned} \tag{2}$$

We will focus on the case where the dividends $\{y_t\}$ are independently identically distributed (IID).

Let the supply z_s of shares be fixed for each date t, normalized to one. In equilibrium, the supply of bonds is zero for society as a whole, thus, $b_t = 0$ in equilibrium. Thus, equilibrium $\{p_t\}$ must solve,

$$p_t u'(w_y - p_t) = E_t[(p_{t+1} + y_{t+1})v'(w_o + p_{t+1} + y_{t+1})]. \tag{3}$$

As shown by Brock (1990), depending upon the shape of u and v, there can be many rational expectations solutions to (3). We concentrate on stationary ones here. Put

$$\begin{aligned} A(p) &= pu'(w_y - p) \\ B(p) &= E_t[(p + y_{t+1})v'(w_o + p + y_{t+1})]. \end{aligned} \tag{4}$$

Since $\{y_t\}$ is IID, $B(p)$ does not depend upon time t. Assume $u' > 0, u'' < 0$. The $A(p)$ is increasing in p. Therefore if we restrict our search for solutions $\{p_t\}$ of (3) to deterministic sequences p_t, these must satisfy,

$$A(p_t) = B(p_{t+1}), \qquad \text{i.e. } p_t = F(p_{t+1}). \tag{5}$$

The last follows because $A' > 0$ implies A is invertible. Notice that $B(p)$ can have many shapes, so cycles, sunspots, and many kinds of equilibria are possible for different B functions. Let $p^* = F(p^*)$ be a fixed point. Brock (1990) shows that there are usually at least two fixed points, one negative (even though $y > 0$) and one positive. He also shows that there can be many such fixed points. Assume the positive fixed point is unique. We call this the "fundamental price" p^*.

Notice also that if A cuts B from below at p^* it looks like "up bubbles" are possible in (5). Simply take an initial condition to the right of p* and solve (5) forward. However once $p_t > w_y$ it cannot be equilibrium because $p_t.z$ can not exceed the wealth in the economy, w_y. This simple argument shows that up bubbles can not be equilibria.

Notice that for a learning scheme where $p^e_{t+1} = G(p_{t-1}, ...)$ is an expectation of p_{t+1} that depends upon past observed prices, (5) generates a learning dynamics. Methods from the learning literature may be adapted and applied to this dynamics. We are interested in generalizing BH heterogeneous beliefs dynamics to this OG setup. Let b_{ht} and z_{ht} denote the demands of type h for the bond and the risky asset respectively. Expectational type h solves

$$\text{Maximize}\{u(w_y - b_{ht} - p_t z_{ht}) + E_{ht}[v(w_o + (p_{t+1} + y_{t+1})z_{ht} + R_t b_{ht})]\}. \tag{6}$$

Assume all h's agree on R_t. First order conditions for optimal choices of z_{ht}, b_{ht} are given by

$$p_t u'(w_y - b_{ht} - p_t z_{ht}) = E_{ht}[(p_{t+1} + y_{t+1})v'(w_o + (p_{t+1} + y_{t+1})z_{ht} + R_t b_{ht})]. \tag{7}$$

$$u'(w_y - b_{ht} - p_t z_{ht}) = R_t E_{ht}[v'(w_o + (p_{t+1} + y_{t+1})z_{ht} + R_t b_{ht})]. \tag{8}$$

We make an assumption about beliefs that parallels BH:

Assumption 1:

A1 $E_{ht}[(p_{t+1} + y_{t+1})v'(w_o + (p_{t+1} + y_{t+1})z_{ht} + R_t b_{ht})] =$

$$E_t[(p^* + f_{ht} + y_{t+1})v'(w_o + (p^* + f_{ht} + y_{t+1})z_{ht} + R_t b_{ht})],$$

A2 $R_t E_{ht}[v'(w_o + (p_{t+1} + y_{t+1})z_{ht} + R_t b_{ht})] =$

$$R_t E_t[v'(w_o + (p^* + f_{ht} + y_{t+1})z_{ht} + R_t b_{ht})],$$

where f_{ht} is a function of past prices $p_{t-1}, p_{t-2}, ...$, which have been observed at date t when h submits his or her demand function before the market has determined the equilibrium price p_t based upon aggregate demand and aggregate supply being equated at date t.

Notice how this assumption is the exact parallel of BH in a simpler mean variance model. It says that beliefs agree on the general functional form of the cumulative distribution function of $p_{t+1} + y_{t+1}$ but disagree on the form of the shift parameter in expectations about p_{t+1}. I.e. beliefs about $p_{t+1} + y_{t+1}$ are of the form

$$E_{ht}[p_{t+1} + y_{t+1}] = p^* + f_{ht} + E_t[y_{t+1}],$$

where there is agreement on the distribution of y_{t+1}, but disagreement on f_{ht} which is deterministic. Notice that this implies full agreement on the support of y_{t+1}. Below we will discuss an example where we allow disagreement on the support of y_{t+1} and we will show how the presence of PCC's forces a certain amount of agreement on the support of y_{t+1} in order for a PCC equilibrium to exist. The important point is that although existence of a PCC equilibrium may force agreement on supports, it does *not* force full agreement on probability masses over the support.

We are now ready to introduce PCC's. Put $q' = p' + y'$ for next period's values of p and y, $p + y$. Partition the non-negative real line as follows

$$0 < a_1 < a_2 < ... < a_n < \infty. \tag{9}$$

Define disjoint sets S_i as follows,

$$S_1 = [0, a_1), S_2 = [a_1, a_2), ..., S_i = [a_{i-1}, a_i), ..., S_n = [a_{n-1}, \infty). \tag{10}$$

The case $n = 1$ corresponds to $a_0 = 0$, $S_1 = S = [0, \infty)$ by convention, i.e. the case $n = 1$ corresponds to the case where only the risky asset itself is traded. We take $n > 1$ unless otherwise noted. Let p_{it} denote today's price of security i which pays $q'.1_i$ next period, where $1_i = 1[q' \in S_i]$ with $1[A]$ the indicator function for event A which is 1 if A occurs, zero otherwise; we will write p_{0t} for the price of the stock, i.e. the original risky asset, and refer to it as asset 0. Let z_{h0t} denote the demand for asset 0 and z_{hit} the demand for the i-th PCC. After introduction of n PCC's the type h young person's problem is to choose the demand vector $\boldsymbol{z} = (z_{h0t}, z_{h1t}, ..., z_{hnt}, b_{ht})$ to solve

$$\text{Maximize}\{u(w_y - b_{ht} - p_{0t}z_{h0t} - \textstyle\sum_{i=1}^{n} p_{it}z_{hit}) + E_{ht}[v(w_o + q_{t+1}z_{h0t} + \textstyle\sum_{i=1}^{n} q_{i,t+1}z_{hit} + R_t b_{ht})]\}, \tag{11}$$

where $q_{t+1} = p_{t+1} + y_{t+1}$ is a random variable and by definition,

$$q_{i,t+1} = q_{t+1}1[q_{t+1} \in S_i]. \tag{12}$$

The definition of $q_{i,t+1}$ implies the random variable q_{t+1} is a linear combination of $\{q_{i,t+1}\}$, i.e.

$$q_{t+1} = \sum_{i=1}^{n} q_{i,t+1}. \tag{13}$$

Therefore in order to prevent arbitrage portfolios (cf. Magill and Quinzii (1996), Chapter 2 and elsewhere) we must have that, at each date, the price of the stock, i.e. asset 0, equals the sum of the prices of the PCC's, i.e.

$$p_{0t} = \sum_{i=1}^{n} p_{it}. \tag{14}$$

This is easy to see. If (14) were not true, e.g. if p_{0t} were larger than the sum, the agent could borrow z shares of the asset 0, and buy z shares of each PCC i. By (13) this arbitrage results in a net income of zero at $t+1$ but generates positive income at t. It can be scaled up to infinity. Hence p_{0t} must be less than or equal to the sum. A similar argument forces p_{0t} equal to the sum. General arguments of this type and the restrictions they place on the prices of dependent random variables are discussed by Magill and Quinzii (1996). Our case is a very special case of this general discussion.

Let $\omega_{hit} = z_{h0t} + z_{hit}$ be the sum of the demands of asset 0 and the i-th PCC, and use (14) to rewrite (11) as follows,

$$\text{Maximize}\{u(w_y - b_{ht} - \sum_{i=1}^{n} p_{it}\omega_{hit}) + E_{ht}[v(w_o + \sum_{i=1}^{n} q_{i,t+1}\omega_{hit} + R_t b_{ht})]\}. \tag{15}$$

The treatment of Brock and Hommes (1998) corresponds to the special case, $u(.) = 0$ (i.e. there is no utility generated by consumption while young), and $v(c') = E(c') - (a/2)Var(c')$. The return, R, on the bond, b, was exogenously given in BH (1998). It is endogenous here. We are now ready to compare the dynamics of an Adaptive Belief System (ABS) with PCC's, to the dynamics of an ABS without PCC's. Let $\boldsymbol{\omega_{ht}}$ denote the demand vector $(\omega_{h1t}, ..., \omega_{hnt})$ and let $\mathbf{e}$ denote the vector of n ones. Let n_{ht} denote the fraction of agents using belief system h at date t. The heterogeneous agents market equilibrium conditions are

$$\sum_{h=1}^{H} n_{ht} b_{ht} = 0, \qquad \text{(bond market)} \tag{16}$$

$$\sum_{h=1}^{H} n_{ht}\boldsymbol{\omega_{ht}} = z_s \boldsymbol{e}, \qquad \text{(n PCC markets).} \tag{17}$$

Our dynamical system will be completely specified once we specify the fractions n_{ht} and the beliefs E_{ht} about q_{t+1} for each type h. As in BH (1998) the fractions n_{ht} will be updated by evolutionary selection through a discrete choice model, as will be discussed in the example in section 3. The beliefs we shall consider are specified by a cumulative probability function as follows

$$F_{ht}(x) = \text{Prob}\{E_{ht}[q_{t+1}] < x\} = \text{Prob}\{p^* + f_{ht} + y < x\} = F_y(x - p^* - f_{ht}), \tag{18}$$

where F_y is the cumulative probability function of the IID dividend process and

$$f_{ht} = f_h(x_{t-1}, x_{t-2}, ..., x_{t-L}). \tag{19}$$

Recall that $x_s = p_s - p^*$. I.e. we are following BH (1998) in restricting the belief space to consist of common beliefs on the distribution of y and deterministic beliefs on p_{t+1} of the form $p^* + f_{ht}$ where f_{ht} is a function of L lags of past deviations from the "fundamental" p^*. This specification seems adequate for an initial exploration of the impact of PCC's upon ABS dynamics. The special case of no PCC's and exogenously determined R that was treated in BH (1998) is nested within our general framework.

Example

Up to now we have not seen a clear role in our models that PCC's play in forcing agreement of beliefs, if any such force exists. We sketch here an example of two PCC's that clearly exposes such a role. Let $S_1 = (-\infty, a_1)$ and let S_2 be the complement of S_1. At date t, let type $h = 1[h = 2]$ believe all support of $q_{t+1} = p_{t+1} + y_{t+1}$ is on $S_1[S_2]$. We consider subcases as follows. First, suppose the model is OG as in (1) above with budget sets for young and old as follows

$$\begin{aligned} w_y &= p_{0t}z_{h0t} + p_{1t}z_{h1t} + p_{2t}z_{h2t} + c_t \\ &= (p_{1t} + p_{2t})z_{h0t} + p_{1t}z_{h1t} + p_{2t}z_{h2t} + c_t \\ &= p_{1t}\omega_{h1t} + p_{2t}\omega_{h2t} + c_t, \end{aligned} \tag{20}$$

$$c_{t+1} = w_o + q_{t+1}1[q_{t+1} \in S_1]\omega_{h1t} + q_{t+1}1[q_{t+1} \in S_2]\omega_{h2t}. \tag{21}$$

Notice that we do not have a riskless bond here. Clearly, since type 1 believes there is no support of q on S_2, if $p_2 > 0$ she will set $\omega_{h2t} < 0$ believing that she has nothing to repay when old. This way she can increase consumption while young to any desired level. This kind of move creates infinite supply of PCC #2 for type 2 and vice versa. The requirement that budget sets be bounded as well as existence of equilibrium will force agreement on supports of the random variable q_{t+1} for this case where utility is increasing in consumption while young.

Second, consider the same case but where there is zero utility for consumption while young. In this case type one might start to indulge in the same operation to increase ω_{h1t} to infinity by decreasing ω_{h2t} to negative infinity. However if $v(.)$ is risk averse (e.g. $v(.)$ is mean-variance) type one will not find it optimal to send ω_{h1t} to plus infinity. In this case we may locate sufficient conditions on $v(.)$ for finite demands even though the two types do not agree on the support of q_{t+1}.

Third, suppose we now add a risk free bond, but still assume there is zero utility from consumption while young. Consumption at t+1 when old is now given by

$$c_{t+1} = w_o + R_t b_{ht} + q_{t+1} 1[q_{t+1} \in S_1]\omega_{h1t} + q_{t+1} 1[q_{t+1} \in S_2]\omega_{h2t}. \tag{22}$$

Type one can now send ω_{h2t} to negative infinity at date t and purchase bonds long to receive $R_t b_{ht}$ while old. Type one believes that

$$q_{t+1} 1[q_{t+1} \in S_2] = 0, \tag{23}$$

so she believes it will cost her zero to pay back the "loan" $\omega_{h2t} < 0$ at date $t+1$. Obviously if $R_t > 0$ she can consume an infinite amount with no variance at date $t+1$ by scaling up this operation.

We have said enough at this point to convince the reader that trading of multiple securities quite easily forces agreement on supports of future price and earnings random variables. However, this alone does not automatically force agreement on the exact values of those random variables. These kind of examples are closely related to the classical conditions for boundedness of budget sets and existence of temporary equilibria in the literature reviewed by Grandmont (1982).

3 A Simple Example

In this section, we work out a special case with 3 belief types and 2 PCC's to allow direct comparison of the evolutionary dynamics with previous work without the presence of PCC's. Consider the case $u(.) = 0$ so that utility of consumption while young is zero. Let $V[c'] = E[c]' - (a/2)Var[c']$. As we said above we are now in the mean-variance setting of BH (1998). Neglecting terms that do not affect the optimal demand $\boldsymbol{\omega}$ and using the budget equation $w_y = b_{ht} + p_{0t}z_{h0t} + \sum_{i=1}^{n} p_{it}z_{hit}$ for the young the optimization problem for the young may now be written as follows,

$$E_{ht}[\sum_{i=1}^{n} A_{hit}\omega_{hit}] - (a/2)Var[\sum_{i=1}^{n} A_{hit}\omega_{hit})], \tag{24}$$

where $A_{hit} = q_{i,t+1} - R_t p_{it}$. Optimization produces the optimal demand

$$\boldsymbol{\omega_{ht}} = (1/a)V_{ht}^{-1}[\boldsymbol{m}_{ht} - R_t \boldsymbol{p_t}], \tag{25}$$

where $\boldsymbol{m}_{ht}$ is an n-vector with i'th element $E_{ht}[q_{i,t+1}]$, $\boldsymbol{p_t}$ is the n-vector of the prices p_{it} of the PCC's and V_{ht} is the covariance matrix whose elements are

$$s_{ij} = \mathrm{Cov}(q_{i,t+1}, q_{j,t+1}). \tag{26}$$

In general, this covariance matrix as well as the variance Var in (24) depends upon the belief type h. However, for analytical tractability, we will make the simplifying assumption that all belief types h use the same covariance matrix $V_{ht} = V$ of the fundamental belief types. This simplifying assumption is similar to the assumption made in BH 1998 that beliefs on the variance σ^2 of $q_{t+1} = p_{t+1} + y_{t+1}$ is the same for all belief types. Equilibrium in each PCC market requires that

$$\sum_{h=1}^{H} n_{ht}\boldsymbol{\omega_{ht}} = z_s \boldsymbol{e}. \tag{27}$$

Inserting the optimal demand (25), with $V_{ht} = V$, into the equilibrium equation (27) and multiplying by the risk aversion coefficient a yields

$$\sum_{h=1}^{H} n_{ht} V^{-1}[E_{ht}[\boldsymbol{q_{t+1}}] - R_t \boldsymbol{p_t}] = a z_s \boldsymbol{e}. \tag{28}$$

Multiplying (28) with the covariance matrix V and solving for the vector of PCC equilibrium prices with the gross rate of return of the risk free asset fixed at $R_t = R$ yields

$$R\boldsymbol{p_t} = \sum_{h=1}^{H} n_{ht} E_{ht}[\boldsymbol{q_{t+1}}] - a z_s V \boldsymbol{e}. \tag{29}$$

Once we specify how the fractions n_{ht} evolve over time, we have a well specified dynamical system for each array of PCC's. But let us first make an important observation how the dynamics of the n PCC prices in (29) are related to the dynamics of the equilibrium price of the underlying asset 0. We claim that summing up the n market equilibrium equations for PCC's in (29) exactly yields an equilibrium equation for the asset 0, given by

$$Rp_{0t} = \sum_{h=1}^{H} n_{ht} E_{ht}[p_{t+1} + y_{t+1}] - a\sigma^2 z_s. \tag{30}$$

This follows immediately from the following three observations: (i) the price of the asset 0 is the sum of the prices of the n PCC's; (ii) by definition (13) it follows that $\sum_{i=1}^{n} E_{ht}[q_{i,t+1}] = E_{ht}[q_{t+1}] = E_{ht}[p_{t+1} + q_{t+1}]$; (iii) the sum of all elements of the covariance matrix V equals $\sigma^2 \equiv \sigma^2_{p_{t+1}+y_{t+1}}$. At first sight, it thus seems that summing up all equilibrium equations for the PCC's in (29) exactly leads to the BH-model (30) with one risky asset and no PCC's. But this is *not* true in general, since the fractions n_{ht} in the world with PCC's in (29) in general are *not* the same as the fractions in a world (30) without PCC's.

Let us now discuss evolution of belief types over time. BH 1997 have proposed an endogenous evolutionary updating of trading or forecasting strategies by a discrete choice model, where the fractions are given by

$$n_{ht} = \frac{exp(\beta U_{h,t-1})}{Z_{t-1}}, \qquad Z_{t-1} = \sum_{h=1}^{H} exp(\beta U_{h,t-1}), \tag{31}$$

where $U_{h,t-1}$ is the evolutionary fitness measure and Z_{t-1} is a normalization factor in order for the fractions n_{ht} to add up to 1. The crucial feature of (31) is that the higher the fitness of trading strategy h, the more traders will select strategy h. The parameter β in (31) is called the *intensity of choice*, measuring how sensitive the mass of traders is to selecting the optimal prediction strategy. Discrete choice models can be derived from a random utility model, where all agents observe the fitness measure with an error, applying a law of large numbers. The intensity of choice β is inversely related to the variance of the noise term. The extreme case $\beta = 0$ corresponds to the case of infinite variance noise, so that differences in fitness cannot be observed and all fractions (31) will be fixed over time and equal to $1/H$. The other extreme case $\beta = +\infty$ corresponds to the case without noise, so that the deterministic part of the fitness can be observed perfectly and in each period, *all* traders choose the optimal forecast. An increase in the intensity of choice β represents an increase in the degree of rationality w.r.t. evolutionary selection of trading strategies.

A natural candidate for evolutionary fitness is accumulated *realized profits*, which in the BH world without PCC's is given by

$$U_{ht} = (p_{0t} + y_t - Rp_{0,t-1})\frac{E_{h,t-1}[p_{0t} + y_t - Rp_{0,t-1}]}{a\sigma^2} + wU_{h,t-1}, \tag{32}$$

where $0 \leq w \leq 1$ is a *memory* parameter measuring how fast past realized fitness is discounted for strategy selection. In a world with n PCC's, realized profits for type h are given by

$$\pi_{ht} = \sum_{i=1}^{n} \omega_{hit}(q_{i,t} - Rp_{i,t-1}). \tag{33}$$

A natural fitness measure in a heterogeneous world with n PCC's is therefore

$$U_{ht} = \pi_{ht} + wU_{h,t-1}. \tag{34}$$

We are now ready to explore a specific example.

Example: Three Belief Types, two PCC's

Let there be two PCC's with $S_1 = p_t + y_t < p^* + B$ and S_2 equal to the complement of S_1. We may vary B as a bifurcation parameter as well as other parameters. The covariance matrix is 2x2 with elements

$$s_{ij} = \mathrm{Cov}(q_{t+1}1[q_{t+1} \in S_i], q_{t+1}1[q_{t+1} \in S_j]). \quad (35)$$

We assume all belief types at each date t are of the form

$$E_{ht}[q_{t+1}] = f_{ht} + E_t[y_{t+1}], \qquad f_{ht} = p^* + f_h(x_{t-1}, ..., x_{t-L}) \quad (36)$$

where there is agreement on the cumulative distribution function of y_{t+1} but there may be disagreement on f_{ht} which is of the form

$$f_{ht} = p^* + f_h(x_{t-1}, ..., x_{t-L}). \quad (37)$$

In order to get some understanding of the impact of introducing PCC's we study a simple case with three very simple belief types (cf. BH (1998, p. 1258, Example 4.2.1). Type 1 is fundamentalist with $f_{1t} \equiv 0$, type 2 is constant upward bias with $f_{2t} = b_2 > 0$, and type 3 is (opposite) downward bias with $f_{3t} = b_3$; we will focus on the symmetric case where $b_2 = -b_3 = b > 0$. The main questions that will be addressed here by numerical simulations are: *What is the impact on bifurcation values in the no PCC case when we add the two PCC's?*, and in particular *Do bifurcations towards instability occur "sooner" when PCC's are present or "later"?*

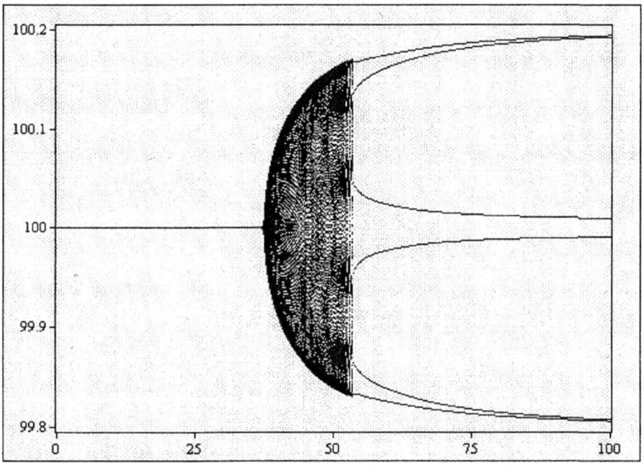

Fig. 1. BH-model with 3 types and no PCC's. Bifurcation diagram w.r.t. the intensity of choice β, $0 \leq \beta \leq 100$, with the other parameters fixed at $R = 1.01$, $\bar{y} = 1$, $z_s = 0$, $b_2 = 0.2$, $b_3 = -0.2$ and $w = 0$. A Hopf bifurcation of the fundamental steady state $p^* = 100$ occurs at $\beta = 37.5$.

Figure 1 shows a bifurcation diagram w.r.t. the intensity of choice parameter β for the three types world without PCC's. A Hopf bifurcation of

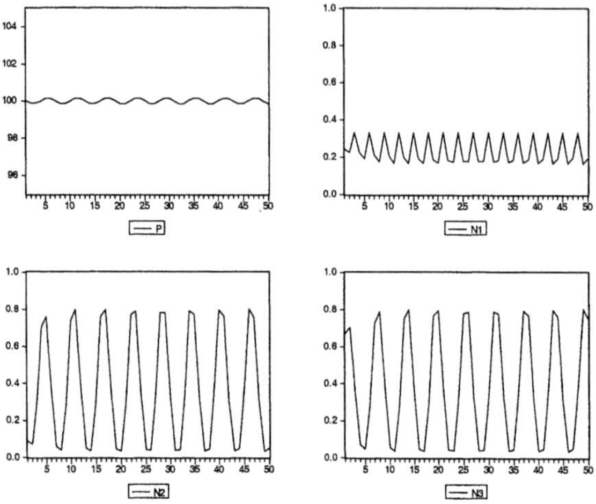

Fig. 2. BH-model with 3 types and no PCC's; (quasi-)periodic time series for $\beta = 50$, $R = 1.01$, $\bar{y} = 1$, $z_s = 0$, $b_2 = 0.2$, $b_3 = -0.2$ and $w = 0$. Fluctuations in asset prices p_t and fraction n_{1t} of fundamentalists are relatively small; fluctuations in fractions n_{2t} of optimists and fractions n_{3t} of pessimists are relatively large.

the fundamental steady state $p^* = 100$ occurs at $\beta = 37.5$. For $\beta < 37.5$ the fundamental steady state is stable; for $\beta > 37.5$ the fundamental steady state is unstable and periodic and quasi-periodic asset price fluctuations arise, as illustrated in Figure 2. For $\beta > 55$ a stable 6-cycle arises. For the given belief parameters and fundamental parameters, fluctuations in asset prices p_t are relatively small. Fluctuations of fractions n_{1t} of fundamentalists are relatively small compared to fluctuations in fractions n_{2t} of optimists and fractions n_{3t} of pessimists.

Figure 3 shows the same bifurcation diagram w.r.t. the intensity of choice parameter β for the three types world with PCC's. These simulations suggests that the fundamental steady state is *unstable* for all β-values. For $0 < \beta < 25$ prices converge to a stable 2-cycle as illustrated in Figure 4; for $\beta > 25$ prices converge to a stable 6-cycle as illustrated in Figure 5. This numerical example suggests that in this 3-type world, the introduction of PCC's is *destabilizing.* Another remarkable fact suggested by Figures 4 and 5 is that the fluctuations in asset prices p_t are relatively small and of the same order of magnitude as in the case without PCC's, whereas fluctuations of prices p_{1t} and p_{2t} of the two PCC's are much larger than the fluctuations in the asset price p_t. This seems in accordance with real markets, where volatility of prices of derivatives is typically higher than volatility of the prices of the underlying asset.

Finally, we investigate the role of the memory parameter w in the evolutionary fitness measure. It is sometimes argued that more memory in the fitness measure should stabilize price fluctuations and force prices to the

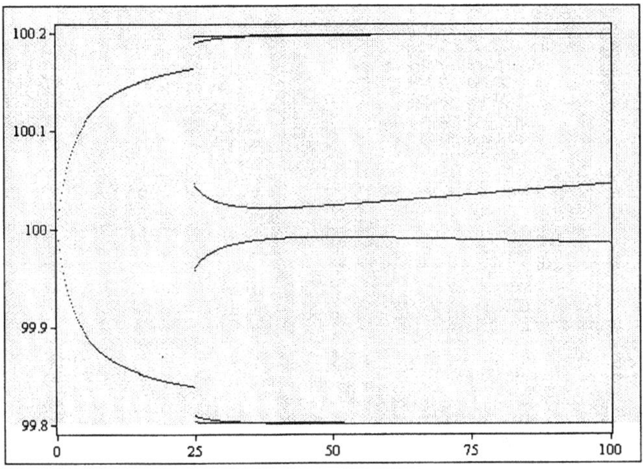

Fig. 3. BH-model with 3 types and 2 PCC's. Bifurcation diagram w.r.t. the intensity of choice β, $0 \leq \beta \leq 100$, with the other parameters fixed at $R = 1.01$, $\bar{y} = 1$, $z_s = 0$, $b_2 = 0.2$, $b_3 = -0.2$, $B = 0$ and $w = 0$. For all positive β-values the fundamental steady state seems to be unstable.

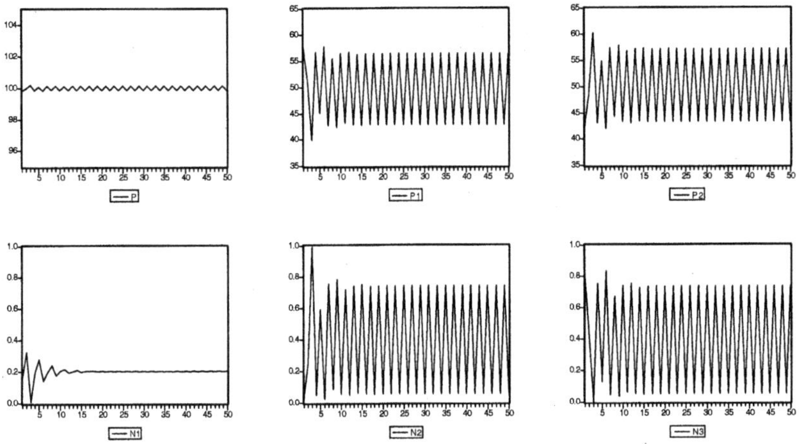

Fig. 4. BH-model with 3 types and 2 PCC's. Convergence to a 2-cycle for $\beta = 10$ with the other parameters fixed at $R = 1.01$, $\bar{y} = 1$, $z_s = 0$, $b_2 = 0.2$, $b_3 = -0.2$, $B = 0$ and $w = 0$. Fluctuations in asset prices p_t are relatively small and of the same order of magnitude as in the case without PCC's. Fluctuations of prices p_{1t} and p_{2t} of the two PCC's are much larger than fluctuations in the asset price p_t. Fluctuations of the fraction n_{1t} of fundamentalists are relatively small; fluctuations in fractions n_{2t} of optimists and fractions n_{3t} of pessimists are relatively large.

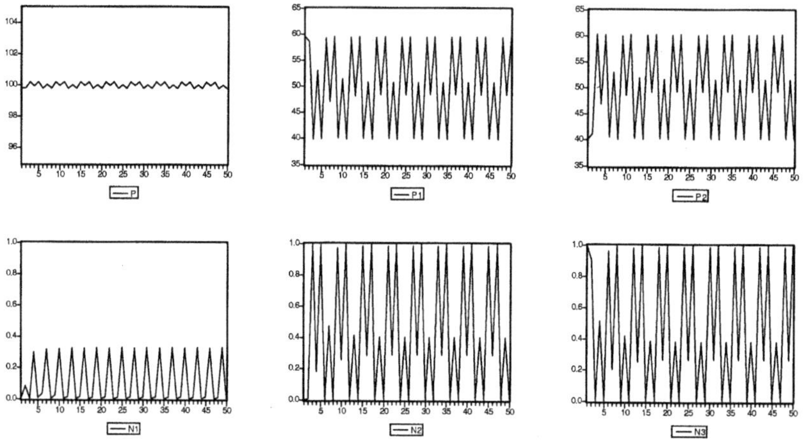

Fig. 5. BH-model with 3 types and 2 PCC's. Convergence to a 6-cycle for $\beta = 50$ with the other parameters fixed at $R = 1.01$, $\bar{y} = 1$, $z_s = 0$, $b_2 = 0.2$, $b_3 = -0.2$, $B = 0$ and $w = 0$. Fluctuations in asset prices p_t are relatively small and of the same order of magnitude as in the case without PCC's. Fluctuations of prices p_{1t} and p_{2t} of the two PCC's are much larger than fluctuations in the asset price p_t. Fluctuations of the fraction n_{1t} of fundamentalists are relatively small; fluctuations in fractions n_{2t} of optimists and fractions n_{3t} of pessimists are relatively large.

fundamental steady state. BH (1999) have presented a 2-type example, with costly fundamentalists versus trend followers, where this is *not* true and an increase in memory can actually destabilize price fluctuations. Figure 6 shows a bifurcation diagram w.r.t to the memory parameter w in the three type world with two PCC's. For $w = 0$, without memory, the fundamental steady state is unstable and the system has a stable 2-cycle as could be seen already for $\beta = 10$ in the bifurcation diagram of Figure 3. As the memory parameter w increases, price fluctuations become more complicated and chaotic price fluctuations arise for w close to 1. Figure 7 shows a strange attractor for $w = 0.99$ with corresponding chaotic time series shown in Figure 8. Our numerical simulations thus suggest that the introduction of PCC's may *destabilize* asset price fluctuations. In future work we hope to get more analytical insight into the exact economic mechanism leading to this numerical observation. But based on our numerical simulations, the main intuition may be this. Consider a "correct bull", i.e. a type who was bullish at date $t-1$ in using biased belief type $b_h = 0.2$ to forecast prices at t and turned out to be more right than the rest of the community in the sense that prices turned out to be above the fundamental. If there were no PCC's this correct bull would have been constrained to invest the same amount in PCC_1 and PCC_2 whereas the availability of PCC's allowed the correct bull to take a larger position in the upside PCC_2 based on his upside belief as well as possibly

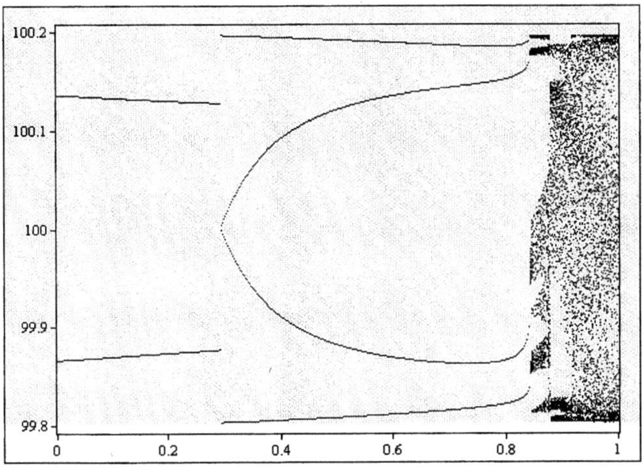

Fig. 6. BH-model with 3 types and 2 PCC's. Bifurcation diagram w.r.t. the weight factor w, $0 \leq w \leq 1$, of the evolutionary fitness measure, with the other parameters fixed at $\beta = 10$, $R = 1.01$, $\bar{y} = 1$, $z_s = 0$, $b_2 = 0.2$, $b_3 = -0.2$, $B = 0$ and $w = 0$. An increase of memory leads to a 'rational route to randomness'.

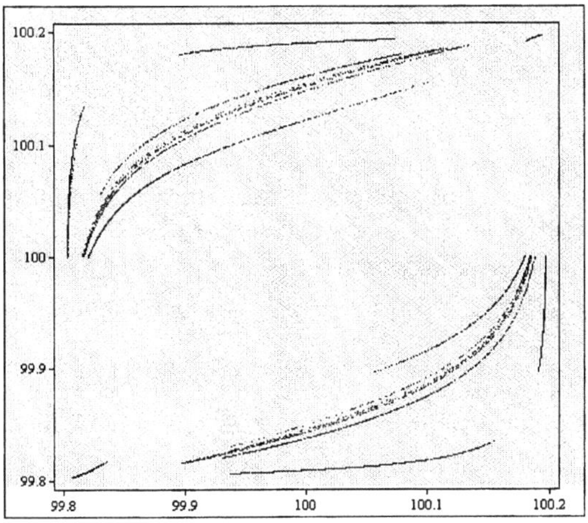

Fig. 7. BH-model with 3 types and 2 PCC's. A strange attractor for high memory $w = 0.99$, with the other parameters fixed at $R = 1.01$, $\bar{y} = 1$, $z_s = 0$, $b_2 = 0.2$, $b_3 = -0.2$, $B = 0$ and $w = 0$.

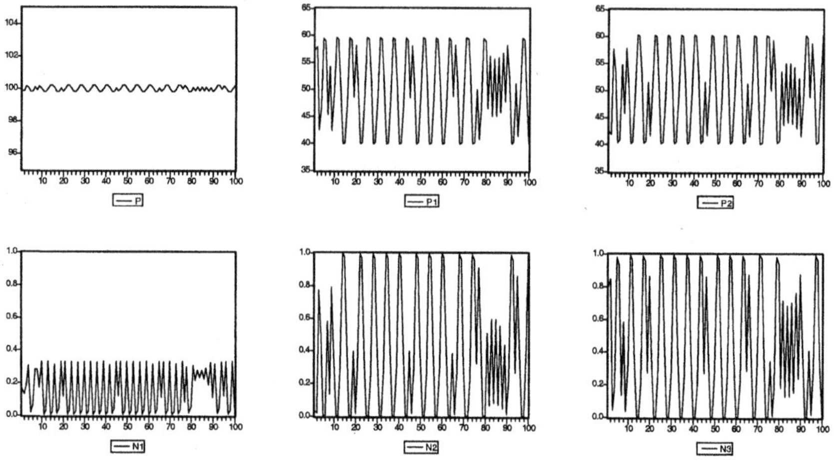

Fig. 8. BH-model with 3 types and two PCC's; chaotic time series corresponding to the strange attractor in Figure 7 for high memory parameter $w = 0.99$, with the other parameters fixed at $R = 1.01$, $\bar{y} = 1$, $z_s = 0$, $b_2 = 0.2$, $b_3 = -0.2$, $B = 0$ and $w = 0$. Fluctuations in asset prices p_t are relatively small, whereas the amplitude of fluctuations of prices p_{1t} and p_{2t} of the two PCC's is much larger. Fluctuations of the fraction n_{1t} of fundamentalists are relatively small compared to fluctuations in fractions n_{2t} of optimists and fractions n_{3t} of pessimists.

borrowing PCC_1 by going negative in it. It seems plausible that this extra freedom would tend to make profits turn out to be larger for correct bulls (and similar reasoning would apply to correct bears). When profit differences are larger, an evolutionary world where strategy selection is based upon past performance becomes more *unstable.*

If this admittedly loose speculation is right, at the minimum, the question of the contribution of the presence of PCC's and other derivatives to the stability or instability of the markets is shown to be a subtle one indeed. Of course one should investigate this same issue in the context of noisy rational expectations models where derivatives can play an additional "Hayekian/Grossmanian" role in transmitting information (Grossman (1989), de Fontnouvelle (2000)) before drawing any general conclusions. But we dare to speculate further that introduction of PCC type securities into a de Fontnouvelle dynamic noisy rational expectations framework may allow de Fontnouvelle's analog of "correct information bulls" discussed here to use PCC's to take larger positions and thus garner larger profits when correct. If indeed PCC's can be used to garner larger profits on the part of a belief type when it is on the "correct side" of the market, this will generate larger profits for that type, attracting copycats at a faster rate and, hence, possibly contribute to less stability of the markets rather than more stability of the markets.

4 General Comments and Discussion

This paper has sketched an approach to studying the impact of introducing additional securities into the ABS framework of BH (1998) upon the evolutionary dynamics studied by BH. This was done in an attempt to deal with one type of criticism of general evolutionary approaches with heterogeneous beliefs. Here are some other criticisms. First, the device of attaching higher probability of playing a strategy (e.g. trading on belief h at date t) based upon how well it performed in the past relative to the other available strategies runs the risk of reproducing the rather dumb type of "cobweb" behavior which was criticized by the original writers on rational expectations. Increasing β makes agents more responsive to past differences in relative performance which accentuates unstable "cobweb-like" behavior. To put it another way, the use of β as a tuning parameter that stands as a surrogate for a "dial of rationality" may be misplaced. A possible remedy might be to increase the strategy space to include condition-action beliefs of the form, "if h did well last period, believe h' this period, thinking most other traders will be believing and acting on h". One can easily add strategies of this form, attach performance indices to each of these new strategies based upon past profits that would have been garnered by such strategies, introduce a discrete choice model over this enlarged set of strategies and proceed as in BH (1998) but with a larger space. Of course there is no limit to the number of conditioning of this type that are placed before each action. But presumably the larger the number of such condition-action strategies that lie in the belief space the more plausible it might be to use β as a proxy for a "dial of rationality." One could also endogenize β, which may be a slow variable. Traders may become more sensitive to selecting the optimal prediction strategy when they are dissatisfied, that is at a low level of realized profits. This relates to the literature on bounded rationality that assumes that people economize on their cognitive activities when they are satisfied.

Second, we have seen a general tendency for a belief to do well if it puts the trader on the opposite side of the market from where the masses are and are moving towards. I.e. if most of the traders are bearish [bullish] today, a trader will do well to be on the bullish [bearish] side today, provided that the masses are not even more bearish [bullish] tomorrow. This is so because relative to the dividends that will be captured next period the price of the asset today is cheap [expensive today so it can be turned into a device to implement a cheap loan by borrowing it today, cashing it out today, investing the proceeds in some other asset, closing out the position tomorrow by paying out the relatively cheap dividends and repurchasing the asset tomorrow when, hopefully, it will be relatively cheap]. Any parameter change that magnifies a trader's move in the right direction (e.g. a decrease in risk aversion, a decrease in perceived variance) will tend to produce increased profits. This reasoning suggests that if PCC's can be used to reduce perceived variance, their presence will lead to more aggressive postures by traders. This force may cause the presence of PCC's to lead to more complex dynamics rather than less complex dynamics.

Finally, a third criticism revolves around what would happen if the time scale of the analysis was more appropriate to real world markets. I.e. might not patient traders with many time periods "smooth out" any dynamics so that in the end, in analogy with Levine and Zame (2000), much of the potential for endogenous complex dynamics vanishes?

Another major criticism of BH type theory is the lack of attention to wealth dynamics of traders who do a superior job of switching across the space of beliefs to garner profits over time. Wealthier traders should loom larger in asset price formation than "average" traders. Formidable analytical challenges must be faced because of the book keeping requirements that require tracking traders according to their switching histories. This makes the "state space" of the underlying dynamical system multiply up in dimension as time proceeds. An analytically tractable compromise is this. At each date t, one could attach a performance index W_{ht} to a belief h that gives the total wealth accumulated up to date t by trading on belief h from $s = 0$ to $s = t$. These wealth-based performance indices could be used in place of the utilities in the discrete choice systems of BH (1998). This would correspond to the performance records seen in actual mutual fund advertisements where an initial investment of size W_0 at $s = 0$ is tracked by the accumulated wealth at each date t. This record is typically displayed by a graph of W_t against t. The BH (1998) model treats investors as using a discrete choice model to choose at each date t, amongst $h = 1, 2, ..., H$ such "mutual funds" (e.g. "beliefs").

The fractions of belief types in BH (1998) possess dynamics which adds dimensions to the minimal characterization of the "state vector" of the economy at date t. This creates tractability problems if there are very many types, say four or more. However Large Type Limit theory developed by Brock, Hommes, and Wagener (2001) drastically reduces the dimension of the state space and removes many of the tractability problems. Hence, at this date two types of evolutionarily adaptive belief systems are analytically tractable: (i) Those with a very small number of types, (ii) Those with a very large number of types. Much more work within the themes *Equilibrium, disequilibrium* and *dynamics* of this Festschrift needs to be done for a better understanding of adaptive evolutionary systems and their empirical and experimental relevance to economics and finance.

References

1. Brock, W.A. (1990), Overlapping generations models with money and transactions costs, In: Friedman, B., Hahn, F., (eds.), *Handbook of Monetary Economics, Volume I*, Amsterdam: North Holland, pp. 263-298.
2. Brock, W.A., (1997), Asset Price Behavior in Complex Environments, in: Arthur, W.B., Durlauf, S.N., and Lane, D.A., eds., *The Economy as an Evolving Complex System II*, Addison-Wesley, Reading, MA, 385–423.

3. Brock, W.A. and de Fontnouvelle, P., (2000), Expectational diversity in monetary economies, *Journal of Economic Dynamics and Control* 24, 725-759.
4. Brock, W.A., and Hommes, C.H., (1997) A rational route to randomness, *Econometrica* 65, 1059-1095.
5. Brock, W.A., and Hommes, C.H., (1998), Heterogeneous beliefs and routes to chaos in a simple asset pricing model, *Journal of Economic Dynamics and Control* 22, 1235-1274.
6. Brock, W.A., Hommes, C.H. and Wagener, F.O.O. (2001), Evolutionary dynamics in financial markets with many trader types, *CeNDEF working paper* 01-01, University of Amsterdam.
7. Brock, W.A. and Kleidon, A., (1992) Periodic market closure and trading volume: a model of intraday bids and asks, *Journal of Economic Dynamics and Control* 16, 451-489.
8. Evans, G.W., and Honkapohja, S., (2001) *Learning in macroeconomics*, Princeton University Press.
9. de Fontnouvelle, P., (2000), Information dynamics in financial markets, *Macroeconomic Dynamics* 4, 139-169.
10. Gaunersdorfer, A. and Hommes, C.H., (2000), A nonlinear structural model for volatility clustering, *CeNDEF working paper* 00-02, University of Amsterdam.
11. Grandmont, J.-M. (1982), Temporary general equilibrium theory, In: Arrow, K.,J., and Intrilligator, M.D., (eds.), *Handbook of Mathematical Economics, Volume II*, Amsterdam: North Holland, Chapter 19, pp. 879-922.
12. Grossman, S.J., (1989), *The informational role of prices*, MIT Press, Cambridge.
13. Guesnerie, R. (2001), *Assessing rational expectations: sunspot multiplicity and economic fluctuations*, MIT Press: Cambridge, Mass.
14. Hommes, C.H., (2001), Financial markets as nonlinear adaptive evolutionary systems, *Quantitative Finance 1*, 149-167.
15. Kurz, M., (ed.), (1997), *Endogenous economic fluctuations: studies in the theory of rational beliefs*, Springer-Verlag: New York.
16. Levine, D. and Zame, W., (2000), Does market incompleteness matter?, Department of Economics, UCLA.
17. Magill, M. and Quinzi, M. (1996), *Theory of incomplete markets*, MIT Press: Cambridge, Mass.
18. Magill, M. and Shafer, W., 1991, Incomplete markets, In: Hildenbrand, W. and Sonnenschein, H., (eds.), *Handbook of Mathematical Economics, Volume IV*, Amsterdam: North Holland, Chapter 30.

Why Are There Nowadays only a Few Articles on Matrices in Economic Journals?

Dave Furth

Deptartment of Economics, Faculty of Law, University of Amsterdam

Abstract. Shortly after World War II, many articles on matrices were published in economic journals. This chapter is an incomplete review of some of the applications of the theory of matrices to the stability of the equilibria of both: exchange economies and Cournot oligopolies. When in the seventies an '(almost) everything is possible' result was derived, there was a shift in interest from linear(ized) models towards non–linear ones and economists were not longer interested in stability properties. When economists are not longer interested in stability, they do not write articles on matrices anymore. That is why nowadays there are only a few articles on matrices in economic journals.

Keywords: General Equilibrium Theory, Cournot Oligopoly, Stability

1 Introduction

Shortly after World War II, many articles on matrices were published in economic journals. It was also around that time that Samuelson (1941,1942) described the tâtonnement process as a set of differential equations, while Arrow and Debreu (1954) and McKenzie (1954) gave a rigorous proof of the existence of an equilibrium in an exchange economy. The main questions in economic theory at that time were:

- *Does an equilibrium always exist?*
- *If an equilibrium exists, is it unique?* and:
- *If an equilibrium exists, is it stable?*

Only in Chicago and its suburbs, there still live economists believing the answer to each of these questions is: *"Yes."*

In the above stability refers to the stability, in the sense of Liapunov, of a stationary point of some adjustment process, in general equilibrium theory mostly related to the adjustment of prices in the Walrasian tâtonnement process. A stationary point of a dynamical system is stable whenever the Jacobi matrix of the dynamical system,[1] calculated at the stationary point, is a stable matrix. In mathematics the conditions for a stable matrix are the Routh–Hurwitz conditions. Unfortunately these conditions are difficult

[1] Many authors call the Jacobi matrix: 'the Jacobian'. I use the term 'Jacobian' only for the determinant (when it exists) of the Jacobi matrix.

to interpret in an economic context. Therefore economists have developed their own characterizations for stable matrices. Let me mention a few of the most remarkable articles on this subject: Samuelson (1941,1942), Metzler (1945), Debreu and Herstein (1953), Morishima (1952), Arrow and Hurwicz (1958), Hahn (1958), Negishi (1958), Arrow, Block and Hurwicz (1959) and McKenzie (1960).

Theocharis (1959) was "On the stability of the Cournot solution on the oligopoly problem." Here the questions were asked with respect to the stability of the equilibrium of a Cournot oligopoly model. Among the reactions on this article there were papers by Fisher (1961), McManus and Quandt (1961), Hahn (1962), Bishop (1962), and Okuguchi (1964). There is also a paper by Seade (1980) on instability of the Cournot solution.

Except for the paper of Seade, all these papers deal with properties of matrices, that explain the stability of an equilibrium. The equilibrium of an exchange economy is a zero of the excess demand function. Stability follows from the stability of the Jacobi matrix of this function. So economists want to know whether or not this excess demand function is 'well behaved'. Sonnenschein (1972, 1973) was one of the first to question this. Debreu (1974) and Mantel (1974) proved that *any* continuous function, that (i) is homogeneous of degree zero in prices, (ii) fulfils Walras' Law and (iii) becomes unbounded when prices go to zero, may be the excess demand function of an exchange economy. Similar results were obtained for oligopoly by Corchón and Mas–Colell (1996).

It follows from these results, that pathological cases, for instance the one given by Scarf (1960), can not be excluded on economic grounds. This means that there is room for 'chaos' and 'strange attractors' in general equilibrium theory, see for instance Deneckere and Pelikan (1986), Bala and Majumdar (1992), Weddepohl (1995), Goeree, Hommes and Weddepohl (1997) and the second chapter of Tuinstra (1999), as well as in oligopoly theory, see Rand (1978), Dana and Montrucchio (1986), Puu (1991), Bischi, Mammana and Gardini (2000) and the fifth chapter of Tuinstra (1999).

The set–up of this chapter is as follows. First in section 2, I deal with the stability question for exchange economies. Section 3 does the same for the stability of the Cournot oligopoly model. In the final section, I return to the question: "Why are there nowadays only a few articles on matrices in economic journals?"

2 Stability in an Exchange Economy

An 'Exchange Economy' is defined by:

(i) a set of agents $N := \{1, \ldots, n\}$ with $n \in \mathbb{N}$, $n \geq 2$;
(ii) for each agent $i \in N$ a bundle of consumption goods $\omega^i \in \mathbb{R}^\ell_+$;
(iii) for each agent $i \in N$ a *utility function* $u^i \colon \mathbb{R}^\ell_+ \to \mathbb{R}$.

There are $\ell \in \mathbb{N}$ commodities in this economy. For each agent $i \in \mathbb{N}$: $\mathbb{R}^\ell_+$ is his *consumption set*; and: ω^i are his *initial endowments*.

An 'assignment' for agent $i \in N$ is a vector $x^i = (x^i_1, \dots, x^i_\ell) \in \mathbb{R}^\ell_+$.

A '(feasible) allocation' is a vector $(x^1, \dots, x^n) \in \mathbb{R}^{n\ell}_+$ such that

$$\sum\nolimits_{i=1}^{n} x^i = \sum\nolimits_{i=1}^{n} \omega^i.$$

As it is not the intention to give the most general treatment of an exchange economy, the following (rather restrictive) assumptions will be made.

I. For all $i \in N$ the utility function u^i is continuous, strictly quasi concave and monotone.

II. There is a $M > 0$, such that for all $i \in N$: $0 < \|\omega^i\| < M$.

III. $\sum_{i=1}^{n} \omega^i > 0$.

Let $x^i(0)$ and $x^i(1) \in \mathbb{R}^\ell_+$. Define for $0 < \lambda < 1$: $x^i(\lambda) := (1-\lambda)x^i(0) + \lambda x^i(1)$. The function u^i is 'strictly quasi concave' when for all $0 < \lambda < 1$: $u^i(x(\lambda)) > \min(u^i(x(0)), u^i(x(1)))$. The function u^i is 'monotone' if $x^i(1) \geq x^i(0)$ implies $u^i(x^i(1)) \geq u^i(x^i(0))$. $\|\omega^i\| := \sqrt{\sum_{j=1}^{\ell} (\omega^i_j)^2}$ is the 'Euclidean norm' of the initial endowments. By **II** the initial endowments are bounded. By **III** the total endowments are positive, that is each of the ℓ commodities is actually present in the exchange economy.

Let $p = (p_1, \dots, p_\ell) \in \mathbb{R}^\ell_+$, $p \neq 0$ be a set of prices. Let $p \cdot x := \sum_{j=1}^{\ell} p_j x_j$ be the usual improduct. The *budget set* of agent $i \in N$ is given by: $\mathcal{B}^i(p) := \{x \in \mathbb{R}^\ell_+ \mid p \cdot x \leq p \cdot \omega^i\}$.

Under the assumptions the maximalization problem: $\max\{u^i(x) \mid x \in \mathcal{B}^i(p)\}$ has a unique solution: $x^i = \varphi^i(p)$.

The function φ^i is the *demand function* of agent $i \in N$. His *excess demand* is given by: $\zeta^i(p) := \varphi^i(p) - \omega^i$. 'Total demand' is given by: $\sum_i \varphi^i(p)$ and 'total excess demand' by: $\zeta(p) = \sum_i \zeta^i(p)$.

The following properties follow from our assumptions.

Continuity: ζ is continuous on $\mathbb{R}^\ell_{++}$;

Homogeneity: for all $p \in \mathbb{R}^\ell_+$ and all $\lambda > 0$: $\zeta(\lambda p) = \zeta(p)$;

Walras' Law: for all $p \in \mathbb{R}^\ell_+$: $p \cdot \zeta(p) = 0$;

Boundedness: when for all $k \in \mathbb{N}$: $p(k) \in \mathbb{R}^\ell_{++}$ and when $\lim_{k\to\infty} p(k) = p^* \in \mathbb{R}^\ell_+$ with $p^*_j = 0$, then $\zeta_j(p(k)) \to \infty$.

Continuity follows from the utility function being *strictly* quasi concave. Homogeneity is obvious from the definition of the budget set. As the utility functions are assumed to be monotone, utility is maximized at the boundary of the budget set where $p \cdot x = p \cdot \omega^i$. From this Walras' Law follows. Finally boundedness follows, among other things, from assumption **III**.

A *Walrasian Equilibrium* is:

- a feasible allocation $(\overline{x}^1, \dots, \overline{x}^n) \in \mathbb{R}^{n\ell}_+$ and:
- a price vector $\overline{p} \in \mathbb{R}^\ell_+$,

such that for all $i \in N: \overline{x}^i = \varphi^i(\overline{p})$.

Notice that for an equilibrium $\zeta(\overline{p}) = 0$. Because of 'homogeneity' one may restrict prices to the price simplex:

$$p \in \Delta^{\ell-1} := \{p \in \mathbb{R}^\ell_+ \mid \sum_{j=1}^{\ell} p_j = 1\}.$$

Proposition When for an excess demand function $\zeta\colon \Delta^{\ell-1} \to \mathbb{R}^\ell$ 'continuity', 'homogeneity' and 'Walras' Law' holds, there exists a Walrasian Equilibrium.

This proposition gives an answer to the first of the three questions of the introduction. The Walrasian equilibrium is fully known from the price vector $\overline{p} \in \Delta^{\ell-1}$. When 'boundedness' holds, the prices in equilibrium are positive and the set of equilibrium prices is given by $\{p \in \Delta^{\ell-1} \mid \zeta(p) = 0\}$. When 'boundedness' fails to hold, prices may become zero and therefor some commodities may become free. One could say, this is very uninteresting and leave out these free commodities right from the beginning. But it is possible that there is more than one equilibrium and that different commodities are free in different equilibria. So it is time to investigate the answers to the two remaining questions of the introduction.

First 'stability' will be investigated. In order to study stability, one needs a 'price adjustment process', as stability is only defined with respect to such a process. Here there will be chosen for the following process:

$$\dot{p} = \zeta(p).$$

A dot above a variable indicates the derivative with respect to time t. That is $p = p(t)$ is considered as a function of time: $p\colon \mathbb{R} \to \Delta^{\ell-1}$. The process has been given by a system of differential equations. A solution of these differential equations is the 'adjustment path', or 'orbit': $\{p(t) \in \Delta^{\ell-1} \mid t \in \mathbb{R}\}$. A Walrasian Equilibrium is a rest point of this differential equation where the 'orbit' is just one point. It should be noticed that this is not just some mathematical model. There is a clear economic intuition behind this process that corresponds to a tâtonnement process because prices increase (decrease) when excess demand is positive (negative).

In what follows it is assumed that ζ has continuous derivatives of the first order. Let

$$\mathcal{J}(\zeta(\overline{p})) := \frac{\partial(\zeta_1, \dots, \zeta_\ell)}{\partial(p_1, \dots, p_\ell)}$$

be the Jacobi matrix of the excess demand function, calculated at $\overline{p}$.

An equilibrium $\overline{p} \in \Delta^{\ell-1}$, of the dynamical system $\dot{p} = \zeta(p)$, is called *locally asymptotically stable*, when there exists an open neighbourhood V, in $\Delta^{\ell-1}$, of $\overline{p}$, such that whenever for a certain value $t_0 \in \mathbb{R}$ it holds that $p(t_0) \in V$, then $\lim_{t\to\infty} p(t) = \overline{p}$.

When in this definition, $\Delta^{\ell-1}$ may be chosen as neighbourhood V, the equilibrium is called *(globally) asymptotically stable.*

Before dealing with the stability of an exchange economy, first some results on the stability of matrices are reviewed. In the following $A = (a_{ij})$ will be a $k \times k$–matrix and $h \in \mathbb{R}^k$ is a column vector. A^T is the transpose of the matrix A and h^T is the row vector, that is the transpose of h. A square matrix A is

- *stable*, iff all its eigenvalues have negative real parts;
- *negative definite* (resp. *quasi negative definite*), when for all $h \neq 0$: $h^T A h < 0$ (resp. $h^T(A + A^T)h < 0$);
- a *Hicksian–matrix*, when its $r \times r$ principal minors have sign $(-1)^r$;
- a *Metzler–matrix*, when its diagonal elements are negative and its off diagonal elements are positive.

The following results have been derived in the literature. A is stable if

(i) A is negative definite (resp. quasi negative definite), see Hicks (1939), Samuelson (1941,1942) and Arrow and Hurwicz (1958);

(ii) A has a negative dominant diagonal, that is there exists a $h \in \mathbb{R}^k_+$ such that for $i = 1, \ldots, k$:

$$a_{ii}h_i + \sum_{j=1, j\neq i}^{k} |a_{ij}| h_j < 0,$$

see McKenzie (1960).

(iii) A is a Metzler–matrix and moreover there exists a $h \in \mathbb{R}^k_+$ such that for $i = 1, \ldots, k$:

$$\sum_{j=1}^{k} a_{ij} h_j < 0,$$

see Metzler (1945), Hahn (1958) or Negishi (1958);

When A is a Metzler–matrix, $-A$ is a M(inkowski)–matrix, See Berman and Plemmons (1979, chapter 5). When A is a Metzler–matrix and there exists a $h \in \mathbb{R}^k_+$ such that for $i = 1, \ldots, k$:

$$\sum_{j=1}^{k} a_{ij} h_j < 0,$$

then A is a Hicksian–matrix.

Let $\mathcal{J}(\zeta(\overline{p}))$ be a Metzler–matrix. When the price of one commodity is raised the excess demand for that commodity decreases, while the excess demand for all other commodities increases. Hence all commodities are (gross) substitutes. As this makes the application rather restrictive, Morishima (1952) extended the above result in the following way. Let the set of indices $\{1, \ldots, k\}$ be divided into two disjoint sets, say K and K'. So $K \cup K' = \{1, \ldots, k\}$ and $K \cap K' = \emptyset$. Of the matrix A all diagonal elements are negative. For $i \neq j$ a_{ij} is positive whenever i and j belong to the same subset, while a_{ij} is negative when they belong to different subsets. Applied to the matrix $\mathcal{J}(\zeta(\overline{p}))$ this means that commodities, belonging to the same subset are (gross) substitutes, while commodities, belonging to different subsets are (gross) complements. Let the matrix $D = (d_{ij})$ be defined by:

$$d_{ij} = \begin{cases} 0, & \text{if } j \neq i; \\ 1, & \text{if } j = i \in K; \text{ and:} \\ -1, & \text{if } j = i \in K'. \end{cases}$$

Notice that $D = D^T = D^{-1}$. A is stable if

(iv) DAD is a Metzler–matrix and there exists a $h \in \mathbb{R}^k_+$ such that for $i = 1, \ldots, k$:

$$\sum_{j=1}^{k} d_{ii}a_{ij}d_{jj}h_j < 0.$$

It would be nice if one could prove that the equilibrium $\overline{p} \in \Delta^{\ell-1}$, of the dynamical system $\dot{p} = \zeta(p)$, is locally asymptotically stable, iff the Jacobi matrix $\mathcal{J}(\zeta(\overline{p}))$ is stable, but this is not true. Up to now $p \in \Delta^{\ell-1}$ has been written as a vector from $\mathbb{R}^\ell$. Also $\zeta(p)$ has been considered as a vector in $\mathbb{R}^\ell$. But because of Walras' Law, the mapping $\zeta\colon \Delta^{\ell-1} \to \mathbb{R}^\ell$ is certainly not surjective. It follows that the Jacobi matrix $\mathcal{J}(\zeta(p))$ has at most rank $\ell - 1$, at least one of the eigenvalues of the Jacobi matrix is zero.

Define an economy to be *regular*, if the Jacobi matrix $\mathcal{J}(\zeta(p))$, for all $p \in \Delta^{\ell-1}$, has rank $\ell - 1$.

When $p \in \Delta^{\ell-1}$, one has $\sum_{i=1}^{\ell} p_i = 1$ and $\sum_{i=1}^{\ell} \dot{p}_i = 0$. Let

$$\Pi^{\ell-1} := \{(p_1, \ldots, p_{\ell-1}) \in \mathbb{R}^{\ell-1}_+ \mid 0 \leq \sum_{i=1}^{\ell-1} p_i \leq 1\}$$

and $\tau\colon \Pi^{\ell-1} \to \mathbb{R}^\ell$ be the map that maps $(p_1, \ldots, p_{\ell-1}) \in \Pi^{\ell-1}$ onto $(p_1, \ldots, p_{\ell-1}, 1 - \sum_{i=1}^{\ell-1} p_i) \in \Delta^{\ell-1}$. Consider the projection $proj\colon \mathbb{R}^\ell \to \mathbb{R}^{\ell-1}$, with $proj(p_1, \ldots, p_\ell) = (p_1, \ldots, p_{\ell-1})$. Notice that τ is the inverse of $proj$ restricted to $\Delta^{\ell-1}$.[2]

Let $p \in \Pi^{\ell-1}$, define for $i = 1, \ldots, \ell-1$: $\overline{\zeta}_i(p) := \zeta_i(\tau(p))$, then $\overline{\zeta}\colon \Pi^{\ell-1} \to \mathbb{R}^{\ell-1}$ is a vector field.

Let $p \in \Pi^{\ell-1}$, for a regular economy the matrix $\mathcal{J}(\overline{\zeta}(p))$ has maximal rank, i.e. $\ell - 1$. Except for the eigenvalue zero, the remaining $\ell - 1$ eigenvalues of $\mathcal{J}(\zeta(\tau(p)))$ are the same as the eigenvalues of $\mathcal{J}(\overline{\zeta}(p))$.

Proposition Consider a regular economy. The equilibrium $\overline{p} \in \Delta^{\ell-1}$, of the dynamical system $\dot{p} = \zeta(p)$, is locally asymptotically stable, iff the matrix $\mathcal{J}(\overline{\zeta}(\tau^{-1}(\overline{p})))$ is stable.

[2] In differential topology, the mapping $proj$ is called a 'chart' or a 'system of coordinates'. Its inverse τ is called a 'parametrization'. See Furth (1986, Appendix B).

After the question of stability, next the question of uniqueness of the equilibrium is treated. Let

$$\zeta\colon \Delta^{\ell-1} \to \mathbb{R}^{\ell}$$

again be the excess demand function of an economy. For each $p \in \Delta^{\ell-1}$, $\zeta(p)$ is a vector in $\mathbb{R}^{\ell}$. Hence ζ defines a *vector field* on $\Delta^{\ell-1}$. Remember that solving the differential equation $\dot{p} = \zeta(p)$ is equivalent to finding a curve in $\Delta^{\ell-1}$, such that in each point of the curve, the vector of the vector field is tangent to the curve.

A point $p \in \Delta^{\ell-1}$ is called a *critical point* of the vector field ζ when $\zeta(p) = 0$. A critical point p of the vector field ζ is called *non–degenerate*, when the matrix $\mathcal{J}(\overline{\zeta}(\tau^{-1}(p)))$ is non–singular, that is $\det(\mathcal{J}(\overline{\zeta}(\tau^{-1}(p)))) \neq 0$.

Let $\overline{p}$ be a (non–degenerate) critical point of the vector field ζ. That is $\overline{p}$ is a Walrasian equilibrium. A point $x \in \Delta^{\ell-1}$ belongs to the *stable manifold* $W^s(\overline{p})$ of $\overline{p}$ (resp. the *unstable manifold* $W^u(\overline{p})$ of $\overline{p}$) whenever there is a sequence $(t_n)_{n=-\infty}^{n=\infty} \subset \mathbb{R}$, with $p(t_0) = x$ and $t_n \to \pm\infty$ whenever $n \to \pm\infty$, and such that $\lim_{n\to\infty} p(t_n) = \overline{p}$ (resp. $\lim_{n\to-\infty} p(t_n) = \overline{p}$). In this $p(t)$ is a solution curve of the differential equation. The dimension of the stable manifold is equal to the (Morse) coindex of $\overline{p}$, that is the number of eigenvalues of the Jacobi matrix $\mathcal{J}(\overline{\zeta}(\tau^{-1}(\overline{p})))$, with negative real parts. The dimension of the unstable manifold is equal to the (Morse) index of $\overline{p}$, that is the number of eigenvalues of this matrix with positive real part. Notice that because the critical point $\overline{p}$ is non–degenerate, there is no *centre manifold*, corresponding to zero eigenvalues.

Debreu (1970) showed that regular economies have a finite number of equilibria. Moreover he showed that the regular economies form an open and dense subset of all economies. Hence one may assume that

- the vector field ζ has a finite number of non–degenerate equilibria;

$$\Sigma := \{p \in \Delta^{\ell-1} \mid \zeta(p) = 0\} = \{p^{(1)}, \dots, p^{(m)}\};$$

- for all $i, j = 1, \dots, m$ the stable manifold $W^s(p^{(i)})$ and unstable manifold $W^u(p^{(j)})$ have a normal intersection, that is

$$\dim(W^s(p^{(i)}) \cap W^u(p^{(j)})) = \dim W^s(p^{(i)}) + \dim W^u(p^{(j)}) - (\ell - 1).$$

Under these conditions, the vector field ζ is a Morse–Smale vector field, see Smale (1960,1961). Assume further that

- the vector field ζ has no *cycles*, that is no closed integral curves; and
- the vector field ζ points inwards at the boundary of $\Delta^{\ell-1}$.

Suppose all these conditions hold. Define c_j as the number of $p \in \Sigma$ such that the matrix $\mathcal{J}(\overline{\zeta}(\tau^{-1}(p)))$ has $j(= 0, 1, \dots, \ell - 1)$ eigenvalues with positive real parts. Then the following Morse inequalities hold

$$c_0 \geq 1$$
$$c_1 - c_0 \geq -1$$
$$\vdots$$
$$c_{\ell-1} - c_{\ell-2} + \cdots + (-1)^{\ell-1} c_0 \geq (-1)^{\ell-1}$$

In fact the last inequality is an equality:

$$c_{\ell-1} - c_{\ell-2} + \cdots + (-1)^{\ell-1} c_0 = (-1)^{\ell-1}$$

From $c_0 \geq 1$ it follows that there is at least one stable equilibrium. When there is more than one stable equilibrium: $c_0 > 1$, then $c_1 \geq c_0 - 1 > 0$. That is there should be at least one unstable equilibrium. From this again follows that when the equilibrium is unique, it must be stable. Finally, the above equality we may write as

$$c_0 - c_1 + c_2 - c_3 + \cdots + (-1)^{\ell-1} c_{\ell-1} = 1.$$

This may be rewritten as

$$c_0 + c_2 + c_4 + \cdots = 1 + c_1 + c_3 + c_5 + \cdots$$

It follows that the total number of equilibria is odd:

$$c_0 + c_1 + c_2 + \cdots + c_{\ell-1} = 1 + 2(c_1 + c_3 + c_5 + \cdots).$$

When 'boundedness' does not hold, the vector field ζ may point outwards at the boundary of Δ^ℓ. However the (Morse) coindex and the Morse inequalities may be modified in that case and one may find boundary equilibria, where as was noticed above, some of the commodities are free.

Also when the vector field has a finite number of isolated cycles as solutions, the Morse inequalities can be modified, see Smale (1973). However this is not possible for the following example of Scarf (1960), where the vector field has one equilibrium, that is not asymptotically stable as a solution and further only cycles.

Example

For this example there are three agents and three commodities, that is $\ell = 3$. The vector field ζ on Δ^2 is defined as follows.

$$\zeta_1(p) = -\frac{p_2}{p_1+p_2} + \frac{p_3}{p_1+p_3};$$
$$\zeta_2(p) = -\frac{p_3}{p_2+p_3} + \frac{p_1}{p_1+p_2};$$
$$\zeta_3(p) = -\frac{p_1}{p_1+p_3} + \frac{p_2}{p_2+p_3}.$$

The equilibrium is $\overline{p} = (\frac{1}{3}, \frac{1}{3}, \frac{1}{3})$.

Notice that in this equilibrium the (3×3) Jacobi matrix $\mathcal{J}(\zeta(\overline{p}))$ has an eigenvalue zero and two conjugated, purely imaginary eigenvalues: $\pm\frac{3}{4}\sqrt{3}i$. The purely imaginary eigenvalues are also the eigenvalues of the 2×2–matrix $\mathcal{J}(\overline{\zeta}(\tau^{-1}(\overline{p})))$.

The example is degenerated. To see that take a (small) open neighbourhood V of the line segment, connecting $\overline{p}$ with the point $(0,0,1)$. Construct a so called 'bump function' ψ. ψ is a smooth function form Δ^2 onto the closed interval $[0,1]$, that takes the value 1 on the line segment connecting $\overline{p}$ with the point $(0,0,1)$ and the value 0 outside V. Now perturb the vector field ζ, by adding $\varepsilon\psi(p)$ to the third component. ε is a number that is small in absolute value. The perturbation changes the solutions of the differential equation only in V, but does not change the solutions outside V. So suppose we are moving on a cycle, an integral curve outside V. After some time we will enter the neighbourhood V. Inside V the vector field is perturbed. When $\varepsilon < 0$, we will move towards a cycle that is closer to the equilibrium $\overline{p}$, but when $\varepsilon > 0$, we will move away from $\overline{p}$. We leave V on a cycle that is closer to (resp. further away from) $\overline{p}$. After some time we arrive again in V. So although $\overline{p}$ was a *centre* of the differential equation, the perturbation of the vector field turns it into a *sink*, when $\varepsilon < 0$, or a *source*, when $\varepsilon > 0$.

A dynamical system is called *structural stable* if a slight perturbation does not change the 'phase portrait' (that is the set off all orbits) too much. Hence Scarf's example is not structural stable.

Besides the above example, Scarf (1960) deals with another example also. The other example is one with an unstable equilibrium and a stable limit cycle. This other example of Scarf is structural stable and Smale's (1973) modified Morse inequalities, as mentioned above, hold for it.

3 Stability in Cournot Oligopoly

Suppose there are $n(\geq 2)$ firms selling a homogeneous good in an oligopolistic market. At (market) price p (total) demand is $D(p)$. The following assumptions are made:

(i) there is a choke price $\rho > 0$, such that $D(p) = 0$ whenever $p \geq \rho$;
(ii) for $0 \leq p < \rho$: the demand function is of class C^2,[3] demand is positive ($D(p) > 0$) and strictly decreasing ($D'(p) < 0$);
(iii) maximal demand $D(0)$ is finite.

As by (ii), demand is strictly decreasing there is an inverse demand function f, that maps the total 'supply' in the market Q onto the *market clearing price* p. Notice that $Q = D(p)$ and therefore $p = 0$ iff $Q \geq D(0)$. As $D(\rho) = 0$, one has $f(0) = \rho$.

[3] A function is of class C^2 if the function itself and its first and second order derivatives are continuous.

It is assumed that each firm $i \in \{1, \dots, n\}$ chooses a quantity $q_i \in [0, D(0)]$, then total supply will be

$$Q = \sum_{i=1}^{n} q_i, \text{ and } Q_{-i} := Q - q_i.$$

Let C_i be the cost function of firm i. It will be assumed for each $i \in \{1, \dots, n\}$ and for $0 \le q \le D(0)$:

(iv) the cost functions are of class C^2, costs are non–negative ($C_i(q) \ge 0$), strictly increasing ($C_i'(q) > 0$) and strictly convex ($C_i''(q) > 0$).

Profits for firm i are

$$\pi_i(q_i; Q_{-i}) = q_i f(q_i + Q_{-i}) - C_i(q_i).$$

A *Cournot–Nash equilibrium* is a vector $(q_1^*, \dots, q_n^*) \in [0, D(0)]^n$ such that for all $i \in \{1, \dots, n\}$ and all $q_i \in [0, D(0)]$ one has

$$\pi_i(q_i^*; Q_{-i}^*) \ge \pi_i(q_i; Q_{-i}^*).$$

In the sequel, we are most interested in *local* Cournot–Nash equilibria. That is the above inequality does not have to hold for all $q_i \in [0, D(0)]$, but only for q's in a neighbourhood $< q_i^* - \varepsilon, q_i^* + \varepsilon >$ of q_i^*. In this ε is a small but positive number.

The first order conditions (FOC) for a (local) Cournot–Nash equilibrium are for all i:

$$\varphi_i(q_i; Q_{-i}) := \frac{\partial \pi_i(q_i; Q_{-i})}{\partial q_i} = f(q_i + Q_{-i}) + q_i f'(q_i + Q_{-i}) - C_i'(q_i) = 0.$$

When $Q_{-i} < D(C_i'(0))$, one has $f(Q_{-i}) > C_i'(0)$. Hence for $q_i = 0$: $\varphi_i(0, Q_{-i}) = f(Q_{-i}) - C_i'(0) > 0$. When $q_i = \rho - Q_{-i}$, one has $\varphi_i(\rho - Q_{-i}; Q_{-i}) = (\rho - Q_{-i}) f'(\rho) - C_i'(\rho - Q_{-i}) < 0$. That is for each i there is at least one solution $q_i(Q_{-i})$, in the open interval $< 0, \rho - Q_{-i} >$, fulfilling the above FOC. As in the previous section, also here one may define an adjustment process: for all $i = 1, \dots, n$

$$\dot{q}_i = \varphi_i(q_i; Q_{-i}).$$

According to this adjustment process, firms adjust their quantities, in order to increase their profits. The stable equilibria of this adjustment process are the (local) Cournot–Nash equilibria of this oligopoly model. The equilibrium is stable, when all eigenvalues of the Jacobi matrix

$$\frac{\partial(\varphi_1, \dots, \varphi_n)}{\partial(q_1, \dots, q_n)}$$

have negative real parts.

As in the previous section, also for this model it can be shown that 'regular oligopolies' have a finite number of equilibria. Moreover the regular oligopolies form an open and dense subset of all economies. The vector field $(\varphi_1, \dots, \varphi_n)$ again is a Morse Smale vector field and again the Morse inequalities hold. So when for a regular oligopoly the equilibrium is unique, it is stable. When there is more than one equilibrium, at least one of them is unstable. For details see Furth (1986).

As

$$\frac{\partial \varphi(q_i; Q_{-i})}{\partial q_j} = \begin{cases} f' - C_i'' + f' + q_i f'', & \text{if } j = i; \\ f' + q_i f'', & \text{if } j \neq i, \end{cases}$$

one may write the Jacobi matrix as

$$\frac{\partial(\varphi_1, \dots, \varphi_n)}{\partial(q_1, \dots, q_n)} = \begin{pmatrix} f' - C_1'' & \cdots & 0 \\ \vdots & \ddots & \vdots \\ 0 & \cdots & f' - C_n'' \end{pmatrix} + \begin{pmatrix} f' + q_1 f'' \\ \vdots \\ f' + q_n f'' \end{pmatrix} (1, \dots, 1).$$

Furth and Sierksma (1989) define an M_r–matrix, as a matrix that can be written as the sum of a diagonal matrix and a matrix of rank r. The above Jacobi matrix is clearly an M_1–matrix.

An $n \times n$–matrix is called *nearly diagonal* iff the submatrix obtained from it after deleting the ith row and column is a diagonal matrix.

Theorem Let $T = (t_{ij})$ be an $n \times n$–matrix, that is not *nearly diagonal.* Equivalent are:

(i) T is an M_1–matrix;
(ii) for each $\{i, j, k, l\} \subset \{1, \dots, n\}$ with $\{i, j\} \cap \{k, l\} = \emptyset$ one has

$$t_{ik} t_{jl} = t_{il} t_{jk},$$

and for each $\{i, j_1, l_1\}$, $\{i, j_2, l_2\} \subset \{1, \dots, n\}$ one has:

$$t_{il_1} t_{j_1 i} t_{j_2 l_2} = t_{il_2} t_{j_2 i} t_{j_1 l_1}.$$

Moreover, if T is an M_1–matrix that is not *nearly diagonal*, then the vector $(d_1, \dots, d_n)$ in the decomposition

$$\begin{pmatrix} d_1 & \cdots & 0 \\ \vdots & \ddots & \vdots \\ 0 & \cdots & d_n \end{pmatrix} + \begin{pmatrix} u_1 \\ \vdots \\ u_n \end{pmatrix} (v_1, \dots, v_n)$$

is uniquely determined.

The proof of this theorem can be found in Furth and Sierksma (1989). A similar result was obtained by Cohen, Johnson, Rodman and Woerdeman (1989).

The following theorem shows that for an M_1–matrix, all but one eigenvalue belong to the interval $[d_1, d_n]$.

Theorem Let T be an M_1–matrix:

$$T = \begin{pmatrix} d_1 & \cdots & 0 \\ \vdots & \ddots & \vdots \\ 0 & \cdots & d_n \end{pmatrix} + A,$$

with $d_1 \le \cdots \le d_n$ and $A = (a_{ij})$ an $n \times n$–matrix of rank 1.

When for all i $a_{ii} \ge 0 (\le 0)$, the eigenvalues λ_i of T are all real. When $\lambda_1 \le \cdots \le \lambda_n$ the following holds:

– If for all i $a_{ii} \ge 0$ then

$$d_1 \le \lambda_1 \le d_2 \le \cdots \le d_n \le \lambda_n \le d_n + \text{trace} A$$

and

$$d_1 + \text{trace} A \le \lambda_n;$$

– If for all i $a_{ii} \le 0$ then

$$d_1 + \text{trace} A \le \lambda_1 \le d_1 \le \cdots \le \lambda_n \le d_n$$

and

$$\lambda_1 \le d_n + \text{trace} A.$$

This theorem was proven in Hosomatsu (1969) and Quandt (1976). The proof that the eigenvalues are real can be found in Neudecker (1970). The proof in Furth and Sierksma (1989)[4] uses Weyl's Theorem, see Horn and Johnson (1985, p. 181). As all eigenvalues are real, the differential equation does not have cycles, closed integral curves, as a solution.

The Jacobi matrix of the dynamical system is an M_1 matrix. By the assumptions made, for all i one has $f' - C_i'' < 0$. Assuming further that $f' + q_i f'' \le 0$ it follows that at least $n-1$ eigenvalues of

$$\frac{\partial(\varphi_1, \ldots, \varphi_n)}{\partial(q_1, \ldots, q_n)} = \begin{pmatrix} f' - C_1'' & \cdots & 0 \\ \vdots & \ddots & \vdots \\ 0 & \cdots & f' - C_n'' \end{pmatrix} + \begin{pmatrix} f' + q_1 f'' \\ \vdots \\ f' + q_n f'' \end{pmatrix} (1, \ldots, 1)$$

are negative. This proves the following proposition:

Proposition The unstable manifold of a Cournot–Nash equilibrium of a homogeneous oligopoly, has at most dimension 1.

[4] A generalisation for M_r matrices, with $r > 1$, has been given by Barrett, Olesky and Driessche (1991).

This proposition was first proven by Seade (1980) without using any properties of matrices. Notice that the fact that the diagonal elements of the above Jacobi matrix: $2f' + q_i f'' - C_i'' < 0$, are precisely the second order conditions for a local Cournot–Nash equilibrium.

4 Final Remarks

This chapter may be seen as a short survey on the use of matrices and dynamical systems in economic theory. In the questions about stability and uniqueness, there is an interaction between these subjects. Stability and/or uniqueness of the equilibria of the dynamical systems, followed from properties of the Jacobi matrices of these systems. In many cases the properties may be derived from economic properties.

From the Sonnenschein–Debreu–Mantel result, for perfect competition and the Corchón and Mas–Colell result, for imperfect competition it follows that pathological cases can not be excluded, both in general equilibrium theory and in oligopoly theory. This caused a shift in the interest of economists. For the 'old' theory, economists were interested in linear and/or linearized models, hence in matrices. For the 'new' theory they got more and more interested in non linear models, that include possibilities, such as chaos, chaotic behavior, strange attractors and other non stability aspects. When everything is possible, economists are not longer interested in stability. When economists are not longer interested in linear models and in stability, they do not write articles on matrices anymore. That is why nowadays there are only a few articles on matrices in economic journals.

References

1. Arrow, K.J.H., D. Block and L. Hurwicz (1958): "On the Stability of the Competitive Equilibrium II," *Econometrica*, **27**, 82–109.
2. Arrow, K.J.H. and G. Debreu (1954): "Existence of an Equilibrium for a Competitive Economy," *Econometrica*, **22**, 265–290.
3. Arrow, K.J.H. and L. Hurwicz (1958): "On the Stability of the Competitive Equilibrium I," *Econometrica*, **26**, 522–552.
4. Bala, V. and M. Majumdar (1992): "Chaotic Tâtonnement," *Economic Theory*, **2**, 437–435.
5. Barrett, W.W., D.D. Olesky and P. van den Driessche (1991): "A Note on Eigenvalues of Fixed Rank Perturbations of Diagonal Matrices," *Linear and Multilinear Algebra*, **30**, 13–16.
6. Berman, A. and R.J. Plemmons (1979): *Nonnegative Matrices in the Mathematical Sciences*, New York, etc.: Academic Press.
7. Bischi, G.I., C. Mammana and L. Gardini (2000): "Multistability and Cyclic Attractors in Duopoly Games," *Chaos, Solitons and Fractals*, **11**, 543–564.
8. Bishop, R.L. (1962): "The Stability of the Cournot Oligopoly Solution: Further Comment," *Review of Economic Studies*, **29**, 232–236.

9. Cohen, N., C.R. Johnson. L. Rodman and H.J. Woerdeman (1989): "Ranks of Completions of Partial Matrices," *Operator Theory: Advances and Applications*, **40**, 165–185.
10. Corchón, L. and A. Mas–Colell (1996): On the Stability of Best Reply and Gradient Systems with Applications to Imperfectly Competitive Models," *Economic Letters* , **51**, 59–65.
11. Cournot, A.A. (1838): *Recherches sur les Principes Mathématiques de la Théorie des Richesses*, Paris: L. Hachette.
12. Dana, R.A. and L. Montrucchio (1986): "Dynamic Complexity in Duopoly Games," *Journal of Economic Theory*, **40**, 26–39.
13. Debreu, G. (1970): "Economies with a Finite Set of Equilibria," *Econometrica*, **38**, 387–392.
14. Debreu, G. (1974): "Excess Demand Functions," Journal of Mathematical Economics, **1**, 15–23.
15. Debreu, G. and J. N. Herstein (1953): "Non–Negative Square Matrices," *Econometrica*, **21**, 597–607.
16. Deneckere, R. and S. Pelikan (1986): "Competitive Chaos," *Journal of Economic Theory*, **40**, 40–56.
17. Fisher, F.M. (1961): The Stability of the Cournot Oligopoly Solution: The Effects of the Speeds of Adjustments and Increasing Returns," *Review of Economic Studies*, **28**, 125–135.
18. Furth, D. (1986): "Stability and Instability in Oligopoly," *Journal of Economic Theory*, **40**, 197–228.
19. Furth, D. and G, Sierksma (1989): "The Rank and Eigenvalues of Main Diagonal Perturbed Matrices," *Linear and Multilinear Algebra*, **25**, 191–204.
20. Goeree, J.K., C. Hommes and C. Weddepohl (1997): "Stability and Complex Dynamics in a Discrete Tâtonnement Model," *Journal of Economic Behavior and Organization*, **33**, 395–410.
21. Hahn, F.H. (1958): "Gross Substitutability and the Dynamic Stability in General Equilibrium," *Econometrica*, **26**, 169–170.
22. Hahn, F.H. (1961): "The Stability of the Cournot Oligopoly Solution,"" *Review of Economic Studies*, **29**, 329–331.
23. Hicks, J. (1939): *Value and Capital*, Clarendon Press: Oxford.
24. Horn, R.A. and C.R. Johnson (1985): *Matrix Analysis*, Cambridge University Press, Cambridge.
25. Hosomatsu, Y. (1969): "A Note on the Stability Conditions in Cournot's Dynamic Market Solution, when neither the Actual Market Demand Function nor the Production Level of the Rivals Are Known," *Review of Economic Studies*, **36**, 117–121.
26. Mantel, R. (1974): "On the Characterisation of Aggregate Excess Demand," *Journal of Economic Theory*, **7**, 348–353.
27. McKenzie, L. (1954): "On Equilibrium in Graham's Model of World Trade and other Competitive Systems," *Econometrica*, **22**, 147–161.
28. McKenzie, L. (1960): "Matrices with Dominant Diagonal and Economic Theory," in: Arrow, K.J.H., S. Karlin and P. Suppes(eds.): *Mathematical Methods in the Social Sciences*, Stanford: Stanford University Press, 47–62.
29. McManus, M. and R.E. Quandt (1961): "Comments on the Stability of the Cournot Oligopoly Model," *Review of Economic Studies*, **28**, 136–139.
30. Metzler, L. (1945): "The Stability of Multiple Markets: The Hicks Conditions," *Econometrica*, **13**, 277–299.

31. Morishima, M. (1952): "On the Law of Change of Price–System in an Economy which Contains Complementary Commodities," *Osaka Economic Papers*, **1**, 101–113.
32. Negishi, T. (1958): "A Note on the Stability of an Economy where all Goods are Gross Substitutes," *Econometrica*, **26**, 445–457.
33. Neudecker, H. (1970): "Cournot's Dynamic Market Solution and Hosomatsu's Lemma: An Alternative Proof," *Review of Economic Studies*, **37**, 447–448.
34. Okuguchi, K. (1964): "The Stability of the Cournot Oligopoly Solution: A Further Generalization," *Review of Economic Studies*, **31**.143–146.
35. Puu, T. (1991): "Chaos in Duopoly Pricing," *Chaos, Solitons and Fractals*, **6**, 573–581.
36. Rand, D. (1978): "Exotic Phenomena in Games and Duopoly Models," *Journal of Mathematical Economics*, **5**, 173–184.
37. Samuelson, P.A. (1941,1942): "The Stability of Equilibrium," *Econometrica*, **9**, **10**. 97–120, 1–75.
38. Scarf, H. (1960): "Some Examples of Global Instability of Competitive equilibrium,"" *International Economic Review*, **1**, 157–172.
39. Seade, J. (1980): "The Stability of Cournot Revisited," *Journal of Economic Theory*, **23**, 15–27.
40. Smale, S. (1960): "Morse Inequalities for a Dynamical System," *Bulletin of the American Mathematical Society*, **66**, 43–49.
41. Smale, S. (1961): "On Gradient Dynamical Systems," *Annals of Mathematics*, **74**, 199–206.
42. Smale, S. (1973): "Global Analysis and Economics I: Pareto Optimum and a Generalisation of Morse Theory," in: Peixoto, M.(ed.): *Dynamical Systems: Procedings of the* 1971 *Brasil Dynamical Systems Simposium*, New York, etc.: Academic Press, 531–544.
43. Sonnenschein, H. (1972): "Market Excess Demand Functions," *Econometrica*, **40**, 549–563.
44. Sonnenschein, H. (1973): "Do Walras Identity and Continuity Characterize the Class of Community Excess Demand Functions?" *Journal of Economic Theory*, **6**, 345–354.
45. Theocharis, R. D. (1959): "On the Stability of the Cournot Solution on the Oligopoly Problem," *Review of Economic Studies*, **27**, 133–134.
46. Tuinstra, J. (1999): *Price Dynamics in Equilibrium Models*, Ph–D thesis University of Amsterdam, Tinbergen Institute Research Series: 214.
47. Weddepohl, H.N. (1995) "A Cautious Price Adjustment Mechanism: Chaotic Behavior," *Journal of Economic Behavior and Organization*, **27**, 293–300.

Learning in Networks

Sanjeev Goyal

Department of Economics, Queen Mary College, University of London, and Econometric Institute, Erasmus University, Rotterdam

Abstract. In recent years, considerable attention has been paid to the question: what are the circumstances under which an economic equilibrium emerges? This research has highlighted the important role of the interaction pattern among individual entities in shaping the process through which an equilibrium emerges. This paper starts with an example of social learning and technology diffusion to illustrate this role. I then study the formation of the social interaction structure; this leads me to develop a model of network formation. The paper presents results on the static as well as dynamic aspects of network formation.

Keywords: networks, social learning, interaction structure.

"The peculiar character of the problem of a rational economic order is determined precisely by the fact that the knowledge of the circumstances of which we must make use never exists in concentrated or integrated form, but solely as the dispersed bits of incomplete and frequently contradictory knowledge which all the separate individuals possess.... The problem is thus in no way solved if we can show that all the facts, if they were known to a single mind (as we hypothetically assume them to be given to the observing economist), would uniquely determine the solution; instead, we must show how a solution is produced by the interactions of people, each of whom possesses only partial knowledge." (von Hayek, 1945; page 519.)

1 Introduction

Modern economics studies the implications of purposeful behaviour on the part of individuals for aggregate outcomes. The two principal theoretical approaches to this study are general equilibrium theory and game theory. Claus Weddepohl has made important contributions to both these approaches and I feel honoured to be asked to contribute to this volume.

In these approaches, the notion of economic order is conceptualized in terms of a state of equilibrium. In the past few decades there has been a great deal of research on the subject of how equilibrium emerges in an economy. Central to this research has been the idea of *learning*: this is the process by which individual entities starting with incomplete information about their environment, and about each other, gradually acquire information and in response adapt their behaviour. The process of learning arises naturally in many settings. For example, consumers typically do not know the quality of a new brand in the market. Another example is the choice of credit cards: consumers and merchants are usually not fully informed about which credit card the other players in the market are adopting. In the consumers example, the uncertainty is about some technological facts, while in the second example the uncertainty is of a strategic nature. Faced with such uncertainty, individuals try and learn from their own experience; moreover, since experiments are often expensive and time consuming, they also try and gather information from the experience of others.[1]

A notable feature of recent work on learning has been the explicit formulation of the *structure of interaction* between individual entities. This is in contrast to the traditional formulation which has focused on centralized models with anonymous individuals. For example, in the classical model of general equilibrium, individuals are supposed to be anonymous and the exchange process impersonal. Every individual submits his demands to a fictitious central auctioneer who derives a set of prices which is then commonly observed by everyone. In game theory too, research has for the most part proceeded under similar assumptions; generally we assume public observability of relevant data and in large games, matching between players is random and anonymous.

A natural way to think about the structure of interaction is in terms of networks. A network consists of a set of points and the links between them. Networks arise in a wide variety of situations. The set of points may be individual consumers and the links represent the flow of information about various brands between them. Alternatively, consider a group of scientists. The set of points is the group of scientists and the links between them represent collaborative work. We wish to understand how the network structure affects the outcome of the learning process?

2 Network Structure and Technological Change

Technological change is an important driving force in economic growth and development. An important aspect of this change is the process by which new inventions and practices spread in an economy. I now present an example to illustrate how network structure influences the diffusion of a new technology.

Consider a group of consumers each of whom faces a similar decision problem: at regular intervals, they make a choice between two brands, an old brand (or old technology) whose value is perfectly known, and a new

[1] The idea of learning was implicitly present in Cournot (1838). This book also contains perhaps the first discussion of the notion of equilibrium. For recent work on the theory of learning see Fudenberg and Levine (1998) and Kalai and Lehrer (1993).

brand, whose value is unknown.[2] There are two possibilities with respect to the relative qualities of the two brands: either the new brand is superior or it is inferior to the old brand. The old brand yields a payoff of 1 and 0 with equal probability given by a $1/2$. The quality of the new brand is unknown, it could be of High quality or of Low quality: in the former case, the new brand yields a payoff of 1 with probability $3/4$, and payoff of 0 with probability $1/4$. In the latter case, the new brand yields a payoff of 1 with probability $1/4$, and a payoff of 0 with probability $3/4$. Hence, the new brand is the optimal action if true quality of the new brand is High, while the old brand is the optimal action if the true quality of the new brand is Low. We shall assume that the true quality of the new brand is High and hence it is the best option for the consumers. However, consumers do not know the true quality of the new brand. They believe that there is a probability $p \in (0, 1)$, that the true quality of the new brand is High. Suppose that the consumer chooses an action that gives him the highest payoff. It is then easily calculated that at any point in time the best course of action for an individual consumer is to choose the new brand if the belief p is above the cut-off value – in this case $1/2$ – and choose the old brand otherwise.

We locate the consumers in a simple communication network: suppose that consumers live along a street and talk to their immediate neighbors about their experience and also observe the experience of a common set of people, whom we call the the Royal Family.[3] We shall suppose, for the sake of illustration, that there are 5 people in this royal family. Regularly, a consumer chooses a brand and gets a payoff. In addition, he also observes the experiences of his neighbors and the royal family. A payoff of 1 with the new brand is more likely if the true quality is High, while the payoff of 0 is more likely if the true quality is Low. Thus upon observing a payoff from the new brand, a consumer can revise his views about the relative quality of the new brand. This in turn might lead him to revise his choice between the two brands, over time. We would like to know if this process of learning leads the consumers to choose the new brand in the long run.

To illustrate our point simply, we assume that the initial belief of each consumer is slightly in favor of the new brand. This implies that to begin with, every consumer chooses the new brand. Yet, we show that it is possible that society will switch to the old brand and stick to it forever after. How does this happen?

Consider the situation where every member of the royal family is unlucky and receives a payoff of 0. This happens with positive probability, $(1/4)^5$. In this case, each person in the society observes a minimum of 5 negative outcomes, while the maximum number of good outcomes he can observe is 3.

[2] This section is based on Bala and Goyal (1998). For greater details and a technically more accurate account the reader is urged to consult the original paper. For related work on social learning, see Ellison and Fudenberg (1993).

[3] For early work on interaction structures see Schelling (1975).

Thus there is a minimum amount of residual negative information for each consumer in this society. If the initial beliefs of consumers are only slightly in favor of the new brand, then this negative information will be sufficient to persuade all of them to switch to the old brand. Once everyone does so, there is no further information being generated in the society concerning the relative qualities of the two brands, and so the society is stuck with the old brand. This is clearly sub-optimal.

In the example above, one reason for incomplete learning, i.e., the choice of sub-optimal brand in the long run, is the structure of communication. This structure allows the negative experience of a few people (the 'royal family') to overwhelm the locally gathered positive information of everyone in the society. This insight motivates the notion of *locally independent* consumers. A pair of consumers is said to be locally independent if they have no neighbors in common. Local independence allows for consumers to be *relatively insulated* from each other. This enables them to choose different actions and learn about the potential of these actions, before information about other actions reaches them. This in turn permits a reasonable amount of information to be generated concerning the different actions. This is crucial if the society is to be prevented from switching out of actions prematurely. These considerations lead us to argue that a loosely connected society is perhaps best suited for adopting new technologies.

In the context of our above example, this general argument has the following interesting implication. Suppose the network structure is modified so that each consumer observes the two individuals immediately next to him only. i.e., there is no royal family. A moment's reflection tells us that there are a large number of locally independent consumers in such a society. For instance, consumers $2, 5, 8$ form a set of locally independent consumers.

$$----\overbrace{(1)-(2)-(3)}^{N(2)}-\overbrace{(4)-(5)-(6)}^{N(5)}-\overbrace{(7)-(8)-(9)}^{N(8)}----$$

Our arguments imply that this society will choose the new brand, in the long run.

To summarize: we have argued that the architecture of the social communication network can critically influence the process of learning in a society. In particular, it can determine whether a society will eventually choose the best technology or get stuck with the inferior technology. As a final remark we note that this society is derived from the previous society by deleting a large number of links between ordinary people and the royal family. This suggests that deleting information links in a society can sometimes increase the likelihood of society moving into the socially desirable outcome!

3 Network Formation

In the previous section, I argued that the network of communication between consumers has a decisive influence on whether society adopts the optimal action in the long run. This leads me to ask: if consumers can decide whom to contact to gain additional information, and if making contacts is costly, then what will be the communication network that will emerge in a society? Will it be socially optimal? This motivates an investigation of the process of network formation.

We consider a setting in which each individual is a source of information which others can tap via the formation of costly pair-wise links.[4] Our focus is on information which is *non-rival.* We suppose that a link with another agent allows access, in part and in due course, to the benefits available to the latter via his own links. Thus individual links generate externalities whose value depends on the level of decay/delay associated with indirect links. A distinctive aspect of our approach is that the costs of link formation are incurred only by the person who initiates the link. This allows us to model the network formation process as a non-cooperative game, where an agent's strategy is a specification of the set of agents with whom he forms links. The links formed by agents define a social network.

The link that agent i forms with agent j yields benefits solely to agent i. The information flow across persons is assumed to be *frictionless*: if an agent i is linked with some other agent j via a sequence of intermediaries, $\{j_1, \ldots, j_s\}$, then the benefit that i derives from j is insensitive to the number of intermediaries. Apart from this, we allow for a general class of individual payoff functions: the payoff is strictly increasing in the number of other people accessed (directly or indirectly) and strictly decreasing in the number of links formed.

We shall suppose that $N = \{1, \ldots, n\}$ is a set of consumers (henceforth referred to as players or agents), and let i and j be typical members of this set. Each person is assumed to possess some information of value to himself and to other agents. He can augment his information by communicating with other people; this communication takes resources, time and effort and is made possible via the setting up of *pair-wise* links. A *strategy* of a player $i \in N$ is a (row) vector $g_i = (g_{i,1}, \ldots, g_{i,i-1}, g_{i,i+1}, \ldots, g_{i,n})$ where $g_{i,j} \in \{0,1\}$ for each $j \in N \backslash \{i\}$. We say agent i has a *link* with j if $g_{i,j} = 1$. A link between agent i and j allows for one-way flow of information; the link $g_{i,j} = 1$ enables agent i to access j's information, but not vice-versa.[5] The set of all strategies

[4] This part of the paper is based on Bala and Goyal (2000). For related work on network formation, see Jackson and Wolinsky (1996). Networks have been studied extensively by sociologists; influential references include Burt (1992), Coleman (1966), and Granovetter (1974). The recent work in economics on network formation departs from the sociological tradition in that it studies link formation – and endogenous networks – as a strategic game among fully rational players.

[5] For example, i could access j's website, or read a paper written by j.

of agent i is denoted by $\mathcal{G}_i$. The set $\mathcal{G} = \mathcal{G}_1 \times \cdots \times \mathcal{G}_n$ is the space of pure strategies of all the agents.

We can depict a strategy profile $g = (g_1, \ldots, g_n)$ in $\mathcal{G}$ as a *directed network*. The link $g_{i,j} = 1$ is represented by an *edge* starting at j with the arrow-head pointing at i. Figure 1 below provides an example with $n = 3$ agents. Here agent 1 has formed links with agents 2 and 3, agent 3 has a link with agent 1 while agent 2 does not link up with any other agent. Note that there is a one-to-one correspondence between the set of all directed networks with n vertices and the set $\mathcal{G}$.

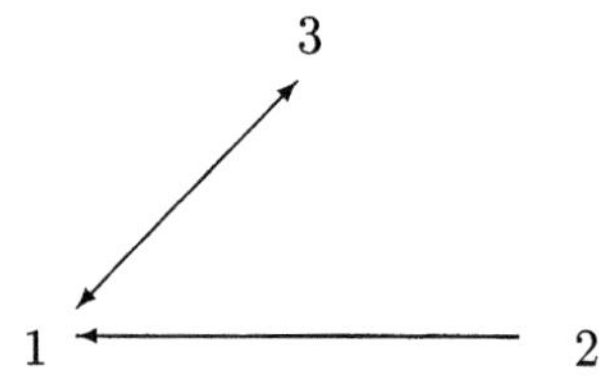

Figure 1

Define $N^d(i; g) = \{k \in N \mid g_{i,k} = 1\}$ as the set of agents with whom i maintains a link. We say there is a *path* from j to i in g either if $g_{i,j} = 1$ or there exist distinct agents $j_1, \ldots, j_m$ different from i and j such that $g_{i,j_1} = g_{j_1,j_2} = \cdots = g_{j_m,j} = 1$. For example, in Figure 1 there is a path from agent 2 to agent 3. The notation "$j \xrightarrow{g} i$" indicates that there exists a path from j to i in g. Furthermore, we define $N(i; g) = \{k \in N \mid k \xrightarrow{g} i\} \cup \{i\}$. This is the set of all agents whose information i accesses either through a link or through a sequence of links. We shall typically refer to $N(i; g)$ as the set of agents who are observed by i. We use the convention that $i \in N(i; g)$, i.e. agent i observes himself. Let $\mu_i^d : \mathcal{G} \to \{0, \ldots, n-1\}$ and $\mu_i : \mathcal{G} \to \{1, \ldots, n\}$ be defined as $\mu_i^d(g) \equiv |N^d(i; g)|$ and $\mu_i(g) \equiv |N(i; g)|$ for $g \in \mathcal{G}$. Here, $\mu_i^d(g)$ is the number of agents with whom i has formed links while $\mu_i(g)$ is the number of agents observed by agent i.

Denote the set of non-negative integers by Z_+. Then define $\Phi : \mathcal{Z}_+^2 \to \mathcal{R}$ to be a function such that $\Phi(x, y)$ is strictly increasing in x and strictly decreasing in y. Define each agent's payoff function $\Pi_i : \mathcal{G} \to \mathcal{R}$ as

$$\Pi_i(g) = \Phi(\mu_i(g), \mu_i^d(g)). \tag{3.1}$$

Given the properties we have assumed for the function Φ, $\mu_i(g)$ can be interpreted as providing the "benefit" that agent i receives from his links, while $\mu_i^d(g)$ measures the "cost" associated with maintaining them. The payoff function in (3.1) implicitly assumes that the value of information does not depend upon the number of individuals it has passed through, i.e. that there is no information decay or delay in transmission.

For a given network $g \in \mathcal{G}$, let g_{-i} denote the network obtained when all of agent i's links are removed. A strategy profile $g^* = \{g_1^*, g_2^*, g_3^*, ..., g_n^*\}$ is a Nash equilibrium if for every $i \in N$, $\Pi_i(g_i^*, g_{-i}^*) \geq \Pi_i(g_i, g_{-i}^*)$. A *strict* Nash network is one where each agent gets a strictly higher payoff with his current strategy than he would with any other strategy.

Given a network g, a set $C \subset N$ is called a *component* of g if for every distinct pair of agents i and j in C we have $j \stackrel{g}{\longrightarrow} i$, (equivalently, $j \in N(i; g)$) and there is no strict superset C' of C for which this is true. A component C is said to be *minimal* if C is no longer a component upon replacement of a link $g_{i,j} = 1$ between two agents i and j in C by $g_{i,j} = 0$, ceteris paribus. A network g is said to be *connected* if has a unique component. If the unique component is minimal, g is called *minimally connected.* A network is said to be *empty* if $N(i; g) = \{i\}$ and it is called *complete* if $N^d(i; g) = N \backslash \{i\}$ for all $i \in N$. We denote the empty and the complete network by g^e and g^c, respectively. A *wheel* network is one where the agents are arranged as $\{i_1, \ldots, i_n\}$ with $g_{i_2,i_1} = \cdots = g_{i_n,i_{n-1}} = g_{i_1,i_n} = 1$ and there are no other links. The wheel network is denoted g^w.

Our first result highlights a general property of Nash networks when agents are symmetrically positioned vis-a-vis information and the costs of access: in equilibrium, either there is no social communication or every agent has access to all the information in the society. Formally, *a Nash network is either empty or minimally connected.* The intuition for this is as follows: Consider a non-empty Nash network, and suppose that agent i is the agent who observes the largest number of agents in this network. Suppose i does not observe everyone. Then there is some agent j who is not observed by i and who does not observe i (for otherwise j would observe more agents than i). Since i gets value from his links, and payoffs are symmetric, j must also have some links. Let j deviate from his Nash strategy by forming a link with i alone. By doing so, j will observe strictly more agents than i does, since he has the additional benefit of observing i. Since j was observing no more agents than i in his original strategy, j increases his payoff by his deviation. This contradiction implies that i must observe every agent in the society. We then show that every other agent will have an incentive to either link with i or to observe him through a sequence of links, so that the network is connected. If the network is not minimally connected, then some agent could delete a link and still observe all agents, which would contradict Nash.[6]

Connectedness is, however, a permissive requirement: for example, with one-way flows a society with 6 agents can have upwards of 20,000 Nash net-

[6] In recent work on social learning and local interaction, connectedness of the society is a standard assumption (see Bala and Goyal, 1998; Ellison and Fudenberg, 1993). Our results may be seen as providing a foundation for this assumption.

works representing more than 30 different architectures.[7] This multiplicity of Nash equilibria motivates an examination of a stronger equilibrium concept. If an agent has multiple best responses to the equilibrium strategies of the others then this may make the network less stable as the agent may be tempted to switch to a payoff-equivalent strategy. This leads us to study the nature of networks which can be supported in a strict Nash equilibrium.

We find that the refinement of strictness is very effective in our setting: *the only strict Nash architectures are the wheel and the empty network.* This result exploits the observation that in a network, if two agents i and j have a link with the same agent k, then one of them (say) i will be indifferent between forming a link with k or instead forming a link with j. We know that Nash networks are either connected or empty. This argument implies that in the one-way flow model a non-empty strict Nash network has exactly n links. Since the wheel is the unique such network, the result follows.

While these findings restrict the set of networks sharply, the coordination problem faced by individuals in the network game is not entirely resolved. In a society with n people there are $(n-1)!$ networks corresponding to the wheel architecture. Thus agents have to choose from among these different equilibria. This leads us to study the process by which individuals learn about the network and revise their decisions on link formation, over time.

We use a version of the best-response dynamic to study this issue.[8] The network formation game is played repeatedly, with individuals making investments in link formation in every period. In particular, when making his decision an individual chooses a set of links that maximizes his payoffs given the network of the previous period. Two features of our model are important: one, there is some probability that an individual exhibits *inertia*, i.e., chooses the same strategy as in the previous period. This ensures that agents do not perpetually miscoordinate. Two, if more than one strategy is optimal for some individual then he *randomizes* across the optimal strategies. This requirement implies, in particular, that a non-strict Nash network can never be a steady state of the dynamics. The rules on individual behavior define a Markov chain on the state space of all networks; moreover, the set of absorbing states of the Markov chain coincide with the set of strict Nash networks of the one-shot game.

Our analysis establishes that the dynamic process converges to a limit network: *for any number of agents and starting from any initial network, the dynamic process converges to a wheel or to the empty network, with probability 1.*[9] The proof exploits the idea that well-connected people generate

---

[7] Two networks have the same architecture if one network can be obtained from the other by permuting the strategies of agents in the other network.

[8] Dynamic network formation with far-sighted players is an open field for further research.

[9] The rate of convergence of the dynamics is exponential (since they constitute a Markov chain). This general property however allows for a wide range of out-

positive externalities. Fix a network g and suppose that agent i is the person who accesses the largest number of people in g, directly or indirectly. Consider an agent j who is not critical for agent i, i.e. agent i is able to access all these people even if agent j deletes all his links. Allow agent j to move; he can form a single link with agent i and access all the different individuals accessed by agent i. Thus if forming links is at all profitable for agent j then one best-response strategy is to form a single link with agent i. This strategy in turn makes agent j well-connected. We now consider some person k who is not critical for j and apply the same idea. Repeated application of this argument leads to a network in which everyone accesses everyone else via a single link, i.e. a wheel network.[10]

We would also like to examine the social efficiency of communication in equilibrium networks. To this end, we define a welfare measure $W(g)$, as the sum of payoffs of all agents. A network g is *efficient* if $W(g) \geq W(g')$ for all $g' \in \mathcal{G}$. Note that the wheel is the only network with n links which is connected. With this observation in place, it is reasonably easy to show that, in addition to being a limit point of the dynamics, the *wheel is also the unique efficient architecture* for a large class of payoff functions.

4 Concluding Remarks

I started by noting that the concept of equilibrium is central to economic theory. In recent years, there has been considerable effort devoted to understanding how and under what circumstances does an economic equilibrium emerge? This research has also illustrated how the structure of interaction matters for the nature as well the possibility of equilibrium. In this paper I started with an example of social learning to illustrate the implications of interaction structure. Theoretical and empirical work on social learning as well as in many other fields, such as human capital formation, crime, and social coordination, has identified similar effects. These findings have in turn led to research on the determinants of the interaction structure itself. In the second

comes; to get a better sense of this issue we have carried out simulations of the dynamic process. These simulations suggest that the rate of convergence is quite rapid. For the details of these findings refer to Bala and Goyal (2000).

[10] The above results are obtained for a benchmark model with no *frictions*. The introduction of decay/delay complicates the model greatly and it is difficult to obtain a complete characterization of equilibrium networks. The main difficulty lies in specifying the agents' best response correspondence. Loosely speaking, in the absence of decay a best response consists of forming links with agents who are connected with the largest number of other individuals. With decay, however, the distances between agents also becomes relevant, so that the entire structure of the network has to be considered. Existing results therefore focus on the case of low levels of decay, where some properties of best responses can be exploited. A detailed exposition of this work lies outside the scope of this survey; the reader is urged to refer to the original paper.

part of the paper, I have presented a simple model of network formation to illustrate that potential of this new line of research. This model shows that individual incentives sharply define the range of sustainable network structures. The model I presented is very simple and the sharpness of the results raises a host of questions, which are being pursued currently.[11] I would like to conclude by noting that the concept of equilibrium again plays a central role in this research programme.

References

1. Bala V. and S. Goyal (1998), Learning from Neighbours, *Review of Economic Studies*, 65, 595-621.
2. Bala, V. and S. Goyal (2000), A Non-Cooperative Model of Network Formation, *Econometrica*, 68, 1181-1231.
3. Burt, R. (1992), *Structural Holes: The Social Structure of Competition*, Harvard University Press, Cambridge, MA.
4. Coleman, J. (1966), *Medical Innovation: A Diffusion Study*, Second Edition, Bobbs-Merrill, New York.
5. Cournot, A. (1838), *Recherches sur les principes mathematiques de la theorie des richesses*, Paris. M. Riviera & Cie. *Researches into the Mathematical Principles of Wealth, New York. A.M.Kelly.*
6. Ellison, G., and D. Fudenberg (1993), Rules of Thumb for Social Learning, *Journal of Political Economy*, 101, 612-643.
7. Fudenberg, D. and D. Levine (1998), *Learning in Games*, M.I.T. Press, Cambridge.
8. Goyal, S. and J.L Moraga (2001), R& D Networks, *Rand Journal of Economics*, forthcoming.
9. Granovetter, M. (1974), *Getting a Job: A Study of Contacts and Careers*, Harvard University Press, Cambridge MA.
10. Hayek, F. (1945), The Use of Knowledge in Society', *American Economic Review*, 35, 519-30.
11. Jackson, M. and A. Wolinsky (1996), A Strategic Model of Economic and Social Networks", *Journal of Economic Theory,* 71, 1, 44-74.
12. Kalai, E. and E. Lehrer (1993), Rational Learning leads to Nash Equilibrium, *Econometrica*, 61, 5, 1019-1047.
13. Kranton, R and D. Minehart (2001), A Theory of Buyer-Seller networks, *American Economic Review*, forthcoming.
14. Schelling, T. (1975), *Micromotives and Macrobehaviour*, Harvard University Press, Cambridge.

[11] In recent years, there has been a virtual explosion in the research on network formation. There are different strands in this work; theoretical work explores the architecture of equilibrium and efficient networks in general models, while applied work explores the implications of networks for economic activity in industrial organization, public economics and labour markets and other fields (for some recent applied work see, e.g., Goyal and Moraga (2001) and Kranton and Minehart (2001).

Imitation of Cooperation in Prisoner's Dilemma Games with Some Local Interaction

Maarten C.W. Janssen*

Department of Economics/Micro, Erasmus University, Rotterdam

Abstract. In this paper I study conditions for the emergence of cooperative behavior in a dynamic model of population interaction. The model has finitely many individuals located on a circle. The pay-off of each individual is partly based on the (local) interaction with neighbors and partly on (uniform) interaction with the whole population. The dynamics is driven by imitative behavior. I show that for a large class of parameters cooperation will emerge if the population is large; if the population is small, defection will prevail in the long run. The result contrasts with conventional wisdom which says that the larger the population, the less likely cooperation will be.

Keywords: cooperation, prisoner's dilemma, evolutionary game theory, local interaction

1 Introduction

Different approaches exist to explaining cooperative behavior in the Prisoner's Dilemma Game. With perfectly rational agents cooperation is an equilibrium outcome if the chance that the game is repeated another time is sufficiently large. With boundedly rational agents in evolutionary game theory, cooperation may emerge if the interaction between agents is non-uniform, i.e., the chance that cooperative types meet each other is higher than the chance that they meet defective types. One such a case is when interaction is between siblings (see, e.g., Bergstrom and Stark, 1993). Another is when interaction is only between neighbors as in Eshel, Shaked and Samuelson (1998). It is commonly thought in both approaches that due to increasing anonymity cooperation is less likely to be observed in large populations than in smaller ones.[1] In this paper I present a dynamic model that tells us that this is not necessarily so.

* The author thanks Daisy Kersemakers, Harry Garretsen, Sanjeev Goyal, Larry Samuelson, Avner Shaked, seminar participants at the University of Groningen and participants of the Summer workshop on Dynamic Games at the University of Warwick for helpful comments.

[1] An exception to this rule is the paper by Kandori (1992) showing how in large communities rational agents may choose to follow social norms that punish non-cooperative behavior.

I consider the interaction between members of two different populations. The two populations have lived in isolation of each other and for reasons not analysed here, one population has evolved in the past towards cooperative behavior and the other towards defective behavior. Now that we see an increase in the interaction between different societies, I want to study the question whether the interaction between members of the two populations results in one type of behavior driving out the other or in the two types of behavior coexisting next to each other. This type of analysis has been done for coordination games in Goyal and Janssen (1997).

In the model each individual agent interacts with anyone in the population, but more so with their neighbors. Given some population configuration each individual receives a certain pay-off. The pay-off is a weighted average of the pay-offs under strict uniform and local interaction. The implications of *two types of imitation dynamics* are examined. Under *Imitating the Best Agent* (IBA), each and every agent observes the pay-off of each individual in the neighborhood and agents imitate the individual who has received the highest pay-off in the previous period. Under imitating the best behavior (cooperation or defection), agents do *not* observe pay-offs of individual neighbors, but only observe the average pay-off of cooperators and the average pay-off of defectors in the neighborhood and they imitate the behavior that has yielded the highest average pay-off in the neighborhood (IBB). I will look at the development of the number of cooperative agents over time by investigating the behavior of the boundary between the two populations.

Roughly speaking, the main result is that for a wide range of parameter values and for any of the above mentioned dynamic processes there exists a threshold value (depending on the parameter values) of the population size: if the population size is larger than this threshold value cooperation will emerge, if on the other hand the population size is smaller than the threshold value cooperation will vanish. (For some other parameter values coexistence of the behaviors of the two populations is possible). The intuition for the main result can be grasped by comparing the pay-off of a cooperative type who is surrounded by other cooperative agents and the pay-off of a defective type who is surrounded by other defective agents. What matters in both classes of dynamic processes is *not* the *absolute* value of the pay-off, but the pay-off of a certain type *relative* to the pay-offs of others (cf., Vega-Redondo, 1997). The total pay-off difference between the cooperative and the defective type under consideration can be decomposed in a pay-off difference due to local interaction and a pay-off difference due to uniform interaction. The pay-off difference due to local interaction is independent of the size of the population and positive, i.e., the cooperative type receives a larger pay-off out of local interaction than the defective type. The pay-off difference due to uniform interaction itself can be decomposed into two parts. A first part is due to the fact that in a match with a given agent of the population, the defective type always receives a higher pay-off than the cooperative type. This part is also independent of the size of the population. A second part is due to the fact that under uniform interaction, a cooperative type meets relatively more defective agents than the defective agents themselves do. This is because a given agent will be matched with one of the *other* agents in the population (cf., Huck and Oechsler (1995)). This second part also favors the defective agents, but this part of the uniform interaction effect vanishes as the population size becomes large. Given the observations above it is then relatively easy to see that the uniform interaction effect may dominate the local interaction effect for small populations, whereas the reverse may hold true for large populations.

From the above it becomes clear that what is important for the result is *(i)* the presence of some neighborhood effect and *(ii)* some form of non-best reply dynamics. If interactions were uniform in the population, each agent will roughly meet the same agents as the others and in such a world, defection always yields a higher pay-off (as it is the dominant strategy). Also, over time best reply dynamics will result in defection as for any given behavior of a set of opponents, defect is always the best reply.

There is a large and rapidly growing literature in evolutionary game theory explaining the emergence of cooperative behavior in populations. Because of lack of space, I will not give an overview. The paper closest to the present paper is Eshel et al. (1998). They show that under IBB with strict local interaction within a small neighborhood the majority of agents in a population will display cooperative behavior. There are a few differences between their paper and the present one. The present paper is simpler as at any point in time, there is only one group of cooperators and one group of defectors. This is convenient as the dynamic process can be investigated by looking at the boundary between regions. A drawback of this assumption is that we can not allow for the possibility of mutants entering the population. On the other hand, the present paper is more general as different types of imitative behavior are considered and as a combination of local and uniform interaction is studied.

The paper is organized as follows. The next section describes the model that is used. Section 3 presents the results under imitating the best behavior dynamics and Section 4 briefly discusses the imitating the best agent dynamics. Proofs are given in the appendix.

2 The Models

There are N individuals located on a circle. Each individual I has a distinct location denoted by i, i=1,..,N. Individual i's immediate neighbors are i-1 and i+1. Individual 1 and individual N are also each other's immediate neighbors. In each period, each member of the population plays the following Prisoner's Dilemma game with other people in the population, with $b > a > d > c$.

Pay-offs:		Strategy of player 2	
		C	D
strategy of player 1	C	(a,a)	(c,b)
	D	(b,c)	(d,d)

Table 1. *Pay-offs in the Prisoner's Dilemma game*

Time is indexed by $t = 0, 1, 2, \ldots$ At $t = 0$, the population is divided into two groups, one group of cooperators and one group of defectors. In each subsequent period, every agent has to choose either to defect or to cooperate. The pay-off of individual i is composed of two parts. The *local interaction* part measures the pay-off out of interaction with the K immediate neighbors on either side. In most cases I will restrict the analysis for computational convenience to neighborhoods of two agents, one on each side. The *uniform interaction* part measures the pay-off out of interaction with anyone in the population. The average pay-off to individual i can be written as

$$U_i = \alpha \frac{\sum_{k=1}^{K} u(i, i-k) + u(i, i+k)}{2} + (1-\alpha) \frac{\sum_{j \neq i} u(i, j)}{N-1} \tag{1}$$

where U_i is the average pay-off of agent i, α is the relative weight of local interaction in the average pay-off of individual i, $u(i,j)$ is the pay-off of individual i when interacting with individual j and N is the size of the population.

Equation (1) allows for two interpretations. First, one may consider a situation in which each individual meets only one other agent in the population in a certain period. The parameter α can then be interpreted as the probability that the individual will meet someone out of his neighborhood (and within this neighborhood, he will meet someone at random). In this interpretation, U_i is the expected pay-off of individual i and I have to allow agents to observe the *expected* pay-off of other agents (in the neighborhood) to retain a deterministic dynamics. A second and more favored interpretation is one in which agents meet everyone in the population, but meet neighbors more frequently than others. In this interpretation α measures the relative importance of local interaction for the average pay-off of an individual. Note that α is assumed to be independent of the population size.

With two neighbors, one can distinguish three subcategories of the cooperative, respectively defective, behavior depending on the neighbors. I will denote by D^+ and C^+ the individuals who defect, respectively, cooperate and who are surrounded by cooperative agents; D^0 and C^0 represent the individuals who have one cooperating neighbor and one defecting neighbor and D^- and C^- stand for the individuals who are surrounded by defecting individuals. When there is just one group of cooperators and one of defectors, the following statements hold. The '0' category of each type is the individual at the boundary between a cooperative region and a defective region. In any period in time there are maximally two individuals of each of these two '0' categories. The categories D^+ and C^- only come into play if there is one defective, respectively one cooperative, type left over in the population and these categories will de discussed only when appropriate.

What is important for the dynamics of the system is how the four different categories that remain (D^0, D^-, C^+ and C^-) are ranked in terms of average pay-off. Straightforward calculations show that when there are x cooperators in the population he pay-offs of the four categories under consideration are as follows: (Note that with x cooperators and (N-x) defectors, there are N-x-2 agents who get a pay-off of type D^-, etc):

category	number	payoff
D^-	$N-x-2$	$\frac{x(b-d)(1-\alpha)}{N-1}+d$
D^0	2	$\frac{x(b-d)(1-\alpha)}{N-1}+d+\frac{\alpha(b-d)}{2}$
C^+	$x-2$	$a-\frac{(N-x)(a-c)(1-\alpha)}{N-1}$
C^0	2	$a-\frac{(N-x)(a-c)(1-\alpha)}{N-1}-\frac{\alpha(a-c)}{2}$

A few observations with respect to these pay-offs can be made. Obviously, the average pay-off of the '+' categories of each type is larger than the average pay-off of the '0' category and this in turn is larger than the average pay-off of the '-' category. Also, the average pay-off of the D^0 individuals is larger than that of the C^0 individuals. This is because the pay-off out of local interaction is the same for both categories, but the pay-off out of uniform interaction is larger for the D^0 individuals.

Under *imitation dynamics*, it is important to specify the information agents have about the pay-offs of other individuals. In this paper, I assume that agents observe the pay-offs of agents in their neighborhood, but not the pay-offs of agents outside the neighborhood. The basic idea here is that one needs a sufficient number of interactions with another agent to figure out what her pay-off actually is. By imitation I mean that an individual chooses the same action (cooperate or defect) that the type (or individual) has chosen who is imitated. If the neighborhood consists of individuals who take the same action, then this action is by definition the best in the neighborhood. Within this framework I analyze the dynamics stemming from "imitating the best agent in your neighborhood" (IBA) as well as from "imitating the behavior (cooperative or defective) with the highest average pay-off in your neighborhood" (IBB). An individual always considers himself to be part of his neighborhood. So, if $K=1$, there are three individuals in a neighborhood. According to the first dynamics, agents observe the pay-off of each individual in their neighborhood and choose, in the next period, the action the agent with the highest pay-off has chosen in this period. According to the second dynamics, agents only observe the average pay-off of cooperative and defective behaviors in their neighborhood and choose, in the next period, the behavior that has yielded the highest pay-off in this period.

3 Imitating the Best Behavior

In this section I consider a dynamic process based on "imitating the best behavior in your neighborhood" (IBB), where the best behavior is defined as the behavior (cooperative or defective) with the highest average pay-off. For simplicity, I restrict attention to the case where $K=1$. In what follows, I shall focus on the behavior of the boundary between the regions of cooperators and defectors. Note that the neighborhoods of the two individuals at the boundary differ: the neighborhood of the cooperative individual at one side of the boundary consists of one C^+ agent, himself (C^0) and one D^0 agent and that the neighborhood of the defector at the other side of the boundary consist of one C^0 agent, himself (D^0) and one D^- agent. Accordingly, I have to calculate the average pay-offs in the two neighborhoods separately.

To understand the main idea behind the Propositions below, I first concentrate on the special cases in which $a-c=b-d$. The expressions for the average pay-off of the cooperator, resp. defector in the neighborhood of the defective agent (D^0) easily follow from the previous section:

$$\begin{aligned} \bar{U}_{D^0}(C) &= a - \frac{(N-x)(a-c)(1-\alpha)}{N-1} - \frac{\alpha(a-c)}{2} \\ \bar{U}_{D^0}(D) &= \frac{x(b-d)(1-\alpha)}{N-1} + d + \frac{\alpha(b-d)}{4} \end{aligned} \tag{2}$$

where $\bar{U}_{D^0}(C)$ resp. $\bar{U}_{D^0}(D)$ is the average pay-off of cooperators, respectively, defectors in the neighborhood of the D^0 individual. From these expressions it follows that in the special case under consideration $\bar{U}_{D^0}(C) > \bar{U}_{D^0}(D)$ if

$$N\left[\frac{\alpha(a-c)}{4} - (d-c)\right] > (a-d) - \frac{3\alpha(a-c)}{4}. \tag{3}$$

Similarly, the average pay-off of the cooperators, resp. defectors in the neighborhood of the cooperative individual C^0 is given by

$$\bar{U}_{C^0}(C) = a - \frac{(N-x)(a-c)(1-\alpha)}{N-1} - \frac{\alpha(a-c)}{4}$$

$$\bar{U}_{C^0}(D) = \frac{x(b-d)(1-\alpha)}{N-1} + d + \frac{\alpha(b-d)}{2}$$

where $\bar{U}_{C^0}(C)$ resp. $\bar{U}_{C^0}(D)$ is the average pay-off of cooperators, resp. defectors in the neighborhood of the C^0 individual. It is easily seen that in the special case of $a-c=b-d$ the following is true: $\bar{U}_{D^0}(C) > \bar{U}_{D^0}(D)$ if, and only if, $\bar{U}_{C^0}(C) > \bar{U}_{C^0}(D)$ Hence, when condition (3) is satisfied, the D^0 individuals will switch to cooperation and, as the above argument is independent of time and the number of cooperators in the population, cooperation will be the most observed behavior in the long run. In the context of proposition 2 and 3 it is shown in the appendix that the population will either have two defectors or will cycle between one and three defectors in the long run. On the other hand, when condition (3) is violated it is clear that $\bar{U}_{D^0}(C) < \bar{U}_{D^0}(D)$ and $\bar{U}_{C^0}(C) < \bar{U}_{C^0}(D)$ so that defection will prevail in the long run. Condition (3) also reveals that for cooperation to emerge in the long run N has to be larger than a certain cut-off value, which depends on the pay-off parameters and the value of α. This result is formally stated in proposition 1 and graphically illustrated in figure 1.

Proposition 1. Suppose $a-c=b-d$ and $x_0 \geq 2$. There is a time period T such that for all $t > T$,

(i) if $N\left[\frac{\alpha(a-c)}{4} - (d-c)\right] > (a-d) - 3\alpha(a-c)/4$, then $x_t \geq N-3$.

(ii) if $N\left[\frac{\alpha(a-c)}{4} - (d-c)\right] < (a-d) - 3\alpha(a-c)/4$, then $x_t = 0$.

The condition $a-c=b-d$ has two important implications. First, it implies that the ranking of pay-offs of cooperators and defectors at the boundary does not depend on the number of cooperators in the population and, hence, that this ranking does not change endogenously over time. Second, it implies that if cooperation is better on average than defection in the neighborhood of the defector at the boundary than it is also better in the neighborhood

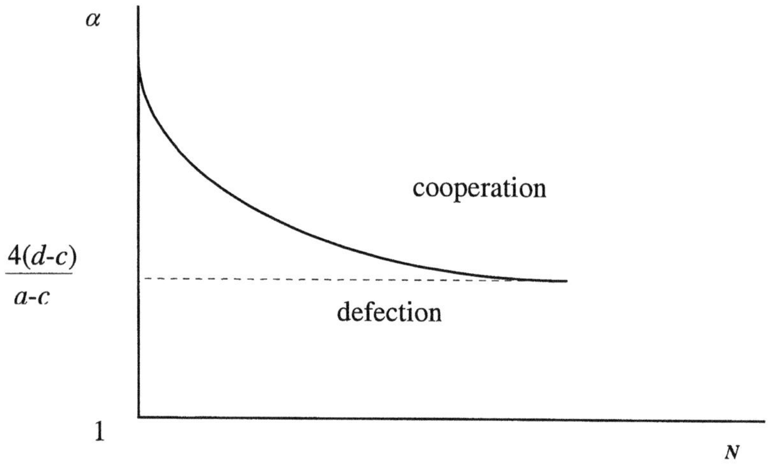

Fig. 1.

of the cooperator at the boundary, and vice versa. The two Propositions below show how the result indicated above generalizes to other values of the pay-off parameters. It turns out that for a given value of α the result generalizes for large and for small values of N. However, for intermediate values of N new phenomena emerge: if $a - c > b - d$, then the initial number of cooperators becomes important in co-determining the long-run outcome of the population (dependence on initial conditions). If, however, $a - c < b - d$, then the population will converge to a stable configuration with a group of cooperators and a group of defectors coexisting next to each other.

To be able to formally state the results for intermediate values of N, I need to introduce three additional parameters, namely x^{1*}, x^{2*} and $x^{1\sim}$. These variables are defined as follows:

$$(a-d)(N-1) = \alpha(N-1)\frac{2(a-c)+(b-d)}{4} + (1-\alpha)[N(a-c) - x^{1*}((a-c)-(b-d))] \tag{4}$$

$$(a-d)(N-1) = \alpha(N-1)\frac{2(b-d)+(a-c)}{4} + (1-\alpha)[N(b-d) - x^{2*}((a-c)-(b-d))] \tag{5}$$

$$(a-d)(N-1) = \frac{\alpha}{4}(N-1)[2(b-d)+(a-c)] \\ +(1-\alpha)[N(a-c) - x^{1\sim}((b-d)-(a-c))] \tag{6}$$

The parameter x^{1*}, resp. x^{2*}, may be interpreted for the case $a-c < b-d$ as the number of cooperators in the population such that the defector respectively cooperator, at the boundary is indifferent between defecting and cooperating. (Note that the number need not be a natural number). A similar interpretation holds true for $x^{1\sim}$ in case $a-c < b-d$.

I am now ready to state the two main propositions.

Proposition 2. Suppose $a-c > b-d$ and $x_0 > 2$. There is a time period T such that

(i) if $[\alpha\frac{2(a-c)-(b-d)}{4} - (d-c)] > (a-d) - \frac{\alpha}{4}[2(a-c)+(b-d)]$ then $x_t \geq N-3$ for all $t > T$,
(ii) if $N[\alpha\frac{2(b-d)-(a-c)}{4} - (b-a)] > (a-d) - \frac{\alpha}{4}[2(b-d)+(a-c)]$ then $x_t = 0$ for all $t > T$,
(iii) if the conditions under (i) and (ii) are *not* satisfied, then

$$\begin{array}{lll} x_t \geq N-3 & \text{for all } t > T & \text{if } x_0 > x^{1*} \\ x_t = 0 & \text{for all } t > T & \text{if } x_0 < x^{2*} \\ x_t = x_0 & \text{for all } t & \text{if } x^{1*} < x_0 < x^{2*}. \end{array}$$

Proposition 3. Suppose $a-c < b-d$ and $x_0 > 2$. There is a time period T such that

(i) if $N[\alpha\frac{2(b-d)-(a-c)}{4} - (b-a)] > (a-d) - \frac{\alpha}{4}[2(b-d)+(a-c)]$ then $x_t > N-3$ for all $t > T$;
(ii) if $N[\alpha\frac{2(a-c)-(b-d)}{4} - (d-c)] > (a-d) - \frac{\alpha}{4}[2(a-c)+(b-d)]$ then $x_t = 0$ for all $t > T$;
(iii) if the conditions under (i) and (ii) are *not* satisfied, then

$$\begin{array}{ll} x_t = 0 & \text{for all } t > T \text{ if } x^{1\sim} < 5 \\ x_t \text{ cycles around } x^{1\sim} \text{ with period 4} & \text{for all } t > T \text{ if } x^{1\sim} > 6. \end{array}$$

Two elements of the stated propositions deserve some further explanation. First, the initial number of cooperators, x_0 has to be larger than or equal to 2 in order for cooperation to have any chance of surviving. Second, the first part of both Propositions should be read as follows: depending on the initial number of defectors (whether it is odd or even) either the population converges to everyone cooperating in the long run or the population converges to a two-period cycle in which the number of defective agents alternates between 1 and 3.

4 Imitating the Best Agent

I have argued in Section 3 that when local interaction is confined to meeting one individual on either side, then at the boundary the individual who defects receives a higher pay-off than the one (at the other side of the boundary) who cooperates. This implies that under IBA with local interaction in a neighborhood of three, the defective agent will keep on playing defect. On the other hand, the cooperative agent at the boundary may (if the D^0 agent receives a higher pay-off than the C^+ agent) or may not (if the D^0 agent receives a smaller pay-off than the C^+ agent) switch to defect. Hence, cooperation is not a feasible long-run outcome in the population at large; only coexistence or defection are. In order to show that this particular result is sensitive to the neighborhood structure and that the general result also holds true under IBA I will now consider local interaction in a neighborhood of five (two on each side). As the main ideas of the analysis are quite similar to the analysis under IBB, I will only consider the case $a - c > b - d$ and concentrate on the condition on the parameters that makes cooperation the predominant long-run behavior in the population.

In the figure below the boundary between a region of cooperators and a defectors is depicted and individuals 1 and 2 have chosen to cooperate in the past and agents 3 and 4 have chosen to defect. I will investigate the long-run behavior of the population by looking at the behavior of the boundary in the next period.

C	C	C	D	D	D
0	1	2	3	4	5

The pay-off of the four individuals is respectively:

$$\begin{aligned} U_1 &= \alpha\frac{3a+c}{4} + (1-\alpha)\frac{(x-1)a+(N-x)c}{N-1} \\ U_2 &= \alpha\frac{2a+2c}{4} + (1-\alpha)\frac{(x-1)a+(N-x)c}{N-1} \\ U_3 &= \alpha\frac{2b+2d}{4} + (1-\alpha)\frac{(xb)+(N-x-1)d}{N-1} \\ U_4 &= \alpha\frac{b+3d}{4} + (1-\alpha)\frac{(xb)+(N-x-1)d}{N-1} \end{aligned} \tag{7}$$

Comparing the pay-offs of individuals 1 and 3 reveals that $U_1 > U_3$, if and only if,

$$\frac{\alpha}{4}[(a-d)+(c-d)+2(a-b)](N-1) > (1-\alpha)[(d-c)N+(a-d)+x(b-d-(a-c))].$$

This inequality holds for all x if it holds for $x = 0$. Rewriting yields

$$N\left[\alpha\frac{3(a-c)-2(b-d)}{4}-(d-c)\right] > (a-d)-\frac{\alpha}{4}[2(a-c)+3(b-d)]. \tag{8}$$

As agent 1 is in 3's neighborhood, agent 3 observes 1's pay-off. Also, 3 observes the pay-off of agents 4 and 5, but these pay-offs are lower than his. Hence, individual 3 will switch to cooperation in the next period if inequality (8) holds and the boundary between the regions of cooperators and defectors in the next period will be between individuals 3 and 4. As the above argument is independent of time, it can be used to prove that in the long run the population consists predominantly of cooperators.[2]

5 Conclusion

In this paper I have looked at the long-run behaviour of a population where individuals imitate best practice in their neighborhood. The main result that is established is that cooperation may emerge if individuals interact sufficiently often within their neighborhood.

Appendix

Proof of Proposition 2

(i) Suppose $a-c > b-d$ In this case the following is true: if $\bar{U}_{D^0}(C) > \bar{U}_{D^0}(D)$, then also $\bar{U}_{C^0}(C) > \bar{U}_{C^0}(D)$. Moreover, it is easily seen that if $\bar{U}_{D^0}(C) > \bar{U}_{D^0}(D)$ for x=0, then it holds for all values of x. Finally, $\bar{U}_{D^0}(C) > \bar{U}_{D^0}(D)$ for x=0 if

$$N\left[\alpha\frac{2(a-c)-(b-d)}{4}-(d-c)\right] > (a-d)-\frac{\alpha}{4}[2(a-c)+(b-d)].$$

Hence, if this condition holds, the D^0 agent at the boundary will switch to cooperation, while the C^0 agent will stick to cooperation. Hence, the boundary between the regions shifts and more agent cooperate. This process continues until either there is no defector left or there is only one D^+ agent left in the population. This agent will be imitated by its neighbors, but they will switch back to cooperation the next period. Hence, for some T, we have that for all $t > T$, $x_t \geq N-3$.

[2] Note that depending on the number of cooperators in the first period in relation to the size of the population, the population may converge to all agents cooperating or to a two-period cycle in which N-1 (or N-2) cooperate in period t and N-5 (or N-6) cooperate in period t+1.

(ii) When $a-c>b-d$, we have that $\bar{U}_{C^0}(D) > \bar{U}_{C^0}(C)$ implies $\bar{U}_{D^0}(D) > \bar{U}_{D^0}(C)$ Moreover, it is easily seen that if $\bar{U}_{C^0}(D) > \bar{U}_{C^0}(C)$ for $x{=}N$, it also holds for any $x<N$. Finally, $\bar{U}_{C^0}(D) > \bar{U}_{C^0}(C)$ for $x{=}N$ if

$$N\left[\alpha\frac{2(b-d)-(a-c)}{4}-(b-a)\right] > (a-d)-\frac{\alpha}{4}[2(b-d)+(a-c)].$$

Hence, when this condition holds, the C^0 agent at the boundary will switch to defection and the D^0 agent will continue to defect. This process of C^0 agents switching to defection continues until all agents defect. (The last cooperative agent may belong to the C^- category and he will switch to defection as well.)

(iii) Suppose that the conditions under (i) and (ii) are both *not* satisfied. For any given x, N and pay-off parameters, there are x^{1*} and x^{2*}, defined by equations (4) and (5). From these equations it is clear that $x^{1*} > x^{2*}$. We also know that $x^{1*} > 0$ and $x^{2*} < N$. I will argue that if $x_0 > x^{1*}$, the analysis under (i) remains valid, if $x_0 < x^{2*}$ the analysis under (ii) remains valid and if $x^{2*} < x_0 < x^{1*}$ there will be coexistence and the boundary remains where it is. Suppose $x_0 > x^{1*}$. It is easily seen that in period 0, $\bar{U}_{D^0}(C) > \bar{U}_{D^0}(D)$. Following the argument under (i) reveals that in period 1 x has increased by 2. However, when $a-c>b-d$ more x favors cooperation, i.e., $\bar{U}_{D^0}(C) > \bar{U}_{D^0}(D)$ in period $1,2,\ldots.$ Hence, the analysis under (i) holds true. Similarly, if $x_0 < x^{2*}$, $\bar{U}_{C^0}(D) > \bar{U}_{C^0}(C)$ in period 0 and the decrease in x favors defection in subsequent periods. Finally, consider $x^{2*} < x_0 < x^{1*}$. In this case in period 0, $\bar{U}_{D^0}(C) < \bar{U}_{D^0}(D)$ and $\bar{U}_{C^0}(C) > \bar{U}_{C^0}(D)$. Hence, nobody will switch to another behavior in period 1 or in any subsequent period. This implies coexistence of the two regions.

Proof of Proposition 3

(i) If $a-c<b-d$, then $\bar{U}_{C^0}(C) > \bar{U}_{C^0}(D)$ implies $\bar{U}_{D^0}(C) > \bar{U}_{D^0}(D)$. Also $\bar{U}_{C^0}(C) > \bar{U}_{C^0}(D)$ holds for all x if it holds for $x=N$. The inequality holds for $x=N$ if $(a-d)(N-1) > \frac{\alpha}{4}(N-1)[2(b-d)+(a-c)]+N(1-\alpha)(b-d)$, which can be rewritten as

$$N\left[\alpha\frac{2(b-d)-(a-c)}{4}-(b-a)\right] > (a-d)-\frac{\alpha}{4}[2(b-d)+(a-c)].$$

For the rest of the argument, I refer to the proof of Proposition 1 (i).

(ii) As above: $\bar{U}_{D^0}(D) > \bar{U}_{D^0}(C)$ holds for all x if it holds for $x=0$. Moreover, $\bar{U}_{D^0}(D) > \bar{U}_{D^0}(C)$ implies $\bar{U}_{C^0}(D) > \bar{U}_{C^0}(C)$ and $\bar{U}_{D^0}(D) > \bar{U}_{D^0}(C)$ holds for $x=0$ if

$$N\left[\alpha\frac{2(a-c)-(b-d)}{4}-(d-c)\right] > (a-d)-\frac{\alpha}{4}[2(a-c)+(b-d)].$$

For the rest of the argument, I refer to the proof of Proposition 1 (ii).

(iii) Suppose that the conditions under (i) and (ii) are not satisfied. For any given α and N, there are $x^{1\sim}$ and $x^{2\sim}$ defined by the following equalities:

$$(a-d)(N-1) = \frac{\alpha}{4}(N-1)[2(b-d)+(a-c)]$$
$$+(1-\alpha)[N(a-c) - x^{1\sim}((b-d)-(a-c))]$$

and

$$(a-d)(N-1) = \frac{\alpha}{4}(N-1)[2(a-c)+(b-d)]$$
$$+N(1-\alpha)(a-c) + x^{2\sim}(1-\alpha)[(b-d)-(a-c)].$$

From the above equations it is clear that $x^{1\sim} < x^{2}\ \sim$. Also, I know that $x^{1\sim} < N$ and $x^{2\sim} > 0$, but it may be that $x^{1\sim} < 0$ and $x^{2\sim} > N$. Consider first an arbitrary $x_t > x^{2\sim}$. As in this case $\bar{U}_{C^0}(D) > \bar{U}_{C^0}(C)$ and $\bar{U}_{D^0}(D) > \bar{U}_{D^0}(C)$, $x_{t+1} = x_t - 2$. On the other hand, for an arbitrary $x_t \in [3, x^{1\sim})$, $\bar{U}_{C^0}(C) > \bar{U}_{C^0}(D)$ and $\bar{U}_{D^0}(C) > \bar{U}_{D^0}(D)$ so that $x_{t+1} = x_t + 2$.
Let us then consider an $x_t \in [x^{1\sim}, x^{2\sim}]$ and suppose there is one region with cooperative behavior. It is clear that both the D^0 and the C^0 agent will switch actions in period $t+1$ so that $x_{t+1} = x_t$. At each of the two boundaries on the circle between cooperators and defectors the sequence of categories at $t+1$ is as follows: $C^+ - C^0 - D^+ - C^- - D^0 - D^-$, where the $D^+(C^-)$ agent is the $C^0(D^0)$ agent of period t. It is clear that at $t+2$, both the C^0 and C^- agent of $t+1$ will switch to defection. Hence, taken the two boundaries together $x_{t+2} = x_t - 4$.
Suppose then that $x^{1\sim} > 6$. Consider $x_t > x^{1\sim}$. From the above, it is clear that $x_{t+2} = x_t - 4$. If $x_{t+2} > x^{1\sim}$, then $x_{t+4} = x_{t+2} - 4$. If, on the other hand $x_{t+2} < x^{1\sim}$, then $x_{t+3} = x_{t+2} + 2$. If $x_{t+3} < x^{1\sim}$, then $x_{t+4} = x_{t+2} + 4$ and I have a cycle of period 4 around $x^{1\sim}$. If $x_{t+3} > x^{1\sim}$, then $x_{t+5} = x_{t+3} - 4$. As $x^{1\sim} > 6$, and as x_t is an integer x_{t+5} is either 3 or 4. This implies that $x_{t+6} = x_{t+5} + 2$ and $x_{t+7} = x_{t+6} + 2 = x_{t+3}$, establishing (again) a cycle of period 4. A similar analysis applies if $3 \leq x_t \leq x^{1\sim}$.
Suppose then that $x^{1\sim} < 5$. If $x_t > x^{1\sim}$, then $x_{t+2} = x_t - 4$. If x_{t+2} is equal to 1 or 2, then $x_{t+3} = 0$. If on the other hand, x_{t+2} equals 3 or 4, then $x_{t+3} = x_{t+2} + 2$ and $x_{t+3} > x^{1\sim}$. In this case, however, $x_{t+5} = x_{t+3} - 4$ and x_{t+5} is equal to 1 or 2.

References

1. Bergstrom, T. 1995. On the Evolution of Altruistic Ethical Rules for Siblings. *American Economic Review* **85**: 58-81.
2. Bergstrom, T., and O. Stark, 1993. How Altruism can prevail in an Evolutionary Environment, *AEA Papers and Proceedings* **83**: 149-155.

3. Eshel, I., L. Samuelson and A. Shaked. 1998. Altruists, Egoists and Hooligans in a Local Interaction Model. *American Economic Review.* **88**: 157-179.
4. Goyal, S. and M. Janssen. 1997. Non-Exclusive Conventions and Social Coordination. *Journal of Economic Theory.* **77**: 34-57.
5. Huck, S. and J. Oechssler. 1995. The Indirect Evolutionary Approach to Explaining Fair Allocations. Department of economics, Humboldt University Berlin, Mimeo.
6. Kandori, M. 1992. Social Norms and Community Enforcement. *Review of Eonomic Studies* **59**: 63-80.
7. Vega-Redondo, F. 1997. The Evolution of Walrasian Behavior. *Econometrica* **65**: 375-84.

Bubbles and Long Range Dependence in Asset Prices Volatilities*

Alan Kirman[1] and Gilles Teyssière[2]

[1] GREQAM, EHESS, Université d'Aix-Marseille III, Institut Universitaire de France.
[2] GREQAM, Université Libre de Bruxelles.

Abstract. A model for a financial asset is constructed with two types of agents. The agents differ in terms of their beliefs. The proportions of the two types change over time according to a stochastic process which models the interaction between the agents. Thus, unlike other models, agents do not persist in holding "wrong" beliefs. Bubble-like phenomena in the asset price occur. We consider several tests for detecting long range dependence and change–points in the conditional variance process. Although the model seems to generate long–memory properties of the volatility series, we show that this is due to the switching of regimes which are detected by the tests we propose.

Keywords: Interaction, bubbles, testing, long–memory, heteroskedasticity, change–point.

1 Introduction

Price dynamics are at the heart of economics. Aggregate movements do not correspond to those that might be generated by the maximising behaviour of a single individual. The problem of explaining such dynamics is one to which Claus Weddepohl has made substantial contributions. One way of explaining the complex dynamics is to suggest that the economy or market is trying to adjust to the equilibrium and that whilst it has not converged out of equilibrium phenomena such as rationing play an important role. This is the route that Claus has taken. We propose to take another route here and to follow an approach which has been also developed by Cars Hommes and others. We suggest that what we are observing in the price behaviour are bubbles created by the direct interaction between the agents at the micro level.

* We would like to thank Hans Föllmer with whom the stochastic model was developed, Raj Bhansali, George Evans, Vitór Gaspar, Liudas Giraitis, Clive Granger, Niels Haldrup, Cars Hommes, Piotr Kokoszka, Remis Leipus, Grayham Mizon, Franz Palm, Danny Quah, Mark Salmon, Tom Sargent, Murad Taqqu, Richard Topol, and Michael Woodford for helpful comments and suggestions on a series of related papers. Both authors will blame each other for the remaining mistakes.

It is surprising that the idea of testing for the existence of "price bubbles" in markets has occurred so recently since, although particularly associated with the markets for financial assets, such phenomena have been documented for a wide variety of markets over a considerable period of time. One of the earliest bubbles was that in the price of red mullet in the first century A.D. The red mullet fever is documented by Cicero, Horace, Juvenal and Martial. A survey of other historical bubbles, such as the Tulip, South Sea and Mississippi bubbles, may be found in Garber (2000). More recently there has been a substantial literature on the theoretical basic for and testing of bubbles, see for example Blanchard and Watson (1982), Flood and Garber (1980), Meese (1986), Tirole (1985), West (1988), Woo (1987), Stiglitz (1990), Flood and Hodrick (1990), Avery and Zemsky (1998) and Shiller (2000).

There are two basic problems involved in the discussion of bubbles, on the one hand their definition, and on the other their detection and identification. This paper, after a brief discussion of the literature, will present an economic model which produces price bubbles and will then examine how various of the tests proposed for detecting bubbles perform on the data generated by the model. The latter is based on a simple stochastic process developed by the first author with Hans Föllmer. It is this process which produces the bubble-like phenomena resulting from agents changing their forecasts and which was used as the basis for the analysis in Kirman (1991, 1993). In introducing bubbles we follow Evans (1991), with two differences. Firstly, instead of simply testing data from a stochastic process with bubble-like characteristics, we use data from a model of economic behaviour with interacting agents. Secondly, since this model is characterized by switches from one type of stochastic process to another, we apply several procedures developed by Kokoszka and Leipus (2000), Horváth, Kokoszka and Teyssière (2001) and Kokoszka and Teyssière (2001) for detecting changes in regime in the conditional second moments.

In Kirman and Teyssière (2000), we have shown that this class of models is able to replicate the empirical long–memory properties of asset prices volatilies: while asset prices returns are uncorrelated, their power transformation display strong dependence, the degree of which is common to several financial time series. Interested readers are referred to Beran and Ocker (2001), Cheung (1993), Granger and Ding (1995, 1996), Mandelbrot (1997), Taylor (1986), Teyssière (1997,1998) for references on these empirical properties. In the model by Kirman and Teyssière (2000) this common degree of long–memory was linked to the swing in opinions of interacting agents. This is in line with the works of Mikosch and Stărică (1999) who claimed that the strong dependence in asset prices volatility is the outcome of some form of non stationarity and changing regimes in the conditional variance process. Thus, it is of interest to test whether the long range dependence generated by our model is spurious, and is the outcome of a change–point process.

When discussing markets in general, it is difficult to separate significant swings in prices into their two components, those due to speculation and those due to swings in fundamental characteristics of the economy, some of which may be in principle measurable such as technological change, and others of which such as tastes may be unmeasurable. However, when considering the market for financial assets the situation is somewhat simpler. Theory suggests that the price of a share should, for example, reflect the discounted expected value of future dividends (the fundamentals). Thus, any prolonged departure from the underlying fundamentals could be defined as a bubble as Froot and Obstfeld (1989), for example, suggest.

We will require two features of bubbles which go beyond the simple departure from fundamentals. Firstly, they should "burst" at some time and not be perpetually explosive, see Diba and Grossman (1988), and secondly that they should be endogenous, i.e., not directly produced by exogenous shocks.

The earlier part of the recent literature on bubbles, such as the contributions of Le Roy and Porter (1981), Shiller (1981) and Blanchard and Watson (1982), all came to the view that asset prices were too volatile to be explained by fundamentals alone. Thus it was argued that there was "excess volatility". Meese and Rogoff (1983) came to the same conclusion for exchange rates.

The debate then, however, swung somewhat in the opposite direction. In part it has been suggested that the econometric analysis in the papers mentioned was faulty and in part that the process governing the fundamentals had been misspecified, a good idea of the main issues in this discussion can be obtained from Campbell and Shiller (1987), Flavin (1983), Kleidon (1986), Mankiw, Romer and Shapiro (1985), Marsh and Merten (1986), West (1987, 1988), and Flood and Hodrick (1990), or even that some unobserved fundamentals might have been omitted, see Hamilton and Whiteman (1985).

As Campbell *et al.* (1997) point out much of the heat has now gone out of this debate since it is now recognised that a rejection of a model which discounts future dividends at a constant rate is not a rejection of the efficient market hypothesis and partly because economists have become convinced that expected asset returns are time–varying. However, this does not in any way answer the basic question of this paper, which is simply, can there be bubbles in the data on asset returns and how might they be detected?

Diba and Grossman (1988) claimed that the data for stock prices do not have the explosive characteristics one would expect if bubbles existed. Perron (1989) however, suggested that the unit roots tests commonly used may fail to reject the presence of unit roots, when in fact the underlying process is one with a "broken trend" or a shift in regime; interested readers are referred to Stock (1994) for a survey. Indeed Evans (1991) has found, by testing data from simulating a stochastic process known to contain bubbles, that in general the unit roots hypothesis was not rejected. Thus, whether or not there are bubbles in asset price data is still an open question.

The real question seems, however, not to be quite as simple as that discussed in the general debate on the issue. If asset prices do in general follow fundamentals, but periodically depart from them, then the situation is rather complicated. If fundamentals follow a random walk as theory might suggest in the case of stock prices for example, then for some, maybe substantial, part of the time this is the process that will be realised. The process that is followed at other periods has, of course, to be specified. If the process always returns, in the long run, to the fundamentals, then this is rather different from the situation envisaged by Perron (1989) for example. He argues that there are occasional large exogenous shocks that cause a permanent change in the system but that between these shocks, stock prices are trend stationary.

The sort of model which we are suggesting will be permanently affected by shocks, but these shocks are endogenous. However, its characteristics are determined by two things. The relative time it spends in the fundamental phase and the particular characteristics of the process when it is not in that phase. Unfortunately, the natural characterisation of the non fundamental phase following the literature in finance is one which has a class of processes and not a single one. Thus, Hamilton's (1989) suggestion that the market could be thought of as switching from one stationary process to another is not really applicable.

In particular, it should be clear that the basic point at issue here is not whether the asset price process has unit roots but how far and for how long does it deviate from that process and how can one separate out these deviations? In a certain sense, in the long run, as indeed the word suggests, bubbles do not matter, but their impact in the short run may be very significant.

In the foreign exchange market which we take as an example, roughly two thirds of all turnover consists of spot transactions, see Suvanto (1993), McKinnon (1979) and Kouri (1983). Since dealers have very short horizons, many have to have a closed position at the end of the day, and their customers are sensitive to price changes, it is clear that episodes in which extrapolatory behaviour can take the market away from fundamentals can be very important. Yet most dealers argue that, "in the long run fundamentals matter", see Barrow (1994). The way in which fundamentals eventually pull prices back is through underlying customer flows, see Kouri (1983), but since these in turn are affected by the evolution of current prices the magnitude and duration of deviations are difficult to calculate.

All of this suggests that the testing strategy used up to now to detect the presence of bubble-like episodes is not appropriate, as will become clear.

We shall now present a simple model which has the basic characteristics just outlined.

2 A Simple Theoretical Model

This simple economic model, developed in Kirman (1991) and originally based on experimental evidence on the behaviour of ants, Kirman (1993), captures four essential features which one would like financial markets with bubbles to have. Firstly, there are underlying fundamentals which the price process follows for certain periods. Secondly, agents are heterogeneous in the sense that over time different opinions prevail in the population. Thirdly, unlike most of the literature in which there are several groups of agents, see for example the papers on "Noise Traders" by De Long *et al.*(1989, 1990a, 1990b, 1991) on "Bulls, bears and market sheep" by Day and Huang (1990) and on "herding" by Sharfstein and Stein (1990), none of the agents in this model can be classified as "irrational", in the sense that they cling to beliefs which are revealed to be "wrong" by their observations. Indeed, all the agents in the model behave rationally in the sense that each agent plays her best response given the common knowledge about the structure of the beliefs and behaviour in the model. Bubbles, in the standard sense, occur in the price process as prices depart from fundamentals and then return to them. Finally, bubbles can be both negative and positive. Models which are close in spirit to this sort of model are those of Brock and Hommes (1999) and Föllmer and Schweizer (1993).

We now give a brief description of the model which was developed with Hans Föllmer. The model is then simulated to generate the data to be used for evaluating the standard testing procedures for bubbles and change point or their absence.

2.1 The Model

Agents are faced with a price process P_t for a financial asset and form expectations about tomorrow's prices. There are two different ways[1] of forming expectations and each agent uses one of them. However, the expectations of the individual agents are influenced by random meetings with other agents. Call the two methods of forming expectations the two "opinions" in the model, then if there are N agents, we say that

> The state of the system is defined by the number k of agents holding opinion one, i.e., $k \in \{0, 1, \dots, N\}$.

The stochastic process governing the state evolves as follows. Two agents meet at random and the first is converted to the second's view with probability $(1 - \delta)$. If the meeting is considered as a drawing from an urn with balls of

[1] There could be any finite number of ways of forming expectations and this would not change the nature of the results. Franz Palm suggested that agents might choose a convex combination of the two processes but this could not be handled within the framework developed here.

two different colours, it is obvious that which agents is the "first" and which is the "second" is of no importance since the symmetric event occurs with the same probability. There is also a small probability ε that the first agent will change her opinion independently of whom she meets. This is a technical necessity to prevent the process from being "absorbed" into one of the two states O or N, but can be allowed to go to zero as N becomes large.[2] Indeed in what follows we shall require for the basic results that ε be small.

The process then evolves as follows:

$$\begin{aligned} &k \longrightarrow k+1, \text{ with probability } p(k,k+1) = (1-\tfrac{k}{N})(\varepsilon + (1-\delta)\tfrac{k}{N-1}) \\ &k \longrightarrow k, \qquad \text{with probability } p(k,k) = 1 - p(k,k+1) - p(k,k-1) \\ &k \longrightarrow k-1, \text{ with probability } p(k,k-1) = \tfrac{k}{N}(\varepsilon + (1-\delta)\tfrac{N-k}{N-1}) \end{aligned} \tag{1}$$

The first problem is to look at the equilibrium distribution $\mu(k)$, $k = 0, \ldots, N$ of the Markov Chain defined by (1).

This is important in the economic model since it describes the proportion of time that the system will spend in each state. The distribution is given by

$$\mu(k) = \sum_{l=0}^{N} \mu(l)p(l,k), \tag{2}$$

but given that the process is symmetric and reversible it follows that

$$\mu(l)p(l,k) = \mu(k)p(k,l). \tag{3}$$

From this expression one obtains

$$\frac{\mu(k+1)}{\mu(k)} = \frac{p(k,k+1)}{p(k+1,k)} = \frac{\left(1-\frac{k}{N}\right)\left(\varepsilon + (1-\delta)\frac{k}{N-1}\right)}{\frac{k+1}{N}\left(\varepsilon + (1-\delta)\left(1-\frac{k}{N-1}\right)\right)}, \tag{4}$$

since it is clear from (3) that

$$\mu(k) = \frac{\frac{\mu(1)}{\mu(0)} \cdots \frac{\mu(k)}{\mu(k-1)}}{1 + \sum_{l=1}^{N} \frac{\mu(1)}{\mu(0)} \cdots \frac{\mu(l)}{\mu(l-1)}}. \tag{5}$$

---

[2] This ε can be thought of as the replacement of some old agents in each new period by agents who may hold either opinion (see for example the evolutionary model of Young and Foster (1991)) or by some external shock which influences some people's expectations.

Now the form of $\mu(k)$ will depend, naturally, on the values of ε and δ. It is easy to see that if $\varepsilon < (1-\delta)/(N-1)$ then $\mu(k)$ will indeed be convex. Thus this case, in which the process spends most of its time in the extremes corresponds to the case in which the probability of "self conversion" is small relative to the probability of being converted by the person one meets. Although this probability of conversion is independent of the numbers in each group, which type will actually meet which type depends on the relative numbers in each type at any moment, i.e. on the state of the system. Thus, when one type is in the minority conversion of any individual is much less likely than when the numbers of the two types are fairly equal.

The ε might be considered as being simply a technical artefact, therefore it is worth looking at what happens to the process when N becomes large and ε goes to zero. Consider the asymptotic form of $\mu(k)$ when we choose ε for each N so that $\varepsilon < (1-\delta)/N$. When N becomes large redefine μ as $\mu(k/N)$ and consider the limit distribution as $N \to \infty$. Call this limit distribution, which will be continuous, f. Then one can prove the following[3]

Proposition 1. *f is the density of a symmetric Beta distribution i.e. $f(x) = Cx^{\alpha-1}(1-x)^{\alpha-1}$, where C is a constant.*

This stochastic model of shifts of opinion given here is related to the urn models of Arthur *et al.* (1985) and also to models which have been developed for shifts in voter opinion (see the examples given by Weidlich, cited in Haken (1977), where a similar bimodal distribution is derived). The latter model could also have been taken as the basis for the conversion from one opinion to another here.

2.2 The Market for a Financial Asset

Consider two types of individuals who forecast the value of an asset or, as in the model developed by Frankel and Froot (1986), the value of the exchange rate.

"Fundamentalists" believe that the exchange rate P_t at time t is related to some underlying fundamental $\bar{P}_t$ which might be a constant $\bar{P}$, some long run equilibrium, or might be governed by some dynamic deterministic or stochastic process. Their forecast for the value at the next period, conditional on the information set I_t available at time t, is given by

$$E^f\left(P_{t+1}|I_t\right) = \bar{P}_t + \sum_{j=1}^{M_f} \nu_j (P_{t-j+1} - \bar{P}_{t-j}), \tag{6}$$

where ν_j, $j = 1, \ldots, M_f$ are positive constants, M_f is the memory of the fundamentalists. We assume that the fundamentals $\bar{P}_t$ follow a random walk,

[3] Clearly for any given value of k, $\mu(k/N)$ increases proportionately with N. This proposition was proved by Hans Föllmer and the proof is given in Kirman (1993).

i.e.,

$$\bar{P}_t = \bar{P}_{t-1} + \varepsilon_t, \quad \varepsilon_t \sim N(0, \sigma_\varepsilon^2). \tag{7}$$

Chartists, on the other hand, forecast by simple extrapolation of the past history of prices [4] and hence predict that the next period exchange rate will be given by

$$E^c\left(P_{t+1}|I_t\right) = \sum_{j=0}^{M_c} h_j P_{t-j} \quad \text{with} \quad \sum_{j=0}^{M_c} h_j = 1, \tag{8}$$

where h_j, $j = 0, \ldots, M_c$ are constants, M_c is the memory of the chartists.

The market view or forecast is given by a weighted average of the two forecasts, i.e.,

$$E^m(P_{t+1}|I_t) = k_t E^f(P_{t+1}|I_t) + (1 - k_t) E^c(P_{t+1}|I_t). \tag{9}$$

The weights are determined in Frankel and Froot's model by portfolio managers who effectively choose the weights in such a way as to make the actual outcome constant with the market forecast.

In the model here the weights k_t are determined endogenously, and they will depend on the number of agents who act according to each view. To see this consider the process as taking place in two steps. Firstly, individuals meet each other and form an opinion after these meetings about how prices will change. "Meeting" of course does not mean meeting in the literal physical sense, a better word might be "contact". A foreign exchange dealer is, for example, faced with several screens full of general information and indicative quotes given by other dealers. In addition he has loudspeakers from which he can hear brokers' quotes, and furthermore he is in telephone communication with other dealers and with his clients. Each of these individual quotes or "contacts" may cause him to shift his expectations and thus modify his action.

In addition agents also observe the market as a whole and decide what is the prevailing opinion. This adheres to Keynes's beauty queen idea. If I am interested in forecasting who will win the contest what matters is not what I think but what the others think. Again, pursuing the example of a foreign exchange dealing room, dealers maintain that they focus on specific pieces of information, this corresponds to the "meetings" in the model, but that they are also aware of all the information with which they are faced and this they somehow process to obtain an idea of "market sentiment". Having made their specific and general observations, agents forecast and then act accordingly.

Thus the process can be described as follows:

[4] It would, of course, be interesting to try other more sophisticated forms of extrapolation. Although this would change the dynamics of the model it would not change the basic stochastic alternation between regimes.

(i) Agents meet each other at random and are converted to each others' opinions as defined in the process described above. Consider w_t as the number of individuals at time t who are "fundamentalists" and the remaining $N - w_t$ as "chartists". Allow some fixed number M of meetings to take place at each time.

(ii) Defining $q_t = w_t/N$ each agent now makes an observation of q_t i.e., tries to assess which opinion is in the majority. She observes q_t with some noise. Thus, the signal she receives is

$$q_{i,t} = q_t + \varepsilon_{i,t} \quad \varepsilon_{i,t} \sim N(0, \sigma_\vartheta^2). \tag{10}$$

If now agent i receives a signal $q_{i,t} \geq 1/2$, then she will make a fundamentalist forecast since the majority is doing so. Conversely, if $q_{i,t} < 1/2$ she will forecast as a chartist and act accordingly. The number and proportion of agents who base their demand on fundamental forecasts are therefore given by

$$k_t = \frac{1}{N} \# \left\{ i : q_{i,t} \geq \frac{1}{2} \right\}. \tag{11}$$

The natural question here is whether it is rational for agents to act in this way. One might think that an agent should act according to her own beliefs. However, if she knows the structure of the model and knows herself to be in a minority it is not rational for her to base her demand on her own forecast rather than that of the majority since it is the latter that determines the price. Indeed it is easy to see that, if everybody knows that everybody else is following the majority, the best response is to do so oneself. Thus, a Nash equilibrium will occur if everybody follows majority behaviour.[5] It should be noted that the relationship between q_t and k_t depends crucially on the variance of the noise in the signal about the majority. If this variance is large the process will spend more time close to the equal proportion situation than does the k_t process. It seems reasonable to consider the variance as small since otherwise agents should give more weight to their private information. Having now determined the proportion of agents who forecast as fundamentalists the market forecast is given by

$$E^m(P_{t+1}|I_t) = k_t E^f(P_{t+1}|I_t) + (1 - k_t) E^c(P_{t+1}|I_t). \tag{12}$$

The price on the market, i.e., the market exchange rate if one is considering the market for foreign exchange, is given by

$$P_t = c \Delta P^m_{t+1} | I_t + Z_t, \tag{13}$$

where c is a constant and Z_t is an index of a vector of fundamental variables according to Frankel and Froot (1986). Economists will require, of course, that

[5] The logic is inescapable but there is a problem, of course, of indeterminacy. For a discussion of similar problems see Ellison and Fudenberg (1993).

this model be justified by some underlying model of individual behaviour. Consider the following simple:[6]

Agent i has a utility function given by

$$U^i(W^i_{t+1}) = E(W^i_{t+1}) - \lambda V(W^i_{t+1}), \tag{14}$$

where λ denotes the risk aversion coefficient, $E(.)$ and $V(.)$ denote the expectation and variance operators, and W^i_{t+1}, her wealth at time $t+1$, is given by

$$W^i_{t+1} = (1+\rho_{t+1})P_{t+1}d^i_t + (W^i_t - P_t d^i_t)(1+r). \tag{15}$$

The variables are defined as follows:

- ρ_{t+1} is the dividend in foreign currency paid on one unit of foreign currency;
- P_{t+1} is the exchange rate at time $t+1$;
- d^i_t is the demand by the ith individual for foreign currency;
- r is the interest rate on holdings of domestic currency.

P_{t+1} the exchange rate at time $t+1$ and ρ_{t+1} the foreign dividend are both considered by agents to be random variables. The first two moments of the distribution of P_{t+1}, from the point of view of individual i, are given by[7]

$$E(P_{t+1}) = \Delta P^i_{t+1} + P_t, \quad V(P_{t+1}) = \sigma^2_P, \tag{16}$$

and for ρ_{t+1} by

$$E(\rho_{t+1}) = \rho, \quad V(\rho_{t+1}) = \sigma^2_\rho. \tag{17}$$

Furthermore, assume ρ_{t+1} and P_{t+1} to be independent. From these assumptions

$$E(W^i_{t+1}|I_t) = (1+\rho)E^i(P_{t+1}|I_t)d^i_t + (W^i_t - P_t d^i_t)(1+r), \tag{18}$$

and

$$V(W^i_{t+1}|I_t) = (d^i_t)^2\zeta_t \quad \text{with} \quad \zeta_t = V\left(P_{t+1}(1+\rho_{t+1})\right). \tag{19}$$

Demand d^i_t is found by maximising utility and writing the first order condition

$$(1+\rho)E^i(P_{t+1}|I_t) - (1+r)P_t - 2\zeta_t\lambda d^i_t = 0, \tag{20}$$

i.e.,

$$d^i_t = \frac{(1+\rho)E^i(P_{t+1}|I_t) - (1+r)P_t}{2\zeta_t\lambda}, \tag{21}$$

[6] This example was suggested by Michael Woodford.

[7] Of course, in fact, the variance of P_{t+1} is time dependent, but we can pardon the agents for making this mistake in the second moment if we accept "noise traders" who fail even to observe the correct mean!

where $E^i(.|I_t)$ denotes the expectation of an agent of type i. Let k_t be the proportion of fundamentalists at time t, the market demand is then given by:

$$d_t = \frac{(1+\rho)\left(k_t E^f(P_{t+1}|I_t) + (1-k_t)E^c(P_{t+1}|I_t)\right) - (1+r)P_t}{2\zeta_t\lambda}. \tag{22}$$

Now consider the supply of foreign exchange X_t and recall that agents only differ in their forecasts as to the value of the future exchange rate, then the market is in equilibrium if $X_t = d_t$, which gives

$$(1+r)P_t = (1+\rho)\left(k_t E^f(P_{t+1}|I_t) + (1-k_t)E^c(P_{t+1}|I_t)\right) - 2\zeta_t\lambda X_t. \tag{23}$$

In Gaunersdorfer and Hommes (2000) X_t is set to zero. In that case, we refer to d_t as excess demand. We suppose here that $2\zeta_t\lambda X_t/(1+\rho) = \gamma\bar{P}_t$. If we assume that $M_f = M_c = 1$, then the equilibrium price is given by

$$P_t = \frac{k_t - \gamma}{A}\bar{P}_t - \frac{k_t\nu_1}{A}\bar{P}_{t-1} + \frac{(1-k_t)h_1}{A}P_{t-1}, \tag{24}$$

where

$$A = \frac{1+r}{1+\rho} - (1-k_t)h_0 - k_t\nu_1. \tag{25}$$

Thus, when k_t switches from 0 to 1 and vice–versa, equation (24) defines a change point process in the conditional mean. It is then of interest to test whether this process generates changes in regime in the conditional variance.

3 Simulations and Testing

The series P_t generated by the underlying Markov process described above have the following features:

(i) time varying coefficients,
(ii) unit roots,
(iii) heteroskedastic errors,[8] the conditional heteroskedastic structure having varying coefficients,
(iv) long-range dependence in the power transformation of the returns $|R_t|^\delta$, where $R_t = \Delta\ln(P_t)$ and $\delta > 0$.

In fact, the series $|R_t|^\delta$ share the second moment properties of long-memory processes, i.e., a slow hyperbolic rate of decay of the autocorrelation function, henceforth ACF, $\rho(k) = O(k^{2d-1})$, where $d \in (0, 1/2)$ is the parameter which

[8] These properties are shared by the high frequency data from the London FOREX market, see Goodhart *et al.* (1993).

governs the slow rate of decay of the ACF and then parsimoniously summarizes the degree of long-range dependence of the series. Interested readers are referred to Beran (1994) and Robinson (1994) for a complete reference on long-memory processes. These empirical features are well captured by the class of long-memory ARCH, ARCH(∞), processes introduced by Robinson (1991) and developed by Granger and Ding (1995) and other authors. The general form of an ARCH(∞) process is:

$$R_t = m(R_t) + \sigma_t \varepsilon_t, \quad \varepsilon_t \sim D(0,1), \quad \sigma_t^\delta = \omega + \psi(L) g(\varepsilon_t), \tag{26}$$

where $m(R_t)$ denotes the regression function, $\psi(L) = \sum_{i=1}^{\infty} \psi_i L^i$ is an infinite order lag polynomial the coefficients of which are positive and have asymptotically the following hyperbolic slow rate of decay $\psi_j = O\left(j^{-(1+d)}\right)$, $g(\varepsilon_t)$ is a function of the innovations ε_t including non-linear transformations, see Teyssière (2001), $D(0,1)$ is a distribution with mean equal to zero and variance equal to one, and $\sigma_t^\delta - g(\varepsilon_t)$ is a martingale difference sequence. The memory properties of the class of ARCH(∞) model have been studied by Giraitis, Kokoszka and Leipus (2000), Kazakevičius and Leipus (2000), Kazakevičius, Leipus and Viano (2000), and Giraitis and Surgailis (2001). This long-range dependence in the conditional second moments can alternatively be modeled by the class of long-memory stochastic volatility processes, interested readers are referred to Robinson (2001) for a complete theory on the memory properties of LMSV models.

However, the first moments of the series $|R_t|^\delta$ differ from the ones of a long-memory process: while a long-memory process exhibits local trends, the slope of which is increasing with the parameter d, the series $|R_t|^\delta$ is not trended.[9] For that reason, we term this type of processes as "pseudo long-memory processes".

The parameters of the model are tuned so that these properties are satisfied. As in Kirman and Teyssière (2000), we test that the returns series is $I(0)$ by using the statistics proposed by Lo (1991), Kwiatkowski *et al* (1992) and Giraitis, Kokoszka, Leipus and Teyssière (2001). These statistics are based on the partial sum process $S_k = \sum_{t=1}^{k} (y_t - \bar{y})$ and the assumption that under the null hypothesis of $I(0)$, the standardised partial sum process satisfies a functional central limit theorem. Lo (1991) considered the standardised range of S_k, i.e.,

$$\tilde{Q}_n(q) = \frac{1}{\hat{\sigma}(q)} \left[\max_{1 \le k \le T} S_k - \min_{1 \le k \le T} S_k \right]. \tag{27}$$

Kwiatkowski *et al* (1992) considered the standardised second moment of S_k:

$$KPSS(q) = \frac{1}{T^2 \hat{\sigma}^2(q)} \sum_{k=1}^{T} S_k^2, \tag{28}$$

[9] We are grateful to Clive Granger for pointing out these features.

while Giraitis *et al.* (2001) considered the standardised variance of S_k:

$$V/S(q) = \frac{1}{T^2\hat{\sigma}^2(q)}\left[\sum_{k=1}^{T} S_k^2 - \frac{1}{T}\left(\sum_{k=1}^{T} S_k\right)^2\right], \tag{29}$$

where $\hat{\sigma}^2(q)$ is the Newey and West (1987) heteroskedastic and autocorrelation consistent variance estimator:

$$\hat{\sigma}^2(q) = \hat{\gamma}_0 + 2\sum_{i=1}^{q} \omega_i(q)\hat{\gamma}_i, \quad \omega_i(q) \equiv 1 - \frac{1}{q+1}, \tag{30}$$

where the sample auto-covariances $\hat{\gamma}_i$ at lag i account for the possible short-range dependence up to the qth order. The V/S statistic is less sensitive to the choice of the truncation order q than Lo's (1991) statistic, and is more powerful than the KPSS statistic. We use the same tests for checking that the power transformation of returns display long-memory.

In Kirman and Teyssière (2000), we have shown that this model is able to generate the sort of strong dependence observed in the volatility of asset returns, and that the degree of long–memory is controlled by the swings in opinions, i.e., the evolution of the process governing k_t. It is well known that statistical tests wrongly detect long–range dependence when the true DGP is a change–point process. In particular, according to Mikosch and Stărică (1999), long-range dependence in the volatility process can be spurious and the consequence of the concatenation of short-range dependent GARCH(1,1) processes with changing coefficients. Given that in our model, asset prices P_t are a varying combination of the previous prices P_{t-1} and the fundamentals $\bar{P}_t$ and $\bar{P}_{t-1}$, it is interesting to check whether we are able to detect the changes in the proportion of fundamentalists k_t, i.e., the swings in opinion, which generate the long-memory in the volatility process. There is a substantial literature on change–point analysis, see e.g., Csörgő and Horváth (1997) Besseville and Nikifirov (1993) for recent references, which focuses on conditional mean processes. In our case, we are interested in the conditional variance processes, and we then use the tests for change–point in the conditional variance proposed by Kokoszka and Leipus (2000), Horváth, Kokoszka and Teyssière (2001) and Kokoszka and Teyssière (2001). A recent survey on change–point tests is given in Kokoszka and Leipus (2001). We compare the relative performance of these tests with respect to our non–standard DGP.

Kokoszka and Leipus (2000) proposed a CUMSUM based estimator for change-point in ARCH(∞) processes at unknown time t. This estimator is defined by:

$$\hat{t} = \min\left\{t : |C_t| = \max_{1\leq j\leq T} |C_j|\right\}, \tag{31}$$

where

$$C_t = \frac{t(T-t)}{T^2}\left(t^{-1}\sum_{j=1}^{t} R_j^2 - \frac{1}{T-t}\sum_{j=t+1}^{T} R_j^2\right). \tag{32}$$

Horváth, Kokoszka and Teyssière (2001) proposed several tests for change–point in ARCH sequences, based on the empirical process of standardised residuals. According to Kokoszka and Teyssière (2001), these tests work well for GARCH(1,1) sequences and do have a limit distribution under the null hypothesis, although a bootstrap based inference gives better results than asymptotic inference. The first statistic is a Kolmogorov-Smirnov type statistic. For $1 \leq k \leq T$, define

$$\hat{T}(k,t) = \sqrt{T}\frac{k}{T}\left(1-\frac{k}{T}\right)\left|\hat{F}_k(t) - \hat{F}_k^*(t)\right|, \tag{33}$$

with

$$\hat{F}_k(t) = \frac{1}{k}\#\{i \leq k : \hat{\varepsilon}_i^2 \leq t\}, \quad \hat{F}_k^*(t) = \frac{1}{T-k}\#\{i > k : \hat{\varepsilon}_i^2 \leq t\}, \tag{34}$$

where the $\hat{\varepsilon}_t^2$ denote the sequence of squared standardised residuals of a GARCH(1,1) process fitted on the simulated returns, i.e.,

$$R_t = m + \varepsilon_t, \quad \varepsilon_t \sim N(0, \sigma_t^2), \quad \sigma_t^2 = \omega + \beta\sigma_{t-1}^2 + \alpha\varepsilon_{t-1}^2. \tag{35}$$

The K–S statistic is defined as

$$\hat{M} = \sup_{0 \leq t \leq \infty} \max_{1 \leq k \leq T} |\hat{T}(k,t)|. \tag{36}$$

According to Horváth, Kokoszka and Teyssière (2001), for ARCH(p) models with finite fourth moment the asymptotic distribution of $\hat{M}$ is the same as for Picard's (1985) generalized Kolmogorov–Smirnov statistic.

Horváth, Kokoszka and Teyssière (2001) proposed also a Cramér – von Mises statistic:

$$\hat{B} = \int_0^1 \left\{\frac{1}{T}\sum_{i=1}^{T}[\hat{T}([Ts], \hat{\varepsilon}_i^2)]^2\right\} ds. \tag{37}$$

The distribution function of B can be derived from Blum, Kiefer and Rosenblatt (1961). Kokoszka and Teyssière (2001) proposed other tests for change–point in GARCH(1,1) processes, that are considered in Kirman and Teyssière (2001).

These tests consider the case of a single change–point. Test for multiple change–points in conditional mean processes have been proposed by Lavielle

(1999) and Lavielle and Moulines (2000) among others. Given that the series of k_t switch from one extreme to the other, it is to be expected that the simulated process P_t will have several change–points for large series. Thus, we restrict our analysis to samples of 500 observations to reduce the probability of occurrence of multiple change–points.

The details of the simulation are as follows:

- $T = 500$, (sample size),
- number of simulated series = 10000,
- number of agents, $N = 1000$,
- the number of fundamentalists at the beginning of the process = $N/2$, thus $k_0 = 0.5$,
- $P_0 = 1000$, $\bar{P}_0 = 1050$,
- $h_0 = 0.625$, $h_1 = 0.375$, $\nu_1 = 0.62$,
- $\delta = 0.10$,
- annual foreign interest rate $\rho = 0.07$,
- annual domestic interest rate $r = 0.04$,

The series of error terms used in this simulation have been generated by using the Box-Muller transformation of a sequence of uniform deviates which succeeds Marsaglia's (1996) DIEHARD randomness tests.

A series of standard tests were now run on the simulated data to see if it was possible to reject alternative specifications which have different statistical characteristics from those of the process used in the model. In particular it is interesting to see if an econometrician faced with this data would have been led to reject the random walk hypothesis and whether he would have been able to detect the changes in regimes. Once again, it is important to emphasize that the interest of this exercise is to see what the results are when the test is applied to data which are, in fact, generated by an "inappropriate" process, although this would not be known a priori to the tester.

Table 1 below reports the results of the tests for long-range dependence on the series of simulated returns, absolute and squared returns. The results for the series of returns are close to the ones reported in Giraitis *et al.* (2001) for *i.i.d.* white noise processes of the same sample size. These tests detect long–memory in the simulated series of absolute returns and squared returns. The tests reject the null hypothesis of $I(0)$ more often for absolute returns than for squared returns, in accordance with empirical results on financial series. As with standard long range dependent processes, the V/S statistic appears to be more powerful than the KPSS statistic and less sensitive to the choice of q than Lo's (1991) statistic.

Table 2 below reports the empirical power of the CVM and KS tests for change–point in the conditional variance of the simulated returns. These tests detect a change in regime, although these tests are very sensitive to the magnitude of the change in the unconditional variance, i.e., the power of the test is equal to its size when the unconditional variance is constant.

Table 1. Tests for long-range dependence on the series of simulated returns R_t, absolute simulated returns $|R_t|$, and squared simulated returns R_t^2. Test size 5%.

	R_t: $\Pr(d=0)$			$\|R_t\|$: $\Pr(d>0)$			R_t^2: $\Pr(d>0)$		
q	Lo	KPSS	V/S	Lo	KPSS	V/S	Lo	KPSS	V/S
0	0.9499	0.9485	0.9480	0.9447	0.9054	0.9441	0.9296	0.8917	0.9331
1	0.9556	0.9474	0.9519	0.9281	0.8894	0.9315	0.9043	0.8712	0.9155
2	0.9633	0.9447	0.9537	0.9166	0.8774	0.9228	0.8874	0.8564	0.9049
3	0.9666	0.9441	0.9562	0.9054	0.8661	0.9157	0.8718	0.8417	0.8950
4	0.9699	0.9444	0.9573	0.8924	0.8544	0.9090	0.8557	0.8297	0.8843
5	0.9709	0.9438	0.9594	0.8799	0.8445	0.9002	0.8388	0.8184	0.8749
6	0.9733	0.9428	0.9607	0.8679	0.8325	0.8912	0.8245	0.8067	0.8653
7	0.9756	0.9424	0.9607	0.8575	0.8200	0.8834	0.8103	0.7942	0.8556
8	0.9759	0.9430	0.9615	0.8423	0.8099	0.8732	0.7919	0.7828	0.8450
9	0.9758	0.9417	0.9623	0.8292	0.8009	0.8623	0.7729	0.7728	0.8348
10	0.9770	0.9420	0.9629	0.8149	0.7903	0.8528	0.7551	0.7627	0.8267

Table 2. Tests for change–point in the conditional variance of simulated returns R_t. Probability of a change–point. Test size 5%.

CVM	KS
0.2062	0.2558

Interested readers are referred to Kokoszka and Teyssière (2001) for further details.

Interestingly, the parameters of the estimated GARCH(1,1) process fitted on the series of simulated returns R_t are close to the empirical values of the estimated parameters on real data, i.e., $\beta+\alpha$ is close to one. On average $\hat{\beta}$ is equal to 0.7988 while $\hat{\alpha}$ is equal to 0.0934, i.e., $\hat{\alpha}+\hat{\beta}$ is close to 0.9 and tends to 1 when the sample size increases, e.g., with samples of 1000 observations $\hat{\beta}+\hat{\alpha}=0.964866$. Our model is able to replicate the standard property of occurence of an IGARCH process, i.e., $\alpha+\beta=1$ for large samples. Interested readers are referred to Kirman and Teyssière (2000, 2001) for the large sample properties of this model.

Figures 1 and 2 below give two examples of the test by Kokoszka and Leipus (2000) for detecting a change in regime in the conditional variance process. These pictures display the series of k_t, i.e., the proportion of fundamentalists. The vertical line shows the point where the change in regime has been detected. These examples show that changes in regime are detected, which correspond to a switch in the evolution of the process k_t. Thus, the change–point test by Kokoszka and Leipus (2000) can be used for detecting bubbles in asset prices.

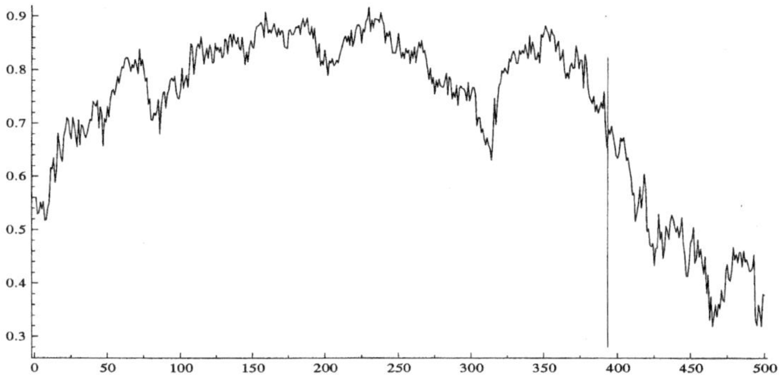

Fig. 1. A change–point has been detected at time $t = 394$

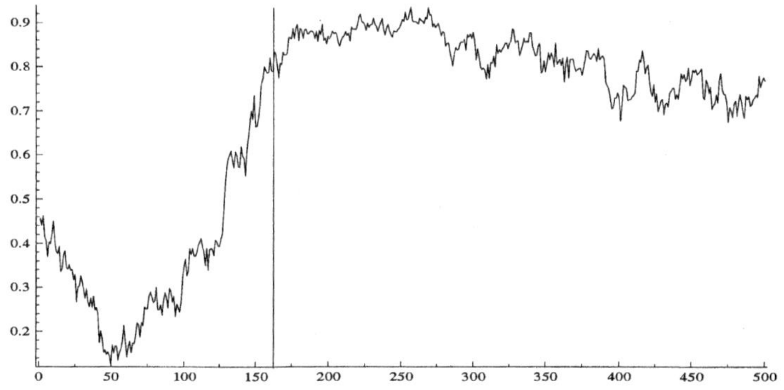

Fig. 2. A change–point has been detected at time $t = 166$

4 Conclusion

In this paper we have used a specific model of the microbehaviour of interacting individuals to generate aggregate price behaviour for the market for a financial asset. The prices are generated by a rather complicated stochastic process and exhibit bubble-like phenomena. Standard tests used to detect bubbles in the past would have led one to deny the existence of these bubbles. What is really happening in the model is that there are switches of regime and therefore that one should use tests to detect these rather than trying to decide if the process over the whole period exhibits the characteristics of a random walk. Such tests which have recently been developed by one of the authors and his collaborators are used here and detect the onset and bursting of bubbles. Furthermore the process generates time series whose properties correspond very well to those of empirical data.

References

1. Arthur, W.B.A. Ermoliev, Y and Kamovski, Y., (1985), Strong laws for a class of path-dependent stochastic processes with applications. In: Arkin, Shiryaev and Wets (eds), *Proceedings of International Conference in Stochastic Optimisation.* Springer Verlag, New York.
2. Avery, C. and Zemsky P., (1998), Multidimensional uncertainty and herd behavior in financial markets, *American Economic Review*, 88, 724–748
3. Barrow, S., (1994), Fundamental currency forecasting: an agnostic view, *Chemical Working Papers in Financial Economics*, 3, September, 8–12.
4. Beran, J., (1994), *Statistics for Long-Memory Processes.* Chapman & Hall.
5. Beran, J. and Ocker, D., (2001), Volatility of stock–market indexes. An analysis based on SEMIFAR models. *Journal of Business and Economic Statistics*, 19, 103–116.
6. Besseville, M. and Nikifirov, I.V., (1993), *Detection of Abrupt Changes: Theory and Applications.* Prentice Hall, Upper Saddle River.
7. Blanchard, O. and Watson, M.W., (1982), Bubbles, rational expectations and financial markets. In: P. Wachtel (ed), *Crisis in the Economic and Financial System*, 295–315. Lexington, MA, Lexington Books,
8. Blum, J.R., Kiefer, J. and Rosenblatt, M., (1961), Distribution free tests of independence based on the sample distribution function, *The Annals of Mathematical Statistics*, 32, 485–498.
9. Brock, W.A. and Hommes, C.H., (1999), Rational animal Spirits. In: P. J. J. Herings, G. van der Laan and A. J. J. Talman (eds), *The Theory of Markets*, 109–137. North Holland, Amsterdam.
10. Campbell, J.Y. and Shiller, R.J., (1987), Cointegration and tests of present value models, *Journal of Political Economy*, 95, 1062–1088.
11. Campbell, J.Y., Lo, A. and MacKinlay, A.C., (1997), *The Econometrics of Financial Markets.* Princeton.
12. Cheung, Y.W., (1993), Long memory in foreign–exchange rates, *Journal of Business and Economic Statistics*, 11, 93–101.
13. Csörgő, M. and Horváth, L., (1997), *Limit Theorems in Change–Point Analysis.* Wiley.
14. Day, R. and Huang, B., (1990), Bulls, bears and market sheep, *Journal of Economic Behavior and Organization*, 14, 299–329.
15. De Long, J.B., Schleifer, A., Summers, L.H. and Waldmann, R., (1989), The size and incidence of the losses from noise trading, *Journal of Finance*, 44, 681–699.
16. De Long, J.B., Schleifer, A., Summers, L.H. and Waldmann, R., (1990a), Noise trader risk in financial markets, *Journal of Political Economy*, 98, 703–738.
17. De Long, J.B., Schleifer, A., Summers, L.H. and Waldmann, R., (1990b), Positive feedback investment strategies and destabilising rational speculation, *Journal of Finance*, 45, 379–396.
18. De Long, J.B., Schleifer, A., Summers, L.H. and Waldmann, R., (1991), The survival of noise traders in financial markets, *Journal of Business*, 64, 1–19.
19. Diba, B.T. and Grossman, H.I., (1988), Explosive rational bubbles in stock prices? *American Economic Review*, 78, 520–530.
20. Ellison, G. and Fudenberg, D., (1993), Rules of thumb for social learning, *Journal of Political Economy*, 101, 612–643.

21. Evans, G.W., (1991), Pitfalls in testing for explosive bubbles in asset prices, *American Economic Review*, 81, 922–930.
22. Flavin, M.A., (1983), Excess volatility in the financial markets: a reassessment of the empirical evidence, *Journal of Political Economy*, **91**, 929–956.
23. Flood, R.P. and Garber, P.M., (1980), Market fundamentals versus price level bubbles: the first tests, *Journal of Political Economy*, 88, 745–770.
24. Flood, R.P. and Hodrick, R.J., (1990), On testing for speculative bubbles, *Journal of Economic Perspectives*, 4, 85–102.
25. Föllmer, H. and Schweizer, M., (1993), A microeconomic approach to diffusion models for stock prices, *Mathematical Finance*, 3, 1–25.
26. Frankel, J.A. and Froot, K.A., (1986), The dollar as an irrational speculative bubble: the tale of fundamentalists and chartists, *Marcus Wallenberg Papers on International Finance*, 1, 27–55.
27. Froot, K.A. and Obstfeld, M., (1989), Intrinsic bubbles: the case of stock prices, *NBER Working Paper no.* 3091.
28. Garber, P.M., (2000), *Famous First Bubbles.* MIT Press.
29. Gaunersdorfer, A. and Hommes, C.H., (2000), A nonlinear structural model for volatility clustering, *CeNDEF Working Paper* 00-02.
30. Giraitis, L. and Surgailis, D., (2001), ARCH–type bilinear models with double long–memory, *Preprint.*
31. Giraitis, L., Kokoszka, P.S. and Leipus, R., (2000), Stationary ARCH models: dependence structure and central limit theorem, *Econometric Theory*, 16, 3–22.
32. Giraitis, L., Kokoszka, P.S., Leipus, R. and Teyssière, G., (2000), Semiparametric estimation of the intensity of long-memory in conditional heteroskedasticity, *Statistical Inference for Stochastic Processes*, 3, 113–128.
33. Giraitis, L., Kokoszka, P.S., Leipus, R. and Teyssière, G., (2001), Rescaled variance and related tests for long-memory in volatility and levels, *Preprint.*
34. Goodhart, C.A.E., Hall, S.G., Henry, S.G.B. and Pesaran, B., (1993), News effects in a high frequency model of the Sterling-Dollar exchange rate, *Journal of Applied Econometrics*, 8, 1–13.
35. Granger, C.W.J. and Ding, Z., (1995), Some properties of absolute returns, an alternative measure of risk, *Annales d'Économie et de Statistique*, 40, 67–91.
36. Granger, C.W.J. and Ding, Z., (1996), Varieties of long-memory models, *Journal of Econometrics*, 73, 61–77.
37. Haken, H., (1997), *Synergetics.* Springer Verlag, Berlin.
38. Hamilton, J.D., (1989), A new approach to the economic analysis of nonstationary time series and the business cycle, *Econometrica*, 57, 357–384.
39. Hamilton, J.D. and Whiteman, C.H., (1985), The observable implications of self–fulfilling expectations, *Journal of Monetary Economics*, 16, 353–373.
40. Horváth, L., Kokoszka, P.S. and Teyssière, G., (2001), Empirical process of the squared residuals of an ARCH sequence, *The Annals of Statistics*, 29, 445–469.
41. Kazakevičius, V. and Leipus, R., (2000), On stationarity in the ARCH(∞) model, *Econometric Theory*, forthcoming.
42. Kazakevičius, V., Leipus, R. and Viano, M.C., (2000), Stability of random coefficients autoregressive conditionally heteroskedastic models, *Preprint.*
43. Kirman, A., (1991), Epidemics of opinion and speculative bubbles in financial markets. In: M. Taylor (ed), *Money and Financial Markets*, chapter 17. Macmillan London.
44. Kirman, A., (1993), Ants, rationality and recruitment, *The Quarterly Journal of Economics*, 108, 137–156.

45. Kirman, A. and Teyssière, G., (2000), Microeconomic models for long-memory in the volatility of financial time series, *Studies in Nonlinear Dynamics and Econometrics*, forthcoming.
46. Kirman, A. and Teyssière, G., (2001), Testing for bubbles and change–point, *Preprint.*
47. Kleidon, A.W., (1986), Variance bounds tests and stock price valuation models, *Journal of Political Economy*, 94, 953–1001.
48. Kokoszka, P.S. and Teyssière, G., (2001), Change-point detection in GARCH models: asymptotic and bootstrap tests, *Preprint.*
49. Kokoszka, P.S. and Leipus, R., (2001), Detection and estimation of changes in regime. Forthcoming in: M. S. Taqqu, G. Oppenheim and P. Doukhan (eds), *Long-Range Dependence: Theory and Applications.* Birkhauser.
50. Kokoszka, P.S. and Leipus, R., (2000), Change-point estimation in ARCH models, *Bernoulli*, 6, 513–539.
51. Kouri, P.J.K., (1983), Balance of payments and the foreign exchange market: a dynamic partial equilibrium model. In: J.S. Bhandari and B.H. Putnam (eds), *Economic Interdependence and Flexible Exchange Rates*, MIT. Press, Cambridge, Massachussets.
52. Kwiatkowski, D., Phillips, P.C.B., Schmidt, P. and Shin, Y., (1992), Testing the null hypothesis of stationarity against the alternative of a unit root: how sure are we that economic series have a unit root? *Journal of Econometrics*, 54, 159–178.
53. Lavielle, M., (1999), Detection of multiple changes in a sequence of dependent variables, *Stochastic Processes and their Applications*, 83, 79–102.
54. Lavielle, M. and Moulines, E., (2000), Least squares estimation of an unknown number of shifts in a time series, *Journal of Time Series Analysis*, 21, 33–59.
55. Le Roy, S.F. and Porter, R.D., (1981), The present value relation: tests based on implied variance bounds, *Econometrica*, 49, 555–574.
56. Lo, A.W., (1991), Long-term memory in stock market prices, *Econometrica*, 59, 1279–1313.
57. Mandelbrot, B.B., (1997), *Fractals and Scaling in Finance: Discontinuity, Concentration, Risk.* Springer Verlag.
58. Mankiw, N.G., Romer, D. and Shapiro, M., (1985), An unbiased reexamination of stock market volatility, *Journal of Finance*, 40, 677–689.
59. Marsaglia, G., (1996), DIEHARD: A Battery of Tests of Randomness. `http://stat.fsu.edu/pub/diehard`.
60. Marsh, T.A. and Merten, R.C., (1986), Dividend variability and variance bounds tests for the rationality of stock market prices, *American Economic Review*, 76, 483–498.
61. Mattey, J. and Meese, R., (1986), Empirical assessment of present value relations, *Econometric Reviews*, 5, 171–233.
62. McKinnon, R., (1979), *Money in International Exchange: The Convertible Currency System.* Oxford University Press.
63. Meese, R.A., (1986), Testing for bubbles in exchange markets: the case of sparkling rates, *Journal of Political Economy*, 94, 345–373.
64. Meese, R.A. and Rogoff, K., (1983), Empirical exchange rate models of the seventies: do they fit out–of–sample? *Journal of International Economics*, 14, 3–24.
65. Mikosch, T. and Stărică, C., (1999), Change of structure in financial time series, long range dependence and the GARCH model, *Preprint.*

66. Newey, W.K. and West, K.D., (1987), A simple positive definite, heteroskedasticity and autocorrelation consistent covariance matrix, *Econometrica*, 55, 703–708.
67. Perron, P., (1989), The great crash, the oil price shock, and the unit root hypothesis, *Econometrica*, 57, 1361–1401.
68. Phillips, P.C.B., (1987), Time series regression with a unit root, *Econometrica*, 55, 277–302.
69. Picard, D., (1985), Testing and estimating change–point in time series, *Advances in Applied Probability*, 17, 841–867.
70. Robinson, P.M., (1991), Testing for strong serial correlation and dynamic conditional heteroskedasticity in multiple regression, *Journal of Econometrics*, 47, 67–84.
71. Robinson, P.M., (1994), Time series with strong dependence. In: C.A. Sims (ed), *Advances in Econometrics, Sixth World Congress*, 47–95. Cambridge University Press.
72. Robinson, P.M., (2001), The memory of stochastic volatility models, *Journal of Econometrics*, 101, 195–218.
73. Schwert, G.W., (1989), Tests for unit roots: a Monte Carlo investigation, *Journal of Business and Economic Statistics*, 7, 147–159.
74. Sharfstein, D.S. and Stein, J.C., (1990), Herd behavior and investment, *American Economic Review*, 80, 465–479.
75. Shiller, R.J., (1981), Do stock prices move by too much to be justified by subsequent changes in dividends? *American Economic Review*, 71, 421–436.
76. Shiller, R.J., (2000), *Irrational Exuberance*. Princeton University Press, Princeton, N.J.
77. Stiglitz, J.E., (1990), Symposium on bubbles, *Journal of Economic Perspectives*, 4, 13–18.
78. Stock, J.H., (1994), Unit roots and trend breaks. In: R.F. Engle and D. McFadden (eds), *Handbook of Econometrics*, 4. North-Holland, Amsterdam.
79. Suvanto, A., (1993), Foreign exchange dealing, *Preprint.*
80. Taylor, S.J., (1986), *Modelling Financial Time Series.* Wiley.
81. Teyssière, G., (2001), Nonlinear and semiparametric long–memory ARCH, *GREQAM DT* 01A12.
82. Teyssière, G., (1998), Multivariate long–memory ARCH modelling for high frequency foreign exchange rates. In: *Proceedings of the HFDF-II Conference*, Olsen & Associates, 1-3 April 1998, Zurich, Switzerland.
83. Teyssière, G., (1997), Modelling exchange rates volatility with multivariate long–memory ARCH processes, in revision for the *Journal of Business and Economic Statistics.*
84. Tirole, J., (1985), Asset bubbles and overlapping generations, *Econometrica*, 53, 1499–1528.
85. West, K.D., (1987), A specification test for speculative bubbles, *Quarterly Journal of Economics*, 102, 553–580.
86. West, K.D., (1988), Bubbles, fads and stock price volatility tests: a partial evaluation, *Journal of Finance*, 43, 639–656.
87. Woo, W.T., (1987), Some evidence of speculative bubbles in the foreign exchange markets, *Journal of Money, Credit and Banking*, 19, 499–514.
88. Young, H.P. and D. Foster, (1991), Cooperation in the short and in the long run, *Games and Economic Behavior*, 3, 145–156.

Nonlinear Dynamics and the Stability of Competitive Equilibria

Jan Tuinstra

CeNDEF, Department of Quantitative Economics, University of Amsterdam

Abstract. In this chapter we review the recent literature on the dynamics of the tâtonnement process, especially from the perspective of the theory of nonlinear dynamics. We show that complicated dynamical phenomena emerge naturally in the tâtonnement process. We also discuss some conceptual drawbacks of the tâtonnement process and directions for future research.

Keywords: Price adjustment processes, general equilibrium models, nonlinear economic dynamics, endogenous fluctuations.

1 Introduction

A fundamental problem in general equilibrium theory is the stability of competitive equilibria. The significance of competitive equilibria for understanding how economies work highly depends upon the existence of a sensible price mechanism that drives the economy to such a competitive equilibrium. The best-known price mechanism in general equilibrium theory is the tâtonnement process, for which convergence to an equilibrium price vector can only be guaranteed under rather restrictive assumptions on the aggregate excess demand functions. Together with the fact that one might question the validity of the tâtonnement process as a realistic price mechanism, these "negative" stability results have caused a decreasing interest in this process, and in price mechanisms in general.

The insight that the mathematical theory of nonlinear dynamical systems is also relevant for many economic models has attracted new attention to the tâtonnement process in the last two decades. Moreover, it has been shown that for reasonable specifications of the fundamentals of the economy the tâtonnement process might lead to endogenous periodic and chaotic fluctuations of prices around an unstable equilibrium price vector. Claus Weddepohl has shown a keen interest in this newly developing field and has made some interesting contributions (e.g. Weddepohl (1995, 1999) and Goeree, Hommes and Weddepohl (1997)). Furthermore, he has inspired others, like myself, to work on these problems.

The purpose of this chapter is to review the literature on the tâtonnement process, especially in light of the theory on nonlinear dynamics. Furthermore, we want to suggest possible directions for future research on price mechanisms. We also highlight some of Claus' contributions.

This chapter is organized as follows. In Section 2 an overview of the literature on the tâtonnement process is given. Section 3 studies two examples of tâtonnement processes in economies with three commodities and shows that different kinds of complicated dynamical behavior can occur in such a simple setting. Section 4 discusses some conceptual problems and points at directions for future research. Section 5 concludes this chapter.

2 The Tâtonnement Process

Consider an exchange economy with m consumers and n commodities. Each consumer i $(i = 1, \ldots, m)$ takes prices as given and solves the optimization problem

$$\max_{x_{i1},\ldots,x_{in}} U_i\left(x_{i1}, \ldots, x_{in}\right) \text{ subject to } \sum_{j=1}^{n} p_j x_{ij} \leq \sum_{j=1}^{n} p_j w_{ij}.$$

Here $U_i\left(.\right)$ is the strictly quasi-concave, strictly monotonic utility function of consumer i, x_{ij} the amount of good j consumed by consumer i, p_j the price of good j and w_{ij} the initial endowment of good j owned by consumer i. Let $x_{ij}\left(\mathbf{p}, \mathbf{w}_i\right)$ be the resulting demand for good j by consumer i, where $\mathbf{p} = \left(p_1, \ldots, p_n\right) \in I\!R_+^n$ is the price vector and $\mathbf{w}_i = \left(w_{i1}, \ldots, w_{in}\right) \in I\!R_+^n$ the vector of initial endowments of consumer i.

Aggregate excess demand for good j is

$$z_j(\mathbf{p}) = \sum_{i=1}^{m} x_{ij}(\mathbf{p}, \mathbf{w}_i) - \sum_{i=1}^{m} w_{ij}, \qquad j = 1, \ldots, n. \tag{1}$$

The excess demand functions satisfy Walras' Law, $\sum_{j=1}^{n} p_j z_j\left(\mathbf{p}\right) = 0$, and are homogeneous of degree zero in prices, i.e. $z_j\left(\theta\mathbf{p}\right) = z_j\left(\mathbf{p}\right), \forall\theta > 0$. The economy is in equilibrium at the equilibrium price vector $\mathbf{p}^*$ if $z_j\left(\mathbf{p}^*\right) = 0$ for all j.

Economists and laymen alike believe that there are economic forces (often referred to as the *law of supply and demand*) driving the economy towards an equilibrium. The tâtonnement process models these forces and is usually associated with a centralized agency (the *auctioneer*) who quotes a price vector and then gathers information about excess demand at the prevailing prices. He then quotes a new price vector which is adjusted from the old one according to the following rule: if there is excess demand for a good, its price goes up; if there is excess supply for a good, its price goes down; if the market for a good is in equilibrium its price remains the same. This process repeats over and over until an equilibrium price vector is reached and all excess demands have vanished. Only then trade is executed, and hence transactions can only occur at equilibrium prices.

The tâtonnement process serves as a first approximation to more realistic adjustment processes. We will discuss some of its drawbacks in Section 4. The tâtonnement process can be formalized as a dynamical system in continuous or in discrete time as follows

$$\frac{dp_j(t)}{dt} = F_j(p_j(t), z_j(\mathbf{p}(t))), \tag{2}$$

$$p_{j,t+1} = p_{jt} + F_j(p_{jt}, z_j(\mathbf{p}_t)), \tag{3}$$

for $j = 1, \ldots, n$, where $F_j(p_j, z_j)$ is a sign-preserving function in z_j that satisfies $F_j(p_j, 0) = 0$.

The continuous tâtonnement process (2) has been extensively studied in the literature. In their seminal paper Arrow, Block and Hurwicz (1959) consider $F_j(p_{j,t}, z_j(\mathbf{p}_t)) = \lambda_j z_j(\mathbf{p}_t)$, that is, price change is proportional to excess demand, and try to prove *global* stability of the equilibrium by constructing an appropriate Lyapunov function. They find that a sufficient condition for global stability is that all goods are gross substitutes, that is: $\frac{\partial z_j}{\partial p_k}(\mathbf{p}) > 0$, $\forall j$, $\forall k \neq j$, $\forall \mathbf{p}$. This condition implies that $\sum_{j=1}^{m} p_j^* z_j(\mathbf{p}) > 0$ for any $\mathbf{p} \neq \mathbf{p}^*$. Thus, the value of the aggregate excess demands weighted with the (unique) equilibrium prices is always positive out of equilibrium. The weighted Euclidean distance between the current price and the equilibrium price vector, that is $V(\mathbf{p}) = \sum_{j=1}^{n} \frac{1}{\lambda_j}(p_j - p_j^*)^2$, can then be taken as a Lyapunov function. We have

$$\begin{aligned}\frac{dV}{dt} &= 2\sum_{j=1}^{n} \frac{1}{\lambda_j}(p_j - p_j^*)\frac{dp_j}{dt} = 2\sum_{j=1}^{n} p_j z_j(\mathbf{p}) - 2\sum_{j=1}^{n} p_j^* z_j(\mathbf{p}) \\ &= -2\sum_{j=1}^{n} p_j^* z_j(\mathbf{p}) < 0,\end{aligned}$$

where the second equality follows from the specification of the adjustment process and the third from Walras' Law. Since the time derivative of $V(\mathbf{p})$ is negative everywhere outside of equilibrium, prices converge to the unique equilibrium price vector.

The condition of gross substitutability is very restrictive, but it cannot be relaxed. Scarf (1960) gives some, otherwise well-behaved, examples for which gross substitutability does not hold and for which the continuous tâtonnement process does not converge, but where prices move over a cycle. Moreover, the possibility of nonconvergence of the tâtonnement process should not be surprising in the light of the so-called Sonnenschein-Mantel-Debreu results (see Debreu (1974), Mantel (1974) and Sonnenschein (1973)). These authors have shown that any set of n continuous functions, defined for relative prices bounded away from zero and satisfying Walras' Law and homogeneity, can be the aggregate excess demand functions of an economy with utility maximizing agents with convex preferences, provided that the number of agents is at least as large as the number of commodities. Therefore no restrictions other than continuity, Walras' Law and homogeneity in prices can be imposed on the aggregate demand functions. This implies that we can expect any type of dynamics to occur in the tâtonnement process.

For the discrete time tâtonnement process (3) the situation is even more complicated since stability of the competitive equilibrium is determined by the properties of the aggregate excess demand functions *and* the specification of the tâtonnement process. Hence, in the discrete time case gross substitutability is not enough to ensure stability (see Uzawa (1959)). However, in Goeree, Hommes and Weddepohl (1997) it is shown that, if price changes are bounded and if the aggregate excess demand functions satisfy gross substitutability, the price will be trapped in a certain neighborhood of the equilibrium. Within this neighborhood different kinds of dynamics are possible.

All this suggests that tâtonnement-like processes only converge for a small set of possible economies. Moreover, it leads to the conjecture that all kinds of (complicated) dynamical phenomena are possible in the tâtonnement process. Indeed, in recent years there have been a number of contributions that show the existence of complicated phenomena in the tâtonnement process, such as periodic, quasi-periodic and chaotic behavior. Among these are Saari (1985) and Weddepohl (1995) who give examples of erratic behaviour in economies with 2 commodities (for which, by price normalization, the tâtonnement process corresponds to a one-dimensional dynamical system). Goeree, Hommes and Weddepohl (1997) and Tuinstra (2000a) investigate tâtonnement processes in economies with 3 commodities (giving rise to two-dimensional dynamical systems). All these papers consider discrete time versions of the tâtonnement process but from our discussion above it should be clear that the same phenomena may arise in continuous time versions of the model.

3 Complicated Dynamics in the Tâtonnement Process

3.1 A Pedagogical Example

In this subsection we study a typical example of a tâtonnement process in a very simple economy. This example features many of the typical dynamical phenomena encountered in the study of nonlinear dynamics.[1] We consider an economy with three commodities. The aggregate excess demand functions of

[1] This example is taken from Tuinstra (2000b), where many other examples of economic models with complicated dynamics can be found.

the first two commodities are given by

$$z_1(p_1, p_2, p_3) = \left(\frac{p_2}{p_1}\right)^{\alpha} + \left(\frac{p_3}{p_1}\right)^{2-\alpha} - 2,$$
$$z_2(p_1, p_2, p_3) = \frac{p_1 + p_3}{p_2} - 2,$$

where $\alpha \neq 0$. The aggregate excess demand for the third commodity follows from Walras' Law. By the Sonnenschein-Mantel-Debreu results these aggregate demand functions are consistent with utility maximization. The unique equilibrium price vector is given by $(p_1^*, p_2^*, p_3^*) = (1, 1, 1)$. We take the third commodity as a numeraire ($p_3 = 1$) and consider the following tâtonnement process in discrete time

$$p_{1,t+1} = p_{1t} + \lambda z_1(p_{1t}, p_{2t}, 1), \qquad (4)$$
$$p_{2,t+1} = p_{2t} + \lambda z_2(p_{1t}, p_{2t}, 1),$$

where $\lambda > 0$ is the speed of adjustment. We are interested in the time path of an arbitrary initial price vector $\mathbf{p}_0$, that is, the sequence of price vectors $\{\mathbf{p}_t\}_{t=0}^{\infty}$ generated by (4). We are particularly interested in whether this time path converges to the competitive equilibrium $\mathbf{p}^*$ or not.

Figure 1 shows some typical orbits of the tâtonnement process (4) in the state space (p_1, p_2) for different values of the parameters α and λ.[2] The pictures in Figure 1 correspond to the long-run behavior of the tâtonnement process (4), and they show that prices do not settle down but that there are perpetual fluctuations around the competitive equilibrium. Such a set of points that captures the (possible) limit behavior of a dynamical system is called an *attractor*. Important features of an attractor are that it is an invariant set (that is, if some orbit is "trapped" in the attractor it will move over the attractor forever) and that it attracts a large set of initial states in some neighborhood. An attractor is in fact a generalization of the concept of a (locally) stable equilibrium. In the case of a stable equilibrium this attractor consists of one point. Other examples of attractors are a stable periodic orbit of period k, where the attractor consists of k different points or a stable closed curve, where orbits move over the closed curve quasi-periodically.

Let us now discuss the different attractors in Figure 1. First, we consider stability of the competitive equilibrium. This can be investigated by considering the eigenvalues of the Jacobian matrix J of the linearization of (4) evaluated at the competitive equilibrium $(p_1^*, p_2^*) = (1, 1)$. Denote by $\mu_1(\lambda, \alpha)$ and $\mu_2(\lambda, \alpha)$ the eigenvalues of J. If $|\mu_1(\lambda, \alpha)| < 1$ and $|\mu_2(\lambda, \alpha)| < 1$ the competitive equilibrium is locally stable. On the other hand, if at least one

[2] These pictures are constructed in the following way. An initial (disequilibrium) price vector is chosen (in this case $p_{10} = 1$ and $p_{20} = 0.99$, which lies close to the equilibrium price vector) and (4) is applied to this initial price vector 6000 times and the last 5000 points (p_{1t}, p_{2t}) are plotted in the Figure.

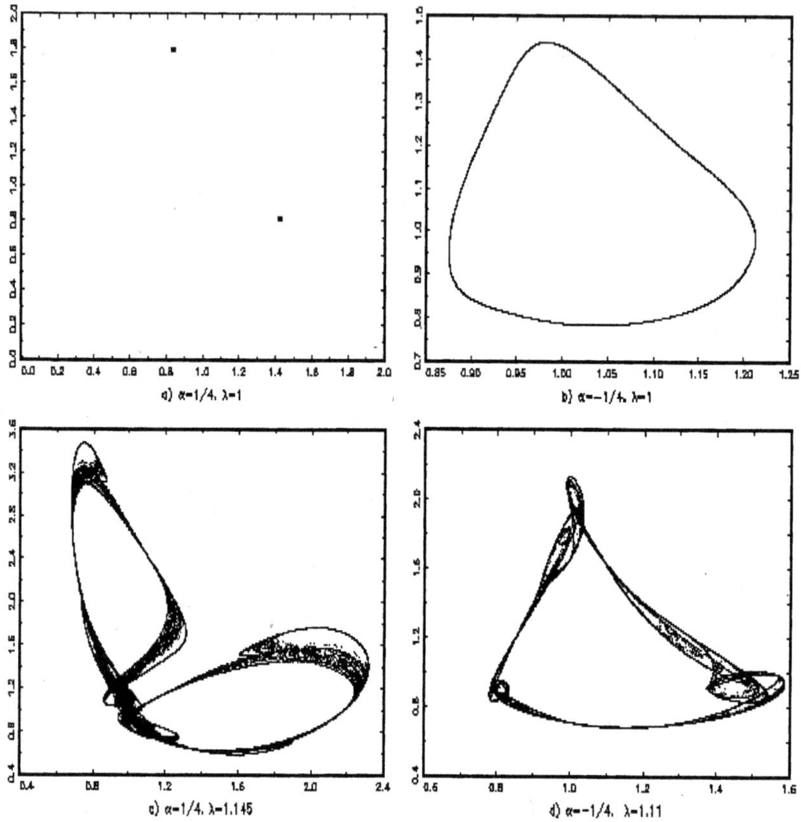

Fig. 1. Attractors for the tâtonnement process (4), with p_1 on the horizontal and and p_2 on the vertical axis. a) (upper left panel) $\alpha = \frac{1}{4}$, $\lambda = 1$; b) (upper right panel) $\alpha = -\frac{1}{4}$, $\lambda = 1$; c) (lower left panel) $\alpha = \frac{1}{4}$, $\lambda = 1.145$; d) (lower right panel) $\alpha = -\frac{1}{4}$, $\lambda = 1.11$.

of these eigenvalues lies outside the unit circle the competitive equilibrium is unstable. As one of the eigenvalues crosses the unit circle as a parameter varies, a so-called *bifurcation* occurs.[3] At such a bifurcation there is a *qualitative change in the dynamical behavior* of the system. For example, the stability of the competitive equilibrium changes and periodic cycles are created. Figure 1 shows two important examples of what can happen at such a *local* bifurcation. Figure 1a shows a period-two cycle, for $\alpha = \frac{1}{4}$ and $\lambda = 1$, that is created in the *flip* or *period-doubling bifurcation* that occurs at $\alpha = \frac{1}{4}$

[3] For a detailed analysis of the different local bifurcations, see for example the introductory textbooks on nonlinear dynamics by Guckenheimer and Holmes (1983) or Kuznetsov (1998).

and $\lambda = \frac{4}{5}$. At these parameter values the eigenvalues are real and the smallest eigenvalue crosses the unit circle at -1 and a period two orbit of the tâtonnement process is created. The tâtonnement process then keeps on fluctuating between two price vectors. Figure 1b shows, for $\alpha = -\frac{1}{4}$ and $\lambda = 1$, a closed curve which is invariant under (4). This attracting closed curve is created in a *Neimark-Sacker bifurcation* at $\alpha = -\frac{1}{4}$ and $\lambda = \frac{16}{17}$. At these values for the parameters, the eigenvalues are complex conjugates and equal to 1 in absolute value.[4]

After the primary bifurcation of the equilibrium an infinite cascade of subsequent bifurcations can occur, culminating in the existence of a *strange attractor*, with a complicated (fractal) geometric structure. Examples are given in Figures 1c and 1d. From these pictures it is clear that the dynamics of the tâtonnement process may behave very erratically. These strange attractors can exhibit *sensitive dependence on initial conditions.* This means that two time series of orbits lying on this attractor and starting arbitrarily close to each other will follow quite different paths. This implies that, even if the deterministic system is perfectly known, long-run predictions for a chaotic system are impossible since they require that the starting conditions are known with infinite precision. The time series generated by a chaotic system is very unpredictable and seems to be generated by a random system. This phenomenon is often referred to as *deterministic chaos.*

3.2 Price Normalization and Symmetry

In the example from the previous subsection we normalized the price of the third commodity to unity. This is not an innocuous assumption. The issue of price normalization in the dynamics of the tâtonnement process is studied in more detail in Tuinstra (2000a); we will briefly discuss some of the main results here. The need to normalize prices stems from the homogeneity of the aggregate excess demand functions (1). This feature of the aggregate excess demand functions implies that, if $\mathbf{p}^*$ is an equilibrium price vector, then so is $\theta\mathbf{p}^*$ for any $\theta > 0$ and we therefore have a *ray* of equilibrium price vectors. None of these equilibria can be locally stable since every neighborhood in $I\!R^n_+$ of such an equilibrium point contains other equilibrium points. This problem is solved by imposing a price normalization rule. However, such a price normalization rule affects the dynamical behavior of the tâtonnement process. For example, for the stability of the competitive equilibrium it matters which commodity is chosen as numeraire.

[4] The Jacobian matrix J, evaluated at the equilibrium price vector $\mathbf{p}^*$, is easily found to be

$$J = \begin{pmatrix} 1-2\lambda & \alpha\lambda \\ \lambda & 1-2\lambda \end{pmatrix}.$$

The eigenvalues of the linearized system are $\mu_{1,2}(\lambda, \alpha) = 1 - 2\lambda \pm \lambda\sqrt{\alpha}$. They are real if $\alpha > 0$ and the smallest of them then is -1 for $\lambda = \frac{2}{2+\sqrt{\alpha}}$. They are complex conjugates when $\alpha < 0$, and then they lie on the unit circle when $\lambda = \frac{4}{4-\alpha}$.

In some cases there is an elegant solution to this price normalization problem. Consider for example the so-called multiplicative tâtonnement process

$$p_{j,t+1} = p_{jt}\left(1 + \lambda z_j\left(\mathbf{p}_t\right)\right), \qquad j = 1, \ldots, n, \tag{5}$$

where $\lambda > 0$ is again the speed of adjustment. Equation (5) states that the rate of change of the price of a good is proportional to its excess demand. An attractive feature of (5) is that it is homogeneous of degree one in prices. If we aggregate over all prices and use Walras' Law we find

$$\sum_{j=1}^{n} p_{j,t+1} = \sum_{j=1}^{n} p_{jt}\left(1 + \lambda z_j\left(\mathbf{p}_t\right)\right) = \sum_{j=1}^{n} p_{jt} + \lambda \sum_{j=1}^{n} p_{jt} z_j\left(\mathbf{p}_t\right) = \sum_{j=1}^{n} p_{jt}.$$

Hence for the multiplicative tâtonnement process the sum of prices will be constant over time. Considering prices at the simplex (that is, price vectors such that $\sum_{j=1}^{n} p_j = 1$) as a normalization rule therefore seems to be a natural choice since it is satisfied automatically and imposes no further restriction on the tâtonnement process (a similar role is played by the sphere, i.e. price vectors with $\sum_{j=1}^{n} p_j^2 = 1$, in the continuous time tâtonnement process $\frac{dp_j(t)}{dt} = \lambda z_j\left(\mathbf{p}\left(t\right)\right)$).

We will now briefly discuss an economy with three commodities and study the resulting tâtonnement dynamics on the simplex. We will consider an economy with three consumers with standard CES preferences, i.e. the utility function of consumer i $(i = 1, 2, 3)$ is $U_i\left(x_{i1}, x_{i2}, x_{i3}\right) = \left(x_{i1}^{\rho} + x_{i2}^{\rho} + x_{i3}^{\rho}\right)^{\frac{1}{\rho}}$ with $\rho < 0$. If we furthermore assume that consumer i has an initial endowment of commodity i of 100 and no initial endowment of the other two commodities the multiplicative price adjustment process becomes

$$p_{j,t+1} = p_{jt}\left(1 + 100\lambda\left[\frac{p_{1t} + p_{2t} + p_{3t}}{p_{1t}^{1-\sigma} + p_{2t}^{1-\sigma} + p_{3t}^{1-\sigma}} p_{jt}^{-\sigma} - 1\right]\right), \qquad j = 1, 2, 3, \tag{6}$$

where $\sigma = \frac{1}{1-\rho}$ is the (constant) elasticity of substitution.

The tâtonnement process (6) turns out to be highly symmetric. We follow Tuinstra (1997, 2000a) in using concepts and results from the literature on symmetric dynamical systems (see Chossat and Golubitsky (1988) and Golubitsky, Stewart and Schaeffer (1988)) for the analysis of this symmetric tâtonnement process. Let us denote dynamical system (6) by $\mathbf{p}_{t+1} = G\left(\mathbf{p}_t\right)$. The mapping G has *symmetry* M if G and M commute, i.e. $MG\left(\mathbf{p}\right) = G\left(M\mathbf{p}\right)$, where M is some permutation matrix. It is easily checked that (6) satisfies *reflection* symmetries, where two prices are interchanged (e.g. $\left(p_1, p_2, p_3\right) \overset{M}{\rightarrow} \left(p_2, p_1, p_3\right)$) and *rotation* symmetries, where all prices are shifted one position (e.g. $\left(p_1, p_2, p_3\right) \overset{M}{\rightarrow} \left(p_3, p_1, p_2\right)$). The equilibrium price

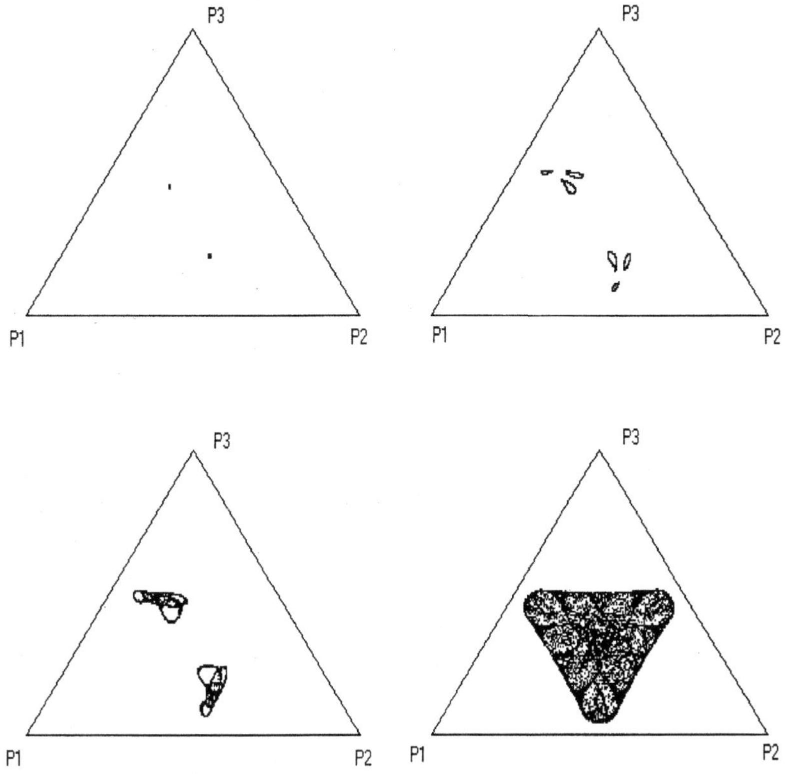

Fig. 2. Attractors for the tâtonnement process (6) with $\sigma = \frac{1}{100}$. a) $\lambda = 2.20$ (upper left panel); b) $\lambda = 2.58$ (upper right panel); c) $\lambda = 2.60$ (lower left panel); d) $\lambda = 2.61$ (lower right panel).

vector, $\mathbf{p}^* = \left(\frac{1}{3}, \frac{1}{3}, \frac{1}{3}\right)$, is invariant under these permutations too. We can therefore also speak about the symmetry of the attractor. Notice that using the simplex as a price normalization rule preserves the symmetry of the tâtonnement process, whereas selecting a numeraire commodity would break it. It therefore seems natural to start the investigation of tâtonnement processes on the simplex, by focussing on symmetric tâtonnement processes first.

Figure 2 shows some of the attractors for tâtonnement process (6) for $\sigma = \frac{1}{100}$ and different values of the speed of adjustment λ. For low values of λ the equilibrium price vector is stable. However, at $\lambda = (50\sigma)^{-1} = 2$ the equilibrium price vector becomes unstable. The period two cycle shown in the upper left panel of Figure 2 is created at this bifurcation. Notice that this period two cycle satisfies a reflection symmetry: interchanging p_2 and p_3 will not change the period two cycle. However, this period two cycle is not invariant under a rotational symmetry: shifting all prices, or rotating the simplex over $\frac{2}{3}\pi$, would give a different period two cycle. By symmetry of (6) these other period two cycles are also attractors of the tâtonnement process. Clearly the bifurcation at $\lambda = 2$ breaks the symmetry of the attractor of the dynamical process. This bifurcation is therefore referred to as a *symmetry breaking period doubling bifurcation* and leads to the existence of three *coexisting* stable period two cycles (moreover, also three unstable period two cycles are created at this bifurcation). When λ increases, each of these period two cycles undergoes the same bifurcation scenario. The upper right and lower left panel of Figure 2 show the attractors for $\lambda = 2.58$ and $\lambda = 2.60$ respectively. Again, these attractors occur in threefold and depending upon the initial price vector, prices can converge to either one of them. As λ becomes even higher, i.e. $\lambda = 2.61$, the three coexisting strange attractors merge and a new, fully symmetric, attractor emerges. This last step of the bifurcation route is called a *symmetry increasing bifurcation*, since the final attractor satisfies rotational symmetry again. In Tuinstra (2000a) it is shown that similar bifurcation scenarios to coexisting strange attractors occur for close-by asymmetric general equilibrium models. The symmetric tâtonnement process can therefore serve as an 'organizing centre' for the global analysis of tâtonnement processes on the simplex.

4 Towards a More Realistic Adjustment Process

From the examples in the previous section it should be clear that the tâtonnement process is likely to lead to complicated price dynamics. Besides the fact that it has poor stability properties there are some conceptual problems with the tâtonnement process that question its validity as a sensible price mechanism (for an extensive account of these problems, see Schinkel (2001)). In this section we identify these problems and suggest some directions for future research on the stability of the competitive equilibrium.

4.1 Trade at Disequilibrium Prices

The tâtonnement process does not allow for trade at disequilibrium prices. Clearly this is an unpalatable assumption and any realistic model of price adjustment would account for disequilibrium trade. In particular, nonconvergence of the tâtonnement process, as encountered in the previous section, implies that trade never takes place. Alternatively, one can assume that trade takes place in each period but that commodities are perishable and that the next period starts with the old allocation of endowments and a new price vector.

Some alternative adjustment processes have been suggested in which trade takes place at "false prices" and where endowments change. Best known among these are the *Edgeworth process* (see Hahn (1962) and Uzawa (1962)) and the *Hahn process* (Hahn and Negishi (1962)). In these processes every trade is a Pareto improvement and convergence to a Pareto efficient allocation occurs under a mild set of conditions. The problem with these processes is that the final allocation, although Pareto efficient, does in general not correspond to a Walrasian equilibrium of the original economy, since trade continually changes the endowments of the consumers.

In recent years Claus Weddepohl has studied price adjustment processes with trade taking place at disequilibrium prices but where each period starts with the same vector of initial endowments. Weddepohl (1999) compares, in a simple production economy, two disequilibrium adjustment processes with the tâtonnement process. The first process is a *fixed price* adjustment process, where a fixed price equilibrium is attained in every period. The *effective* excess demands (i.e. excess demands after rationing) then determine next period's prices. In the second adjustment process each time period is divided into a number of subperiods and trade is assumed to take place *sequentially* instead of simultaneously: in the first subperiod only the market for the first commodity is open and trade on that market takes place. When all trades on this market have been executed, the model proceeds to the second subperiod and the second market opens, and so on. Notice that trade on the first market will have ramifications for excess demand for the goods traded at the following markets. Hence, in between markets agents have to update their demanded consumption bundles. Weddepohl (1999) shows that complicated dynamics arise in these two disequilibrium adjustment processes. In the sequential adjustment process expectations of firms play an important role. Moreover, the time series generated by the sequential adjustment process are qualitatively different from the time series generated by the fixed price adjustment process.

4.2 Price-setting Firms

Another problem with the tâtonnement process, first noticed by Arrow (1959), arises when we consider who is setting the prices. In the perfectly competitive economy described in Section 2 all agents take prices as given in determining their optimal consumption and production level. The traditional solution to this problem is the introduction of an auctioneer who sets prices. However, whereas all decisions of the consumers arise from utility or profit maximization, the behavior of the auctioneer is postulated ad hoc. This seems to be inconsistent with the general equilibrium framework. Whose behavior does the auctioneer represent? It might be argued that a realistic model of price adjustment requires that prices are set by economic agents in such a way that they believe that it maximizes their utility or profit. This type of model falls within the literature on general equilibrium models with monopolistic competition.[5] Since these general monopolistic equilibrium models are typically quite complicated, a first step is to study price adjustment processes in *partial* equilibrium models with monopolistic competition.

The general set-up of this type of model is as follows. Suppose there are n commodities, indexed j, where each commodity is produced by a single firm (also indexed j), which sets the price for this commodity. The demand for each commodity j is given by an *objective demand curve* $D_j(p_1, p_2, \cdots, p_n)$ and is unknown to the firms. Instead, each firm j believes that the demand for its product is given by a *subjective demand curve* $d_j(p_j)$. Beliefs of firms are given by simple *rules of thumb* and do not account for all interactions within the economy. In particular, firms neglect cross-price effects and the structure of the subjective demand curve is in general much simpler (e.g. linear) than the structure of the objective demand curve. Firms revise their beliefs, that is, their subjective demand curve, on the basis of observations on the objective demand curve. The price adjustment process with price setting firms then works as follows. Given their initial beliefs, firms set the price that maximizes their (expected) profit. They sell a certain amount of their commodity against this price, which gives a new observation on the objective demand curve. This new observation may lead them to update their beliefs and revise their perceived optimal price again, and so on. The question is now whether such a price adjustment process converges to an equilibrium.

Kirman (1983) considers a version of this model with two firms and (symmetric) linear objective demand functions. Firms estimate the subjective demand curve by applying a least-squares regression on past combinations of prices and (potential) sales. Numerical simulations show that this adjustment process always converges to an element of the set of *conjectural* equilibria. A conjectural equilibrium consists of beliefs and prices such that firms' expectations with respect to supply possibilities are satisfied. A drawback of this model is that the set of conjectural equilibria is very large. It contains, for example, the Bertrand-Nash equilibrium, the cooperative (cartel) solution and the competitive equilibrium. Moreover, it can easily be seen that any pair of prices can be supported as a conjectural equilibrium. Convergence of Bayesian learning in a generalized version of this model is studied in Schinkel, Tuinstra and Vermeulen (2000).

Tuinstra (2000c) studies a different learning procedure. Firms use only the most recent information on the objective demand curve, in particular: for the last quoted price p_{jt} they learn the quantity they could have sold at that price, that is $D_j(p_{1t}, \cdots, p_{nt})$, which also depends on prices quoted by the other firms. Furthermore, for example through surveying or small market experiments, they learn the sensitivity of the objective demand curve at that price, $D_j^j(\mathbf{p}_t) \equiv \frac{\partial}{\partial p_j} D_j(\mathbf{p}_t)$. Based upon this information they estimate a linear subjective demand curve $d_{jt}(p_j) = a_t - b_t p_j$. The slope of the subjective demand curve has to coincide with the slope of the objective demand curve,

[5] For a comprehensive survey of this literature see Bonanno (1988).

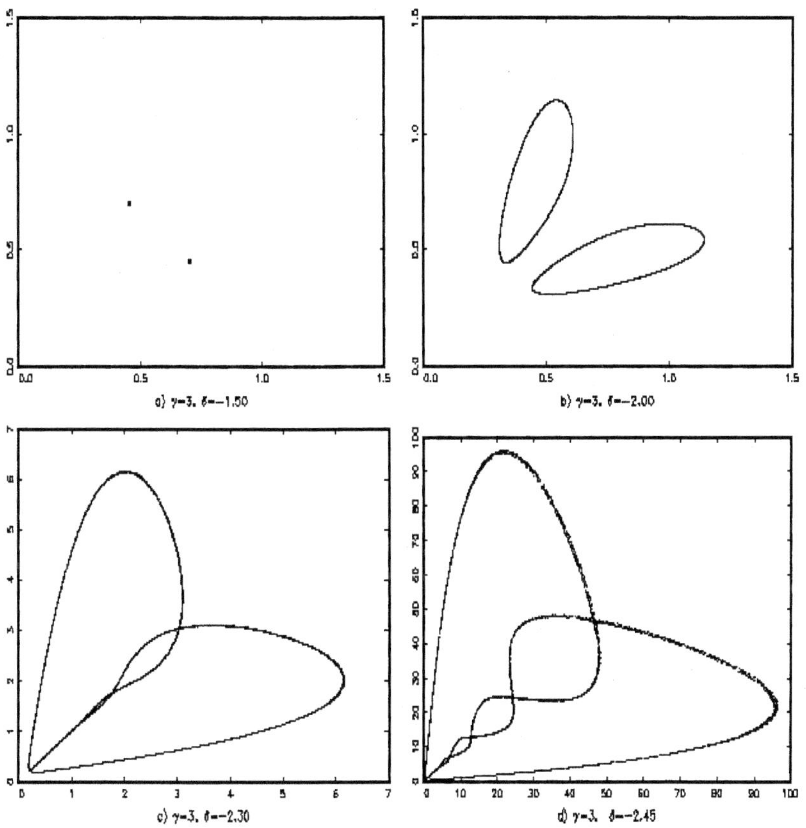

Fig. 3. Attractors for model (7), with $\gamma = 3$. a) (upper left panel) $\delta = -1.50$; b) (upper right panel) $\delta = -2.00$; c) (lower left panel) $\delta = -2.30$; d) (lower right panel) $\delta = -2.45$.

which implies $b_t = -D_j^j(\mathbf{p}_t)$. Furthermore, subjective demand at p_j and objective demand at p_j should be equal, that is $a_t - b_t p_j = D_j(\mathbf{p}_t)$, or $a_t = D_j(\mathbf{p}_t) - D_j^j(\mathbf{p}_t)\, p_{jt}$. The estimated demand function therefore becomes

$$d_{jt}(p_j) = a_t - b_t p_j = \left(D_j(\mathbf{p}_t) - D_j^j(\mathbf{p}_t)\, p_{jt}\right) + D_j^j(\mathbf{p}_t)\, p_j.$$

Given this estimated subjective demand curve the firm sets a new optimal price $p_{j,t+1} = a_{jt}/2b_{jt}$ (for simplicity we abstract from costs of producing), which becomes

$$p_{j,t+1} = \frac{1}{2}p_{jt} - \frac{1}{2}\frac{D_j(\mathbf{p}_t)}{D_j^j(\mathbf{p}_t)}, \qquad j = 1, \ldots, n. \tag{7}$$

Prices will converge to a Bertrand-Nash equilibrium if cross-price effects *and* the curvature of the objective demand functions is small.[6] If these two effects are not so small the price mechanism may feature complicated dynamics. Notice that these conditions are related to the assumption of the firms about demand: they neglect cross-price effects and assume demand is linear.

As an example consider $n = 2$, $D_1(p_1, p_2) = \exp\left[-p_1^{\gamma} p_2^{\delta}\right]$ and $D_2(p_1, p_2) = \exp\left[-p_1^{\delta} p_2^{\gamma}\right]$, where $\gamma > 0$ and $|\delta| < \gamma$. We interpret γ as a measure of the curvature and δ as a measure of the size of the cross-price effect of the objective demand functions. If these effects are not too large, in particular for $\gamma + |\delta| < 4$, the price mechanism (7) will converge to the *Bertrand-Nash* equilibrium $p_1^* = p_2^* = [\gamma]^{-1/(\gamma+\delta)}$.

For $\gamma + |\delta| > 4$ the Bertrand-Nash equilibrium is unstable and endogenous fluctuations emerge. The price mechanism is based upon the conjectures of firms that demand functions are linear and that demand does not depend on the prices of other commodities. For large values of γ and δ the mistakes made by the firms are relatively large (although they may be hard to detect, especially in a noisy environment). These relatively large mistakes drive the complicated dynamics of the price adjustment process. Figure 3 shows some attractors for the dynamical system (7), for $\gamma = 3$ and different values of δ. At $\delta = -1$ the Bertrand-Nash equilibrium undergoes a period doubling bifurcation leading to the creation of a period two cycle (Figure 3a). As δ decreases further, the period-two cycle undergoes a Neimark-Sacker bifurcation leading to an invariant set of two closed curves, with quasi-periodic motion on this invariant set (Figure 3b). For even lower values of δ strange attractors emerge (Figure 3c-d). From this simple example we see that complicated dynamics are also likely to occur in price mechanisms other than the tâtonnement process.

The models presented here are very simple: there is no trade and there are no inventories. Moreover, firms are myopic in the sense that they are only interested in maximizing profit in the current period. However, these models provide a mechanism where prices are set by firms that want to maximize their profits. This is surely an improvement over the tâtonnement process.

5 Concluding Remarks

In this chapter we have reviewed the literature on the tâtonnement process. In general, the tâtonnement process leads to a highly nonlinear dynamical system and we have shown that different types of complicated dynamic behavior are indeed likely to occur in this process. Therefore, the emergence of complicated dynamics in the tâtonnement process is the rule rather than the exception, as it is in many other models from economic dynamics.

[6] Notice that stability of the continuous tâtonnement process depends upon the size of the cross-price effects but not upon the curvature of the excess demand functions.

Another purpose of this chapter has been to identify in which direction one should look for finding a satisfactory price mechanism. However, for many price mechanisms other than the tâtonnement process we should also expect the possibility of complicated dynamics.

The fundamental problem of finding a sensible price mechanism is still unsolved, but it deserves our full attention and hopefully progress will be made in the near future. The quest continues!

References

1. Arrow, K.J. (1959): Towards a theory of price adjustment. In M. Abramovitz et al: *the Allocation of Economic Resources,* Stanford: Stanford University Press.
2. Arrow, K.J., H.D. Block and L. Hurwicz (1959): On the stability of the competitive equilibrium, II. *Econometrica,* **27,** 82-109.
3. Bonanno, G. (1990): General equilibrium theory with imperfect competition. *Journal of Economic Surveys,* **4,** 297-328.
4. Chossat, P. and M. Golubitsky (1988): Symmetry-increasing bifurcation of chaotic attractors. *Physica D,* **32,** 423-436.
5. Debreu, G. (1974): Excess demand functions. *Journal of Mathematical Economics,* **1,** 15-21.
6. Goeree, J.K., C. Hommes and C. Weddepohl (1997): Stability and complex dynamics in a discrete tâtonnement model. *Journal of Economic Behavior and Organization,* **33,** 395-410.
7. Golubitsky, M., I.N. Stewart and D.G. Schaeffer (1988): *Singularities and Groups in Bifurcation Theory,* vol.2, Berlin: Springer-Verlag.
8. Guckenheimer, J. and P. Holmes (1983): *Nonlinear Oscillations, Dynamical Systems, and Bifurcations of Vector Fields,* New York: Springer-Verlag.
9. Hahn, F.H. and T. Negishi (1962): A theorem on non-tâtonnement stability. *Econometrica,* **30,** 463-469.
10. Kirman, A.P. (1983): Mistaken beliefs and resultant equilibria. In: R. Frydman and E. Phelps (eds.), *Individual Forecasting and Collective Outcomes,* Cambridge: Cambridge University Press.
11. Kuznetsov, Yu.A. (1998): *Elements of Applied Bifurcation Theory* (2^{nd}ed.), New York: Springer-Verlag.
12. Mantel, R. (1974): On the characterization of aggregate excess demand. *Journal of Economic Theory,* **7,** 348-353.
13. Saari, D.G. (1985): Iterative price mechanisms. *Econometrica,* **53,** 1117-1131.
14. Scarf, H. (1960): Some examples of global instability of the competitive equilibrium. *International Economic Review,* **1,** 157-172.
15. Schinkel, M.P. (2001): *Disequilibrium Theory. Reflections towards a Revival of Learning.* Ph.D Dissertation Universiteit Maastricht.
16. Schinkel, M.P., J. Tuinstra and D. Vermeulen (2000): Convergence of Bayesian learning in mis-specified models. University of Bilbao working paper 2000.05.
17. Sonnenschein, H. (1973): Do Walras' identity and continuity characterize the class of community excess demand functions? *Journal of Economic Theory,* **6,** 345-354.
18. Tuinstra, J. (1997): A price adjustment process with symmetry. *Economics Letters,* **57,** 297-303.

19. Tuinstra, J. (2000a): A discrete and symmetric price adjustment process on the simplex. *Journal of Economic Dynamics and Control*, **24,** 881-907.
20. Tuinstra J. (2000b): *Price Dynamics in Equilibrium Models: The Search for Equilibrium and the Emergence of Endogenous Fluctuations*, Boston: Kluwer Academic Publishers.
21. Tuinstra, J. (2000c): A price adjustment process in a model of monopolistic competition. CeNDEF Working Paper Series - WP 00-13. University of Amsterdam.
22. Uzawa, H. (1959): Walras' tatonnement in the theory of exchange. *Review of Economic Studies,* **27**, 182-194.
23. Uzawa, H. (1962): On the stability of Edgeworth's barter process. *International Economic Review,* **3,** 218-232.
24. Weddepohl, H.N. (1995): A cautious price adjustment mechanism: chaotic behavior. *Journal of Economic Behavior and Organization,* **27,** 293-300.
25. Weddepohl, H.N. (1999): Price adjustment models in a production economy. *Annals of Operations Research,* **89,** 149-164.

GPSR Compliance

The European Union's (EU) General Product Safety Regulation (GPSR) is a set of rules that requires consumer products to be safe and our obligations to ensure this.

If you have any concerns about our products, you can contact us on ProductSafety@springernature.com

In case Publisher is established outside the EU, the EU authorized representative is:

Springer Nature Customer Service Center GmbH
Europaplatz 3
69115 Heidelberg, Germany

Batch number: 10704035

Printed by Printforce, the Netherlands